Current Topics in Microbiology and Immunology

Volume 414

Series editors

Rafi Ahmed
School of Medicine, Rollins Research Center, Emory University, Room G211, 1510 Clifton Road, Atlanta, GA 30322, USA

Klaus Aktories
Medizinische Fakultät, Institut für Experimentelle und Klinische Pharmakologie und Toxikologie, Abt. I, Albert-Ludwigs-Universität Freiburg, Albertstr. 25, 79104, Freiburg, Germany

Arturo Casadevall
W. Harry Feinstone Department of Molecular Microbiology & Immunology, Johns Hopkins Bloomberg School of Public Health, 615 N. Wolfe Street, Room E5132, Baltimore, MD 21205, USA

Richard W. Compans
Department of Microbiology and Immunology, Emory University, 1518 Clifton Road, CNR 5005, Atlanta, GA 30322, USA

Jorge E. Galan
Boyer Ctr. for Molecular Medicine, School of Medicine, Yale University, 295 Congress Avenue, room 343, New Haven, CT 06536-0812, USA

Adolfo Garcia-Sastre
Icahn School of Medicine at Mount Sinai, Department of Microbiology, 1468 Madison Ave., Box 1124, New York, NY 10029, USA

Akiko Iwasaki
Department of Immunobiology, Yale University School of Medicine, 300 Cedar Street TAC S655B, New Haven, CT 06520, USA

Bernard Malissen
Centre d'Immunologie de Marseille-Luminy, Parc Scientifique de Luminy, Case 906 13288, Marseille Cedex 9, France

Klaus Palme
Institute of Biology II/Molecular Plant Physiology, Albert-Ludwigs-Universität Freiburg, Freiburg, 79104, Germany

Rino Rappuoli
GSK Vaccines, Via Fiorentina 1, Siena, 53100, Italy

Honorary editors

Michael B. A. Oldstone
Department of Immunology and Microbiology, The Scripps Research Institute, 10550 North Torrey Pines Road, La Jolla, CA 92037, USA

Peter K. Vogt
Department of Molecular and Experimental Medicine, The Scripps Research Institute, 10550 North Torrey Pines Road, BCC-239, La Jolla, CA 92037, USA

More information about this series at http://www.springer.com/series/82

R. Luke Wiseman · Cole M. Haynes
Editors

Coordinating Organismal Physiology Through the Unfolded Protein Response

Responsible series editor: Peter K. Vogt

Editors
R. Luke Wiseman
Department of Molecular Medicine
The Scripps Research Institute
La Jolla, CA
USA

Cole M. Haynes
Department of Molecular, Cell,
and Cancer Biology
University of Massachusetts Medical School
Worcester, MA
USA

ISSN 0070-217X   ISSN 2196-9965 (electronic)
Current Topics in Microbiology and Immunology
ISBN 978-3-319-78529-5   ISBN 978-3-319-78530-1 (eBook)
https://doi.org/10.1007/978-3-319-78530-1

Library of Congress Control Number: 2018939014

© Springer International Publishing AG, part of Springer Nature 2018
This work is subject to copyright. All rights are reserved by the Publisher, whether the whole or part of the material is concerned, specifically the rights of translation, reprinting, reuse of illustrations, recitation, broadcasting, reproduction on microfilms or in any other physical way, and transmission or information storage and retrieval, electronic adaptation, computer software, or by similar or dissimilar methodology now known or hereafter developed.
The use of general descriptive names, registered names, trademarks, service marks, etc. in this publication does not imply, even in the absence of a specific statement, that such names are exempt from the relevant protective laws and regulations and therefore free for general use.
The publisher, the authors and the editors are safe to assume that the advice and information in this book are believed to be true and accurate at the date of publication. Neither the publisher nor the authors or the editors give a warranty, express or implied, with respect to the material contained herein or for any errors or omissions that may have been made. The publisher remains neutral with regard to jurisdictional claims in published maps and institutional affiliations.

Printed on acid-free paper

This Springer imprint is published by the registered company Springer International Publishing AG part of Springer Nature
The registered company address is: Gewerbestrasse 11, 6330 Cham, Switzerland

Preface

The endoplasmic reticulum (ER) is a central stress-sensing platform responsible for regulating diverse biologic functions including protein secretion, lipid synthesis, and cellular metabolism. As such, genetic, environmental, and aging-related insults that challenge ER function (i.e., ER stress) can profoundly influence cellular physiology and contribute to the pathogenesis of diverse diseases. To protect cells against ER stress, eukaryotes evolved a complex, integrated stress-responsive signaling pathway called the Unfolded Protein Response (UPR). While initially identified in yeast as a single signaling pathway, the eukaryotic UPR is a complex signaling network consisting of three integrated signaling pathways activated downstream of the ER stress-sensing transmembrane proteins IRE1, PERK, and ATF6. In response to ER stress, these sensors are activated through diverse mechanisms, resulting in a transient attenuation of new protein synthesis and the activation of stress-responsive transcription factors such as XBP1s (activated downstream of IRE1), ATF4 (activated downstream of PERK), and ATF6 (the active transcription factor cleaved off of full length ATF6). Through these mechanisms, the UPR functions to alleviate ER stress and adapt cellular physiology to protect against a given environmental, genetic, or aging-related insult. However, if activation of these pathways prove insufficient to alleviate a severe or chronic ER insult, the UPR promotes pro-apoptotic signaling to induce cell death. Through this combined adaptive and pro-apoptotic activity, the UPR has a critical role in both regulating cellular physiology and dictating cell fate in response ER stress.

Due to its role in coordinating cellular physiology and fate in response to pathologic insults, UPR signaling is inextricably linked to the regulation of diverse biologic functions including secretory proteostasis maintenance, immune cell development, and cellular and organismal metabolism. In addition, alterations in UPR signaling are implicated in the onset and pathogenesis of diverse diseases including cancer, diabetes, and neurodegenerative disorders. The central importance of UPR signaling in regulating both health and disease has led to significant interest in targeting specific aspects of the UPR to intervene in diverse diseases. As we continue to learn more about the UPR in the context of health and disease, we will gain additional understanding into how to best target UPR signaling in the context of a given disease and better appreciate the overall implications of how activating or inhibiting selective UPR signaling pathways influence global organismal physiology.

In this volume, we include chapters from experts in diverse aspects of UPR signaling. The first five chapters are designed to highlight unique roles of UPR signaling in the regulation of different aspects of cellular and organismal physiology such as secretory proteostasis, cell–cell signaling, immune cell development, and metabolism. These chapters provide a broad background in the different mechanisms by which UPR signaling can influence biological functions. In the final three chapters, we describe the contributions of altered UPR signaling in etiologically diverse diseases such as cancer, neurodegenerative disease, and cardiovascular disorders to directly demonstrate the critical role for UPR signaling during different types of pathologic insults. While a comprehensive description of the UPR and all of its functions could fill many volumes, the chapters included in this volume are designed to provide a background in understanding how the UPR can influence diverse physiologic functions in the context of health and disease. Our goal in putting this together was to incorporate diverse aspects of UPR signaling to provide the reader the necessary resources to help them understand the importance of UPR signaling in the context of their own interests and systems. We are confident that readers will enjoy the diverse chapters included this volume and hope that this work will continue to spur new interest in understanding the molecular mechanisms by which the UPR influences cellular and organismal biology.

Worcester, USA Cole M. Haynes
La Jolla, USA R. Luke Wiseman

Contents

Adapting Secretory Proteostasis and Function Through the Unfolded Protein Response 1
Madeline Y. Wong, Andrew S. DiChiara, Patreece H. Suen, Kenny Chen, Ngoc-Duc Doan and Matthew D. Shoulders

Cell Non-autonomous UPR^{ER} Signaling 27
Soudabeh Imanikia, Ming Sheng and Rebecca C. Taylor

The Unfolded Protein Response in the Immune Cell Development: Putting the Caretaker in the Driving Seat 45
Simon J. Tavernier, Bart N. Lambrecht and Sophie Janssens

Mitochondria-Associated Membranes and ER Stress 73
Alexander R. van Vliet and Patrizia Agostinis

Coordinating Organismal Metabolism During Protein Misfolding in the ER Through the Unfolded Protein Response 103
Vishwanatha K. Chandrahas, Jaeseok Han and Randal J. Kaufman

ER Stress and Neurodegenerative Disease: A Cause or Effect Relationship? 131
Felipe Cabral-Miranda and Claudio Hetz

Driving Cancer Tumorigenesis and Metastasis Through UPR Signaling 159
Alexandra Papaioannou and Eric Chevet

ER Protein Quality Control and the Unfolded Protein Response in the Heart 193
A. Arrieta, E.A. Blackwood and C.C. Glembotski

Adapting Secretory Proteostasis and Function Through the Unfolded Protein Response

Madeline Y. Wong, Andrew S. DiChiara, Patreece H. Suen,
Kenny Chen, Ngoc-Duc Doan and Matthew D. Shoulders

Abstract Cells address challenges to protein folding in the secretory pathway by engaging endoplasmic reticulum (ER)-localized protective mechanisms that are collectively termed the unfolded protein response (UPR). By the action of the transmembrane signal transducers IRE1, PERK, and ATF6, the UPR induces networks of genes whose products alleviate the burden of protein misfolding. The UPR also plays instructive roles in cell differentiation and development, aids in the response to pathogens, and coordinates the output of professional secretory cells. These functions add to and move beyond the UPR's classical role in addressing proteotoxic stress. Thus, the UPR is not just a reaction to protein misfolding, but also a fundamental driving force in physiology and pathology. Recent efforts have yielded a suite of chemical genetic methods and small molecule modulators that now provide researchers with both stress-dependent and -independent control of UPR activity. Such tools provide new opportunities to perturb the UPR and thereby study mechanisms for maintaining proteostasis in the secretory pathway. Numerous observations now hint at the therapeutic potential of UPR modulation for diseases related to the misfolding and aggregation of ER client proteins. Growing evidence also indicates the promise of targeting ER proteostasis nodes downstream of the UPR. Here, we review selected advances in these areas, providing a resource to inform ongoing studies of secretory proteostasis and function as they relate to the UPR.

Contents

1 Introduction...	2
2 The UPR in Health and Disease.....................................	4
2.1 Development, Professional Secretory Cells, and Immunity.................	4
2.2 Emerging Functions of the UPR.......................................	6

M.Y. Wong · A.S. DiChiara · P.H. Suen · K. Chen · N.-D. Doan · M.D. Shoulders (✉)
Department of Chemistry, Massachusetts Institute of Technology, 77 Massachusetts Ave,
Cambridge, MA 02139-4307, USA
e-mail: mshoulde@mit.edu

Current Topics in Microbiology and Immunology (2018) 414:1–26
DOI 10.1007/82_2017_56
© Springer International Publishing AG 2017
Published Online: 20 September 2017

	2.3 Dysregulated ER Proteostasis and Disease	7
	2.4 Concept Summary	8
3	Targeting the UPR to Modulate ER Proteostasis	9
	3.1 Stress-Dependent Methods to Modulate the UPR	9
	3.2 Stress-Independent Methods to Modulate the UPR	9
	3.3 Activating the UPR to Address Diseases Linked to Dysregulated ER Proteostasis	12
	3.4 Concept Summary	14
4	Beyond the UPR	15
	4.1 Targeting ATP-Dependent Chaperone Systems in the ER	15
	4.2 Targeting ERAD	17
	4.3 Concept Summary	19
5	Conclusions	19
References		20

1 Introduction

The endoplasmic reticulum (ER) is responsible for secretory proteostasis, involving the coordinated folding, processing, quality control, and trafficking of $\sim 1/3$ of the proteome. Protein folding is a highly complex and error-prone process, requiring a delicate balance between function and risk of aggregation in crowded biological microenvironments where total protein concentrations can range from 100 to 400 mg/mL (Gershenson et al. 2014; Hartl et al. 2011). ER clients, which include secreted, membrane, and lysosomal proteins, face additional challenges, including unique posttranslational modifications (e.g., N-glycosylation) that require specialized cellular machinery (Aebi 2013), oxidative folding processes associated with selective disulfide bond formation (Tu and Weissman 2004), and both spatial and temporal restraints on the completion of folding, modification, assembly, and transport steps. Cells account for this complexity via a diverse array of folding (Hartl et al. 2011) and quality control mechanisms (Smith et al. 2011a), some of which are only recently coming to light. The resulting balance of protein synthesis, folding, and recycling is essential for health. Dysregulated proteostasis in the secretory pathway underpins a diverse array of diseases.

Maintaining secretory proteostasis requires the ability to dynamically respond to challenges such as protein misfolding, often by large-scale remodeling of the ER and the ER proteostasis environment (Walter and Ron 2011). The unfolded protein response (UPR; Fig. 1) is the central stress response pathway involved. The three arms of the metazoan UPR are controlled by the signal transducers IRE1, PERK, and ATF6 (Cox et al. 1993; Harding et al. 1999; Haze et al. 1999; Tirasophon et al. 1998). Activation of these ER transmembrane proteins induces a transcriptional response mediated by three transcription factors, XBP1s (Calfon et al. 2002; Yoshida et al. 2001), ATF4 (Harding et al. 2000; Vattem and Wek 2004), and ATF6f (ATF6-fragment) (Adachi et al. 2008). This coordinated transcriptional response alleviates the burden of protein misfolding in the secretory pathway by upregulating ER chaperone, quality control, and secretion mechanisms (Adachi et al. 2008; Harding et al. 2000; Shoulders et al. 2013b). UPR activation also

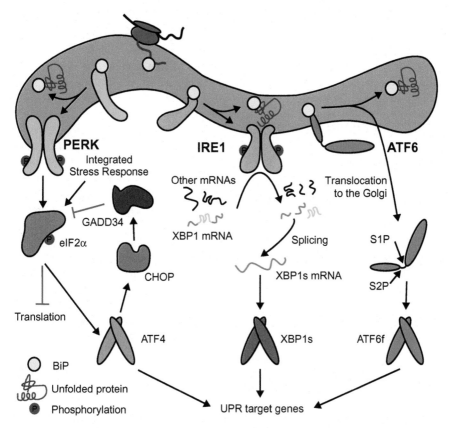

Fig. 1 The unfolded protein response (UPR). Accumulation of misfolding proteins in the endoplasmic reticulum (ER) activates the transmembrane protein UPR signal transducers PERK, IRE1, and ATF6. Dimerization and auto-phosphorylation of PERK and IRE1, or trafficking to the Golgi and subsequent proteolytic processing of ATF6, result in the production of the UPR transcription factors by enhancing translation of ATF4, cleaving and splicing *XBP1* mRNA to yield *XBP1s*, and releasing ATF6f from the Golgi membrane. These transcription factors proceed to the nucleus and remodel the ER proteostasis environment by upregulating chaperones, quality control components, and other UPR target genes to maintain or recover secretory proteostasis. PERK can globally reduce the nascent protein load on the ER via phosphorylation of eIF2α, a pathway that can be similarly induced by the integrated stress response. The RNase domain of activated IRE1 degrades several ER-targeted transcripts and may play a related role

inhibits protein translation to lower the net nascent protein load on the ER, a process mediated primarily by PERK activation and subsequent phosphorylation of eIF2α (Harding et al. 1999), but also influenced by the selective degradation of ER-directed mRNA transcripts by IRE1 (Hollien et al. 2009; Moore and Hollien 2015). If proteostasis cannot be restored, pro-apoptotic mechanisms within the UPR lead to programmed cell death in part through induction of the transcription factor CHOP downstream of PERK.

While extensive research has yielded a relatively well-defined picture of the UPR, the discovery of new regulatory mechanisms continues to shape our understanding of how the UPR relates to ER homeostasis. The ER not only functions as a protein-folding factory, but also participates in calcium storage and lipid biosynthesis (Fu et al. 2011). Along these lines, a recent study highlighted the capacity of IRE1's membrane-spanning domain to activate the protein in response to lipid perturbation even when the luminal protein misfolding stress-sensing domain is deleted using CRISPR/Cas9 (Kono et al. 2017). Moreover, the ER is involved in cellular responses to oxidative stress, metabolic imbalance, and pathogen invasion (Malhotra and Kaufman 2007; Roy et al. 2006; Volmer and Ron 2015). Each of these processes is modulated by the UPR. Thus, despite its name, the UPR is not simply a reaction to protein misfolding, but is instead a fundamental driving force for physiology and pathology. The central roles of the UPR in health and disease have catalyzed the development of methods to modulate the UPR, with the goal of better understanding key regulatory axes and identifying opportunities to influence phenotypic outcomes. Below, we review our current picture of the metazoan UPR in the context of secretory proteostasis, from its connections to health and disease (Sect. 2), to methods for selectively perturbing the UPR and their potential applications in disease therapy (Sect. 3), to efforts to target downstream nodes in ER proteostasis (Sect. 4).

2 The UPR in Health and Disease

Key functional nodes within the ER proteostasis network include chaperones, quality control mechanisms, posttranslational modifiers, and trafficking pathways (Fig. 2). Each of these nodes is dynamically regulated by the UPR to match proteostatic capacity to demand, thereby maintaining balanced levels of protein folding and quality control both during normal cellular function and under stressful conditions.

2.1 Development, Professional Secretory Cells, and Immunity

The UPR plays critical roles during development that have been demonstrated in several model systems (Coelho et al. 2013; Shen et al. 2001). In particular, UPR activation appears to upregulate ER-resident chaperones and signaling pathways, which work collectively to relieve stress and regulate development in differentiating cells (Dalton et al. 2013; Laguesse et al. 2015; Reimold et al. 2001). For example, upon B-cell differentiation into plasma cells, the ER undergoes extensive XBP1-driven expansion (Reimold et al. 2001; Shaffer et al. 2004), in part to

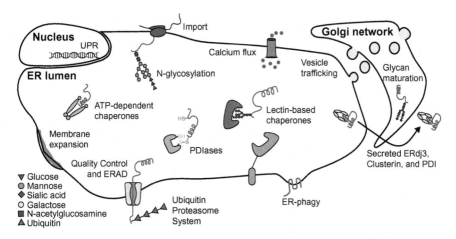

Fig. 2 Representative nodes in the secretory proteostasis network. Diverse proteins and pathways collectively modulate folding, secretion, quality control, and/or degradation of ER clients. ATP-dependent chaperones and PDIases assist in the folding of client proteins, as do lectin-based chaperones such as calnexin and calreticulin. Terminally misfolded proteins are typically cleared by ER-associated degradation (ERAD) via the ubiquitin-proteasome system. ER-phagy can serve as a counterpart to membrane expansion mechanisms, reducing organelle size to regulate ER proteostasis. Meanwhile, calcium flux, vesicle trafficking, and UPR-mediated changes in the chaperone:client balance, import, disulfide bond formation, and N-glycosylation of nascent polypeptides help to maintain or create favorable folding conditions and buffer ER protein-folding capacity. A handful of chaperones, including ERdj3, can accompany proteins to the extracellular space

accommodate high levels of antibody synthesis. B-Cell differentiation in vitro also induces the UPR-regulated proteins XBP1s, BiP, and Grp94. Notably, the process occurs without expression of CHOP or inhibition of protein translation, suggesting that a physiologic UPR need not involve all three UPR arms, in contrast to the case of attenuating acute ER stress-induced protein misfolding. Moreover, induction of XBP1, BiP, and Grp94 transcripts apparently occurs prior to any significant protein-folding load on the ER, suggesting further differences between developmental and stress-associated signaling pathways (Gass et al. 2002). Other professional secretory cells, including pancreatic β-cells, hepatic cells, and osteoblasts, also must sustain high rates of ER client protein synthesis, folding, and secretion, and thus rely on the UPR and its downstream signaling mechanisms for survival and function.

Other work highlights roles of the UPR in cellular responses to pathogen invasion. Binding of unfolded cholera toxin A subunit induces IRE1α ribonuclease activity, but not the canonical UPR involving PERK and ATF6 (Cho et al. 2013). The fragments of endogenous mRNA produced by IRE1α prompt RIG-I to activate NF-κB and interferon signaling. Other work indicates that Toll-like receptors (TLRs) in macrophages promote splicing of XBP1 to optimize the production of proinflammatory cytokines (Martinon et al. 2010), although XBP1s can also be essential in protecting against the effects of prolonged inflammation (Adolph et al.

2013; Richardson et al. 2010). Viral pathogens are also capable of hijacking the UPR to promote proliferation in host cells. For example, IRE1 activity is critical for the replication of at least some strains of the influenza virus (Hassan et al. 2012). In contrast, HSV1 suppresses both PERK and IRE1 signaling: glycoprotein B interacts with the luminal domain of PERK to block kinase activation, the late viral protein γ1 34.5 recruits PP1α to dephosphorylate eIF2α, and the UL41 protein acts as an endoribonuclease to degrade XBP1 mRNA (Zhang et al. 2017). Similarly, recent studies of *Legionella pneumophila*, the organism responsible for Legionnaires' disease, show that the pathogen forestalls a typical ER stress response by repressing translation of a subset of UPR-associated genes to prevent host-cell apoptosis that would otherwise be induced (Hempstead and Isberg 2015; Treacy-Abarca and Mukherjee 2015). The relevant bacterial effector proteins may serve as springboards for biomimetic approaches to modulate UPR pathways.

2.2 Emerging Functions of the UPR

Beyond established roles in development and immunity, new functions for and consequences of the UPR continue to emerge. ER recycling via ER-phagy is critical for ER homeostasis (Bernales et al. 2006), and several constituent biochemical pathways were recently mapped (Fumagalli et al. 2016; Khaminets et al. 2015). A possible role for the IRE1-XBP1s arm of the UPR in inducing such ER-phagy may exist (Margariti et al. 2013). By reducing organelle size and/or disposing of dysfunctional ER regions, UPR induction of selective ER-phagy could serve as a counterpart to membrane expansion mechanisms for resolving ER stress (Schuck et al. 2009).

The discovery that the IRE1-XBP1s axis of the UPR is responsible for cell non-autonomous UPR activation (Taylor and Dillin 2013) is also intriguing. Such cell-to-cell communication of stress is likely to have important biological consequences that merit further investigation. ATF6 activation was recently shown to increase not just expression but also secretion of ERdj3, an ER-localized Hsp40 co-chaperone (Genereux et al. 2015). The consequent co-secretion of ERdj3 with misfolded client proteins may be protective for the origin cell or ameliorate harmful protein aggregation in the extracellular milieu. Stress-induced ERdj3 secretion thus provides a mechanism by which the UPR can modify not just ER but also extracellular proteostasis.

Emerging functions of the UPR described above focus on direct modulation of protein folding and production. In addition to these mechanisms, a role for the ER in regulating the extent of protein posttranslational modifications has emerged. For example, two groups showed that the UPR can modulate hexosamine biosynthesis to promote ER client clearance and prolong life in the face of chronic protein misfolding stress (Denzel et al. 2014; Wang et al. 2014). These studies suggest that UPR activation, and especially the IRE1-XBP1s arm of the UPR, may enhance the extent of client protein N- and O-glycosylation and/or modify oligosaccharyltransferase efficiency. While the consequences require further investigation, N-glycosylation promotes both ER client folding, by providing access to the lectin-based chaperone machinery, and the identification of misfolded proteins for ER-associated degradation (ERAD) via the lectin-based quality control machinery, providing a potential rationale for IRE1-XBP1s enhanced N-glycosylation (Ruiz-Canada et al. 2009).

Surprisingly, the UPR can also remodel the actual molecular architecture of N-glycans added to ER client proteins by modulating their biosynthesis (Dewal et al. 2015). Stress-independent activation of XBP1s changes transcript levels of N-glycan modifying enzymes, leading to altered mature glycan structures on model secreted N-glycoproteins. More work is required to establish the biological relevance and consequences of this phenomenon. However, this newly established connection between N-glycan signatures and the UPR suggests that the UPR may unexpectedly influence processes such as cell–cell interactions, cell–matrix interactions, and trans-cellular communication by actually modifying the molecular structure of secreted ER clients (Dewal et al. 2015).

2.3 Dysregulated ER Proteostasis and Disease

When proteostasis networks function properly, cells maximize production of properly folded, functional proteins. Meanwhile, quality control mechanisms ensure that only folded proteins are transported to their final locations, while production of misfolded and aggregated proteins is minimized (Fig. 3a). The UPR regulates this process by sensing the accumulation of misfolded proteins, whether due to genetic mutations or adverse physiological conditions, and remodeling the ER proteostasis network to resolve emerging problems.

Chronically dysregulated ER proteostasis, unresolved by the UPR, leads to diverse protein misfolding and aggregation-related diseases. For many mutations that destabilize or prevent the folding of a protein, the UPR may in principle have the potential to resolve the proteostatic defect—if it is activated. However, just one mutant protein misfolding in a background of thousands of well-behaved proteins may not always be sufficient to trigger a protective UPR. In other cases, the ER may be overwhelmed by high concentrations of an aggregating mutant protein, leading to chronic ER stress and cellular apoptosis. In either scenario, pharmacologic perturbation of the UPR could be therapeutically useful. Moreover, many cancer cells rely on constitutive activation of pro-survival pathways (in particular the IRE1-XBP1s arm) within the UPR (Chen et al. 2014b; Jamora et al. 1996; Mimura et al. 2012). This observation suggests that UPR inhibition could also prove valuable for diseases that do not stem directly from protein misfolding.

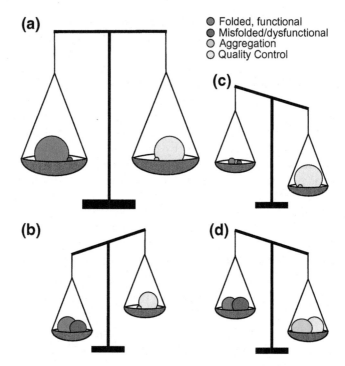

Fig. 3 Proteostasis (im)balances in health and disease. **a** When proteostasis is properly balanced, production of folded, functional client proteins and quality control surveillance are maximized, while misfolding and aggregation are minimized. Maintaining the balance between production of folded, functional proteins and quality control or clearance of misfolded, non-functional proteins is essential for health. In protein misfolding/aggregation disorders linked to defects in ER proteostasis, **b** insufficient quality control, **c** hyperactive quality control and failed folding, and/or **d** failed clearance leading to intracellular protein aggregation and chronic ER stress/UPR activation can all cause a pathologic loss of proteostasis balance. Modulating proteostasis network activities holds potential to resolve such defects

2.4 Concept Summary

Even this cursory survey of the roles of the UPR in health and disease reveals a striking functional spectrum. Beyond the traditional UPR, IRE1, PERK, and ATF6 participate in various physiological processes, upregulating UPR-associated transcripts to facilitate differentiation and sustain the activity of professional secretory cells. Studies on different models of infection demonstrate that UPR arms also can be selectively suppressed or activated in order to bypass cytotoxic stress-signaling pathways and promote cell survival. Such mechanisms are appealing starting points for efforts to better understand and manage cases of chronic ER stress associated with disease. Finally, recent findings suggest new roles for the UPR in regulating extracellular proteostasis and signaling, perhaps via posttranslational alterations in ER client protein molecular structures (Dewal et al. 2015).

3 Targeting the UPR to Modulate ER Proteostasis

The intrinsic functions of the UPR in development and immunity, regulating cell survival and death, and resolving proteostatic stress have motivated extensive method development efforts to uncover small molecule UPR activators and inhibitors. Such tools are enabling detailed dissection of innate UPR function and revealing the therapeutic potential associated with targeting the UPR. This section reviews selected chemical genetic and small molecule approaches to modulate the UPR, and briefly describes insights obtained using such methods in disease model systems.

3.1 Stress-Dependent Methods to Modulate the UPR

Traditional approaches to activate the UPR rely on small molecules that cause ER stress. Such compounds include tunicamycin, which ablates protein N-glycosylation; thapsigargin, which disrupts ER calcium homeostasis; and dithiothreitol, which reduces disulfides. All these methods globally and strongly activate the UPR by inducing extensive protein misfolding and aggregation. While such strategies have proven valuable for mapping UPR signaling pathways, they are less suited for mechanistic work because they induce high and physiologically irrelevant levels of stress. Moreover, these compounds are not helpful for therapeutic proof-of-principle studies, as the cytotoxic and pleiotropic side effects of their use obscure any potentially beneficial effects of UPR activation.

One approach to bypass these issues is to administer very low concentrations of ER toxins to induce only moderate stress (Rutkowski et al. 2006). Alternatively, a chemical genetic strategy was recently pioneered to transiently activate the UPR via moderate stress without inducing apoptosis (Raina et al. 2014). The method involves expressing an ER-targeted HaloTag protein (ERHT) that can be conditionally destabilized by conjugation to a small molecule hydrophobic tag (HyT36). Treatment of ER-targeted HaloTag-expressing cells with HyT36 destabilizes the protein, causing mild but acute accumulation of misfolded proteins in the ER and consequent UPR activation. While the mechanistic details are not yet fully elucidated, the ERHT/HyT36 system provides a valuable tool to study aspects of global UPR activation distinct from the stress-independent, arm-selective techniques described below.

3.2 Stress-Independent Methods to Modulate the UPR

An important element of any strategy to discover and validate small molecule UPR arm activators is verifying that the compounds directly and selectively activate a particular arm of the UPR, without inducing ER protein misfolding or other forms of stress. To address this concern, recent work has delineated comprehensive transcriptional profiling strategies to validate stress-independent, arm-specific UPR activators (Plate et al. 2016). While several small molecule leads have been developed that activate endogenous IRE1 and/or ATF6 (Mendez et al. 2015; Plate et al. 2016), further work will improve potency and selectivity, as well as clarify the mechanisms of action. Selective small molecule activators of endogenous PERK have thus far proven elusive. On the other hand, salubrinal (Boyce et al. 2005), guanabenz (Tsaytler et al. 2011), ISRIB (Sidrauski et al. 2015), and Sephin1 (Das et al. 2015) can modulate the PERK-induced translational block either up or down by altering eIF2α phosphorylation status (Fig. 1). Other explanations for the biological phenotypes induced by guanabenz and Sephin1 may also prove relevant (Crespillo-Casado et al. 2017). Notably, these compounds also modulate overlapping aspects of the integrated stress response, rendering such small molecules neither UPR- nor PERK-specific. For example, salubrinal also targets the constitutively expressed eIF2α phosphatase CReP (Boyce et al. 2005). Finally, while small molecule inhibitors of endogenous IRE1 have been well-characterized (Cross et al. 2012; Volkmann et al. 2011; Wang et al. 2012), compounds that inhibit ATF6 are only now beginning to emerge (Gallagher and Walter 2016).

Given the paucity of small molecules to modulate arms of the UPR in a stress-independent manner, over the past 15 years the field has turned to chemical genetic approaches. One strategy involves fusion of PERK's kinase domain to a tandem-modified FK506-binding domain, Fv2E. Treatment with the small molecule AP20187 induces Fv2E-PERK dimerization, auto-phosphorylation, and downstream PERK signaling (Lin et al. 2009; Lu et al. 2004). The Fv2E fusion strategy was also applied to recapitulate the regulated RNase activity of IRE1 (Back et al. 2006). A complementary strategy based on bump-hole protein engineering employs an I624G amino acid substitution to allow binding of a small molecule ligand in the ATP pocket of IRE1, but not other kinase active sites, resulting in selective, stress-independent activation of IRE1's RNase domain. The addition of the ligand 1NM-PP1 to cells expressing IRE1α^{I624G} results in IRE1-dependent XBP1 mRNA splicing in the absence of ER stress (Han et al. 2008, 2009; Papa et al. 2003).

These PERK and IRE1-targeted chemical genetic strategies allow for uncoupling of UPR arm activation from stress, but require careful engineering of cells to ensure minimal background signaling and robust inducibility. Their application has helped to clarify the consequences of ER stress. For example, chronic PERK activity, but not 1NM-PP1-induced IRE1 activity, is cytotoxic, suggesting a model in which the duration of PERK signaling regulates cell survival versus apoptosis decisions (Lin et al. 2009). Intriguingly, the advent of RNase inhibitors targeting wild-type IRE1, such as KIRA6, revealed that IRE1 can also promote apoptosis through formation of oligomers with hyperactive RNase activity (Ghosh et al. 2014). Thus, cell fate may not necessarily be determined by opposing signals from IRE1 and PERK, but rather by the degree and timescale of activation (whether homeostatic or cytotoxic) experienced by both kinases.

An alternative chemical genetic strategy is to confer small molecule-dependent activity upon the UPR transcription factors, independent of the upstream signal transducers IRE1, PERK, and full-length ATF6. This approach has proven most successful for XBP1s and ATF6f (Fig. 1). In the simplest case, tetracycline (Tet)-regulated expression of the UPR transcription factors ATF6f (Okada et al. 2002) and XBP1s (Lee et al. 2003; Shoulders et al. 2013b) under control of the Tet-repressor domain has been applied in several model systems. A challenge is the requirement for incorporation of the Tet-repressor in target cells and tissues. Moreover, Tet control of protein expression is rarely dose-dependent, resulting in high, non-physiologic levels of the Tet-inducible gene (Shoulders et al. 2013a, b). Such overexpression of UPR transcription factors can cause off-target mRNA upregulation and often induces apoptosis, limiting its applicability.

More recently, destabilized domains (Banaszynski et al. 2006; Iwamoto et al. 2010) were leveraged to achieve orthogonal activation of XBP1s and/or ATF6f in a single cell. Fusing a destabilized variant of the dihydrofolate reductase from *Escherichia coli* (DHFR) to the N-terminus of ATF6f results in a constitutively expressed DHFR.ATF6f fusion protein that is directed to rapid proteasomal degradation. The addition of the small molecule pharmacologic chaperone trimethoprim stabilizes DHFR, preventing degradation and allowing the DHFR-fused ATF6f transcription factor to function (Shoulders et al. 2013b). A related approach was used to control the activity of the XBP1s transcription factor, via fusion to an FKBP12 destabilized domain that can be stabilized by the small molecule Shield-1 (Shoulders et al. 2013a). The method requires minimal cell engineering, can be used in virtually any cell line of interest (including primary cells), and provides for highly dose-dependent control of transcription factor activity within the physiologically relevant regime. Moreover, fusion of destabilized domains to dominant negative versions of the UPR transcription factors can permit small molecule-dependent inhibition of endogenous transcription factor activity (Shoulders et al. 2013a). The advantages of destabilized domains for conferring small molecule control onto UPR and other stress-responsive transcription factors (Moore et al. 2016) have led to their adoption by a number of research groups in studies of stress-responsive signaling.

Chemical genetic strategies provide valuable temporal and ligand concentration-dependent control of UPR arm activity. However, they must be applied with caution to ensure that engineered domains are compatible with the natural function of the target protein. IRE1^{I624G}, for example, does not share the same mechanism of activation as wild-type IRE1, and may exhibit altered kinase activity; similarly, the fusion of destabilized domains to UPR transcription factors may alter their function. Expression levels of engineered proteins, whether dictated by the leakiness of the system or by copy number per cell, must also be optimized. Thus, while such strategies have enabled substantial progress in the UPR field (especially with regard to potential therapeutic benefits of UPR activation in disease model systems, discussed below), they do not ablate the need for highly selective and potent small molecule UPR modulators that function independently of protein misfolding stress.

3.3 Activating the UPR to Address Diseases Linked to Dysregulated ER Proteostasis

Small molecule and chemical genetic methods to modulate the UPR, including those described above, have been extensively applied to test whether UPR-mediated remodeling of the ER proteostasis network is a viable strategy to address ER protein misfolding- and aggregation-related diseases. In particular, various approaches have been used to address dysregulated proteostasis associated with the three types of defects highlighted in Fig. 3.

In the first category of pathologic ER proteostasis defects, protein misfolding/aggregation diseases manifest owing to a combination of underactive quality control and insufficient folding activity (Fig. 3b). The insufficient ER proteostasis environments that characterize such disorders stem from an inability to sense misfolding of an individual mutant protein and/or the permitted secretion of malformed, aggregation-prone proteins into the extracellular milieu.

One example is the transthyretin (TTR) amyloidoses, wherein misfolded or unstable TTR escapes the ER and later aggregates in peripheral tissues (Johnson et al. 2012). TTR is a secreted tetrameric protein whose extracellular disassembly provides the necessary template for oligomers to form (Hammarstrom et al. 2001, 2003). Numerous mutations destabilize the tetramer, accelerating disassembly, aggregation, and disease pathology. Two treatments currently exist: (1) gene therapy via liver transplantation, as the liver is the primary source of TTR, and (2) small molecule pharmacologic chaperones that stabilize secreted TTR tetramers and thereby prevent oligomerization (Johnson et al. 2012).

An alternative, potentially synergistic, possibility is enhancing the stringency of ER quality control, perhaps via UPR-mediated remodeling of the ER proteostasis network to reduce the secretion and promote the degradation of destabilized TTR variants. By lowering misfolding TTR concentrations in the sera, this strategy would likely reduce pathologic oligomer formation. Indeed, stress-independent activation of ATF6 using the chemical genetic DHFR.ATF6f construct (Sect. 3.2) preferentially directs destabilized TTR variants towards ERAD, drastically reducing their secretion (Chen et al. 2014a; Shoulders et al. 2013b). Importantly, the approach does not influence the secretion of stable, wild-type TTR. These results hint at the potential of stress-independent UPR activation to improve cellular capacity to prevent the secretion of misfolded, potentially toxic species. Stress-independent XBP1s and/or ATF6 activation is also effective in preventing the secretion of amyloidogenic light chain (Cooley et al. 2014). Such findings motivated significant efforts to identify small molecule activators of ATF6 that appear very promising and may provide leads for clinical development (Plate et al. 2016).

A second, less intuitive category of ER proteostasis defects arises when cells execute an overly stringent survey of client protein-folding status (Fig. 3c). As in cases of healthy proteostasis, aggregation is minimized. Inconveniently, however, a slow-folding or moderately misfolded ER client is prematurely directed to degradation when, if given sufficient time or assistance, it could have adopted a sufficiently folded, functional state to avert pathology. Such indiscriminate degradation owing to excessively stringent quality control can result in loss-of-function phenotypes. Onset of disease occurs when the amount of protein present and able to function in its natural location falls below the required threshold of essential biological activity. Relaxed quality control may, in these cases, allow such ER clients to adopt sufficiently folded, functional states to traffic to their destinations and thereby ameliorate disease.

One case where hyperactive quality control appears to be associated with pathology is the lysosomal storage disorders. An example is Gaucher's disease, which is caused by mutations that destabilize glucocerebrosidase (GCase). GCase is a hydrolytic enzyme that folds in the ER and is then trafficked to the lysosome, where it degrades encapsulated glycolipids. Misfolding-prone and destabilized GCase variants can be identified by overly stringent ER quality control machinery and targeted for degradation by the proteasome (Ron and Horowitz 2005). Even a slight improvement in trafficking to the lysosome would likely provide therapeutic benefit. Mu and coworkers discovered that global UPR activation by induction of mild ER stress significantly enhances trafficking of the L444P GCase variant to the lysosome in an IRE1- and PERK-dependent manner, with a corresponding decrease in degradation (Mu et al. 2008b). The resulting enhancement in lysosomal GCase activity can be synergistically increased by co-administration of GCase-stabilizing pharmacologic chaperones. More recently, modulation of Ca^{2+} concentrations in the ER and knockdown of specific ER proteostasis network components like FKBP10 and ERdj3, which direct GCase to degradation, were both shown to provide similar benefits (Mu et al. 2008a; Ong et al. 2010, 2013; Tan et al. 2014). Thus, reducing quality control stringency via UPR-mediated and/or more targeted perturbations of the ER proteostasis network may hold potential for diseases associated with hyperactive quality control.

A third category of disease-causing ER proteostasis defects emerges when protein misfolding or aggregation in the ER causes chronically unresolved stress and consequent cellular dysfunction. While the IRE1 and ATF6 arms of the UPR can help cells tolerate such chronic stress, as has been observed in fly models of retinal degeneration (Ryoo et al. 2007; Wu et al. 2007), chronic activation of the PERK arm leads to apoptosis. Thus, long-term, global UPR activation can severely threaten tissue homeostasis (Fig. 3d), engendering a vicious cycle in which protein misfolding caused by mutations, aging, and the like results in cell death, increasing the demand on surviving cells in a given tissue to produce folded, functional protein. The increased protein-folding load placed on the surviving cells results in additional chronic ER stress, continued UPR activation, and further tissue loss. Unresolved chronic ER stress is associated with a wide variety of disorders and diseases, including obesity and neurodegeneration (Lindholm et al. 2006). Strategies to inhibit UPR-mediated apoptosis could be valuable in these situations, as could methods to alleviate chronic stress by enhancing folding or quality control (as in Figs. 3b and c).

UPR-mediated ER proteostasis network remodeling has shown promise for several diseases where protein aggregation is observed in the ER. For example, ATF6 and

PERK activation enhance clearance of mutant, aggregating rhodopsin variants that cause retinitis pigmentosa (Chiang et al. 2012). Similarly, ATF6 activation assists the clearance of the aggregating Z variant of α-1-antitrypsin that normally causes inflammation-induced hepatotoxicity (by a gain-of-function mechanism) and/or proteolytic damage to lung tissue (by a loss-of-function mechanism) (Smith et al. 2011b). Clearance of these protein aggregates from the ER is thus an attractive therapeutic strategy. An alternative approach, instead of activating UPR arms, is to simply lower the net protein load on the ER and allow endogenous mechanisms to resolve the proteostasis defect. This strategy involves application of molecules that function downstream of PERK to translationally attenuate the protein-folding challenge faced by the ER. Such strategies have shown promise in both diabetes- and amyotrophic lateral sclerosis-related model systems (Boyce et al. 2005; Das et al. 2015; Tsaytler et al. 2011).

While the division of disease-causing ER proteostasis defects into the three categories in Fig. 3b–d provides valuable context, few diseases fit seamlessly. For example, lax quality control that permits accumulation of mutant GCase variants may cause chronic, damaging ER stress and/or disrupt general protein trafficking, a mechanism that could well be biologically relevant for Gaucher's disease. Intriguingly, GCase deficiency promotes neurodegeneration in Parkinson's disease, possibly through the loss of stabilizing effects on α-synuclein oligomers or by inducing lysosome dysfunction, providing further evidence that multiple shared mechanisms may underlie and connect otherwise disparate pathologies (Abeliovich and Gitler 2016). Another pertinent example is the collagenopathies, such as osteogenesis imperfecta (Forlino and Marini 2016), which appear to encompass all three types of ER proteostasis defects described above. Collagen is a challenging protein to fold (DiChiara et al. 2016), with both globular and triple-helical regions, abundant and critical posttranslational modifications, and dimensions too large for standard COP-II transport vesicles. Numerous mutations in collagen strands and in components of the collagen proteostasis network cause diverse collagenopathies, depending on the specific type of collagen involved. Such mutations can lead to intracellular collagen accumulation and chronic ER stress, can cause collagen strands to be subjected to excessive quality control and degradation, and/or produce variants that functionally disrupt tissue architecture (Fitzgerald et al. 1999; Mirigian et al. 2016). Whether ER proteostasis imbalances can be remedied in osteogenesis imperfecta and the other collagenopathies remains to be determined, but such pathologies highlight the complexity associated with many protein misfolding-related diseases.

3.4 Concept Summary

Directly targeting particular arms of the UPR via stress-dependent and -independent approaches will require minimizing off-target effects and separating overlapping activities, whether this entails uncoupling an apoptotic, late-stage UPR from a stress-sensing, early-stage UPR, uncoupling IRE1 kinase from RNase activity, or uncoupling UPR activation from ER stress. While the development of improved methods continues, the application of existing tools to various model systems confirms the therapeutic potential of adapting secretory proteostasis through the UPR. Although the details of the energy landscapes of misfolding and aggregation-prone proteins are unique, we note that they often share features in terms of their interactions with the ER proteostasis network. Focusing on commonalities may thus yield broadly applicable rather than disease-specific therapeutic approaches.

4 Beyond the UPR

The preceding discussion emphasizes the substantial promise of UPR modulation to resolve disease-causing ER proteostasis defects. However, global remodeling of the ER proteostasis network by UPR perturbation could have deleterious effects on otherwise well-behaved wild-type proteins or on innate functions of the UPR. Targeted drug delivery to diseased tissues could help sidestep off-target effects, but remains difficult. But while the UPR is the master regulator of the ER proteostasis network (Fig. 2), it is still just one of many hubs (Fig. 4). In many cases, it may be possible to design small molecules that target proteostasis nodes downstream of the UPR's transcription/translation response to resolve proteostasis defects. Such approaches could have a larger therapeutic window owing to more precise manipulation of specific mechanisms. Testing these strategies requires selective chemical biology and genetic tools to manipulate the activities of chaperones, quality control pathways, and the like. In addition to the example of translational attenuation discussed above, this section briefly considers two other nodes downstream of the UPR where targeted investigation has become increasingly possible: the ER's ATP-dependent chaperones and ERAD-mediated quality control.

4.1 Targeting ATP-Dependent Chaperone Systems in the ER

The ATP-dependent chaperones Grp94 (a heat shock protein 90, or Hsp90, isoform), BiP (a heat shock protein 70, or Hsp70, isoform), and their corresponding Hsp40-like co-chaperones and nucleotide exchange factors are central components of the ER proteostasis network (Fig. 2). They play key roles both in assisting client protein folding as well as in recognizing and targeting misfolded proteins for quality control. Given the importance of dysregulated proteostasis in cancer and other diseases, Hsp90 inhibitors have been particularly sought after as therapeutics. A number of potent inhibitors have been discovered (McCleese et al. 2009; Schulte

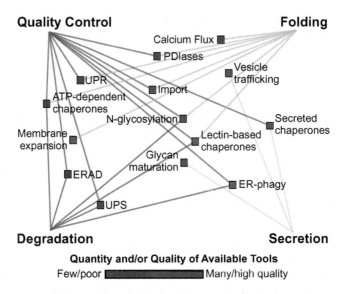

Fig. 4 Targeting functional nodes in the ER proteostasis network. Representative hubs in the ER proteostasis network. *Colored boxes* give a qualitative measure of methods available to modulate each corresponding node: *Red* few and/or poorly characterized tools; *blue* many and/or well-characterized tools. Nodes from (Fig. 2) are aligned with the broader axes of quality control, folding, degradation, and secretion. For many diseases, coordinated regulation of multiple nodes in one or more of the relevant pathways may produce the most therapeutic benefit

and Neckers 1998; Shi et al. 2012; Whitesell et al. 1994) and show promise in cancer model systems. However, the high sequence and structural similarity of Hsp90 isoforms (Chen et al. 2005) poses difficulties for the design of selective small molecule inhibitors. Indeed, the most widely used Hsp90 inhibitors target multiple Hsp90 isoforms. Such promiscuity remains a significant challenge, even among compounds reported to be selective for the ER-resident isoform Grp94 (Liu and Street 2016). Particularly with respect to ER proteostasis, development of selective Grp94 inhibitors is essential to reduce undesirable cytotoxicity and ensure that only the intended proteostasis network node is targeted.

Fortunately, the ATP binding pocket of Grp94 is distinctive amongst the Hsp90 chaperone isoforms (Soldano et al. 2003), and recent progress in developing chemical tools for targeting Grp94 is encouraging. Rigorous structure-based analyses revealed that certain members of a purine-scaffold series bind a Grp94-specific hydrophobic pocket, permitting functional characterization of the effects of Grp94 inhibition in cancer cells (Patel et al. 2013). Other compound classes also show substantial promise, including radamide derivatives like BnIm, the first Grp94-selective inhibitor discovered (Duerfeldt et al. 2012). Improvements on BnIm have yielded even more potent Grp94 inhibitors (Crowley et al. 2016). As such chemical tools have become available, various groups have begun to test the consequences of Grp94 inhibition in protein misfolding-related disease model systems. For example, BnIm-mediated Grp94 inhibition ameliorates excessive quality control of the Ala322Asp variant of GABA$_A$ receptors, whose insufficient trafficking is associated with specific forms of epilepsy (Di et al. 2016). Such results are reminiscent of the ability of UPR activation to reduce excessive quality control of GCase (Mu et al. 2008b). Here, however, targeting an ER proteostasis node downstream of the UPR offers a more specific biological response.

Selective inhibition of the ER's Hsp70 analogue BiP has proven more challenging. Eukaryotes express as many as thirteen Hsp70 isoforms (Kampinga et al. 2009), including BiP in the ER, mtHsp70 in mitochondria, the constitutive cytosolic Hsc70, and the inducible cytosolic isoform Hsp72. Moreover, BiP association with IRE1, PERK, and ATF6 regulates UPR signaling, making it difficult to target the chaperone without also inducing the UPR (Carrara et al. 2015; Pincus et al. 2010). While a few small molecules have been introduced as BiP inhibitors, it is not clear that they act selectively. Nonetheless, selective modulation of BiP activities remains a promising avenue for resolving ER proteostasis defects because it facilitates proper folding of nascent polypeptide chains, targets misfolded proteins for ERAD, and regulates ER stress signaling. The identification of JG-98 as an inhibitor of the Bag3-Hsp72 interaction suggests that targeting protein–protein interactions of co-chaperones with BiP could be a viable alternative strategy for modulating BiP activity in the absence of small molecules that directly and selectively bind to BiP (Li et al. 2013, 2015). The discovery that reversible AMPylation regulates BiP function may likewise yield new opportunities to target BiP activity with selective small molecules (Preissler et al. 2016).

4.2 Targeting ERAD

Quality control mediated by ERAD is another proteostasis node downstream of the UPR where selective modulation offers substantial promise in disease. The process requires coordination of many different proteins in the ER lumen, lipid bilayer, and the cytosol in order to recognize, capture, transport, and degrade aberrant ER client proteins that pose a threat to cell health (Smith et al. 2011a). ERAD can be viewed as a three-step process, involving (1) selection/recognition of misfolded ER clients, (2) retrotranslocation to the cytosol, and (3) proteasomal degradation (Fig. 5). The detailed mechanism involves intricate networks of chaperone–chaperone and chaperone–client interactions, as well as transport machines that we are only beginning to understand in biochemical detail (Christianson et al. 2012; Stein et al. 2014; Timms et al. 2016).

The selection/recognition step of ERAD has thus far proven challenging to target selectively, whereas considerable progress has been made in modulating retrotranslocation and degradation. For example, diverse and potent small molecules are available for proteasome inhibition (Kisselev et al. 2012). Targeting the proteasome has proven therapeutically valuable in multiple myeloma, where treatment with bortezomib or other inhibitors appears to enhance proteotoxic load in the ER,

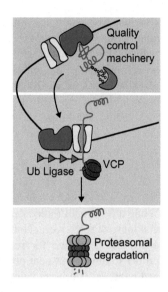

Fig. 5 Advances in understanding and adapting ERAD. Current methods primarily modulate ERAD by influencing transport and/or degradation of misfolded substrates (indicated by *red* and *green boxes*, respectively). Other promising targets for therapeutic intervention include the E3 ligase Hrd1 (*blue box*), which interacts with a broad range of protein substrates via its co-chaperones, and is involved in the selection of terminally misfolded ER clients for ERAD

triggering stress-induced apoptosis (Obeng et al. 2006). However, proteasome inhibition can also be severely toxic for normal cells, and it disrupts protein recycling not just in the ER but also in other subcellular compartments.

The need for more selective modulators of ERAD prompted the search for inhibitors of ER retrotranslocation, and specifically for inhibitors of VCP/p97. VCP/p97 is an ATPase that assists in the extraction of misfolded proteins from the ER for proteasomal degradation, and also participates in other biological processes via its "segregase" activity (Meyer et al. 2012; Rabinovich et al. 2002). A number of VCP inhibitors have been identified via high-throughput screening and optimization efforts, ranging from the promiscuous first-generation compound Eeyarestatin I (Fiebiger et al. 2004; Wang et al. 2008) to more potent second-generation inhibitors like DBeQ (Chou et al. 2011) and the allosteric inhibitor NMS-873 (Magnaghi et al. 2013). Rigorous biophysical characterization of compound selectivity and mechanism of action, including the recent determination of a cryo-EM structure of a VCP-inhibitor complex, will continue to guide inhibitor optimization (Banerjee et al. 2016). Meanwhile, application of existing inhibitors has begun to reveal the potential of VCP inhibition to reduce excessive quality control of misfolding ER client proteins in a manner that is considerably more selective than proteasome inhibition (Han et al. 2015; Wang et al. 2011).

The promising results obtained by perturbing ER proteostasis network nodes downstream of the UPR motivate extensive further investigation in this area.

As highlighted in Fig. 4, we still lack small molecule tools to modulate many of these nodes. The emergence of such tools will open new opportunities for mechanistic studies and disease therapies. In the meantime, Cas9-based methods for transcriptional regulation of individual proteostasis network components in a highly selective, small molecule dose-dependent manner (Maji et al. 2017) should assist the prioritizing of nodes for future method development. The continued advance of high-throughput genetic screening, as recently outlined in two reports that combine CRISPR technology with single cell RNA-seq, also provides opportunities for identifying new targets (Dixit et al. 2016; Adamson et al. 2016).

4.3 Concept Summary

Advances in targeting Hsp90 and ERAD exemplify how much fruitful ground lies beyond IRE1, ATF6, and PERK. Proteomic and high-throughput genetic screening approaches constitute a critical first step in identifying relevant nodes for specific diseases. Where key downstream nodes are known, the next task is to develop tools and modulators of sufficient potency for mechanistic work and drug development. As is the case for Hsp90 and other abundant protein families, researchers must not only identify what isoform is important for a particular process, but also whether and how the relevant isoform can be targeted. Due to high sequence and structural similarities between isoforms, this second question will likely continue to be a significant hurdle to modulating nodes downstream of the UPR. However, recent progress in Grp94 and VCP inhibition offers hope that combined biophysical and biochemical approaches will yield similar advances for isoform-specific targeting of other ER proteostasis network components.

5 Conclusions

As the master regulator of proteostasis in the secretory pathway, the UPR plays critical roles in both health and disease. Various groups have demonstrated UPR participation in organ development and innate immunity; more recently, others have revealed alternate cellular mechanisms for resolving stress, whether in the form of newly discovered extracellular chaperones, membrane expansion, or changes to the oxidative folding environment. The advent of diverse chemical and chemical genetic methods now allows researchers to more precisely control UPR arm activation and inhibition, opening doors to study not only the fundamental biology of the UPR and the secretory pathway, but also the potential of UPR modulation for therapeutic intervention. When selecting from existing tools to modulate or perturb the UPR, researchers must balance specificity with how much engineering or deviation from the natural system is required. The development of improved compounds to target both the UPR and downstream secretory proteostasis network nodes remains an urgent need and, when realized, will surely yield significant new findings. As the field continues to progress, we anticipate not only the elucidation of new functions of the UPR and improved mechanistic understanding, but also the emergence of additional disease-modifying UPR- and ER-targeted therapeutic strategies.

References

Abeliovich A, Gitler AD (2016) Defects in trafficking bridge Parkinson's disease pathology and genetics. Nature 539:207–216

Adachi Y, Yamamoto K, Okada T, Yoshida H, Harada A, Mori K (2008) ATF6 is a transcription factor specializing in the regulation of quality control proteins in the endoplasmic reticulum. Cell Struct Funct 33:75–89

Adamson B et al (2016) A multiplexed single-cell CRISPR screening platform enables systematic dissection of the unfolded protein response. Cell 167:1867–1882

Adolph TE et al (2013) Paneth cells as a site of origin for intestinal inflammation. Nature 503:272–276

Aebi M (2013) N-Linked protein glycosylation in the ER. Biochim Biophys Acta 1833:2430–2437

Back SH, Lee K, Vink E, Kaufman RJ (2006) Cytoplasmic IRE1α-mediated XBP1 mRNA splicing in the absence of nuclear processing and endoplasmic reticulum stress. J Biol Chem 281:18691–18706

Banaszynski LA, Chen LC, Maynard-Smith LA, Ooi AGL, Wandless TJ (2006) A rapid, reversible, and tunable method to regulate protein function in living cells using synthetic small molecules. Cell 126:995–1004

Banerjee S et al (2016) 2.3 Å resolution cryo-EM structure of human p97 and mechanism of allosteric inhibition. Science 351:871–875

Bernales S, McDonald KL, Walter P (2006) Autophagy counterbalances endoplasmic reticulum expansion during the unfolded protein response. PLoS Biol 4:2311–2324

Boyce M et al (2005) A selective inhibitor of eIF2α dephosphorylation protects cells from ER stress. Science 307:935–939

Calfon M, Zeng H, Urano F, Till JH, Hubbard SR, Harding HP, Clark SG, Ron D (2002) IRE1 couples endoplasmic reticulum load to secretory capacity by processing the XBP-1 mRNA. Nature 415:92–96

Carrara M, Prischi F, Nowak PR, Kopp MC, Ali MM (2015) Noncanonical binding of BiP ATPase domain to Ire1 and Perk is dissociated by unfolded protein CH1 to initiate ER stress signaling. eLife 4:e03522

Chen B, Piel WH, Gui LM, Bruford E, Monteiro A (2005) The HSP90 family of genes in the human genome: insights into their divergence and evolution. Genomics 86:627–637

Chen JJ, Genereux JC, Qu S, Hulleman JD, Shoulders MD, Wiseman RL (2014a) ATF6 activation reduces the secretion and extracellular aggregation of destabilized variants of an amyloidogenic protein. Chem Biol 21:1564–1574

Chen X et al (2014b) XBP1 promotes triple-negative breast cancer by controlling the HIF1α pathway. Nature 508:103–107

Chiang WC, Hiramatsu N, Messah C, Kroeger H, Lin JH (2012) Selective activation of ATF6 and PERK endoplasmic reticulum stress signaling pathways prevent mutant rhodopsin accumulation. Invest Ophthalmol Vis Sci 53:7159–7166

Cho JA et al (2013) The unfolded protein response element IRE1α senses bacterial proteins invading the ER to activate RIG-I and innate immune signaling. Cell Host Microbe 13:558–569

Chou TF et al (2011) Reversible inhibitor of p97, DBeQ, impairs both ubiquitin-dependent and autophagic protein clearance pathways. Proc Natl Acad Sci USA 108:4834–4839

Christianson JC et al (2012) Defining human ERAD networks through an integrative mapping strategy. Nat Cell Biol 14:93–105

Coelho DS, Cairrão F, Zeng X, Pires E, Coelho AV, Ron D, Ryoo HD, Domingos PM (2013) XBP1-independent IRE1 signaling is required for photoreceptor differentiation and rhabdomere morphogenesis in Drosophila. Cell Rep 5:791–801

Cooley CB, Ryno LM, Plate L, Morgan GJ, Hulleman JD, Kelly JW, Wiseman RL (2014) Unfolded protein response activation reduces secretion and extracellular aggregation of amyloidogenic immunoglobulin light chain. Proc Natl Acad Sci USA 111:13046–13051

Cox JS, Shamu CE, Walter P (1993) Transcriptional induction of genes encoding endoplasmic reticulum resident proteins requires a transmembrane protein kinase. Cell 73:1197–1206

Crespillo-Casado A, Chambers JE, Fischer PM, Marciniak SJ, Ron D (2017) PPP1R15A-mediated dephosphorylation of eIF2α is unaffected by Sephin1 or Guanabenz. eLife e26109

Cross BCS et al (2012) The molecular basis for selective inhibition of unconventional mRNA splicing by an IRE1-binding small molecule. Proc Natl Acad Sci USA 109:E869–E878

Crowley VM et al (2016) Development of glucose regulated protein 94-selective inhibitors based on the BnIm and radamide scaffold. J Med Chem 59:3471–3488

Dalton RP, Lyons DB, Lomvardas S (2013) Co-opting the unfolded protein response to elicit olfactory receptor feedback. Cell 155:321–332

Das I, Krzyzosiak A, Schneider K, Wrabetz L, D'Antonio M, Barry N, Sigurdardottir A, Bertolotti A (2015) Preventing proteostasis diseases by selective inhibition of a phosphatase regulatory subunit. Science 348:239–242

Denzel MS et al (2014) Hexosamine pathway metabolites enhance protein quality control and prolong life. Cell 156:1167–1178

Dewal MB, DiChiara AS, Antonopoulos A, Taylor RJ, Harmon CJ, Haslam SM, Dell A, Shoulders MD (2015) XBP1s links the unfolded protein response to the molecular architecture of mature N-glycans. Chem Biol 22:1301–1312

Di XJ, Wang YJ, Han DY, Fu YL, Duerfeldt AS, Blagg BS, Mu TW (2016) Grp94 protein delivers γ-aminobutyric acid type A (GABAA) receptors to Hrd1 protein-mediated endoplasmic reticulum-associated degradation. J Biol Chem 291:9526–9539

DiChiara AS, Taylor RJ, Wong MY, Doan ND, Rosario AM, Shoulders MD (2016) Mapping and exploring the collagen-I proteostasis network. ACS Chem Biol 11:1408–1421

Dixit A et al (2016) Perturb-Seq: dissecting molecular circuits with scalable single-cell RNA profiling of pooled genetic screens. Cell 167:1853–1866

Duerfeldt AS et al (2012) Development of a Grp94 inhibitor. J Am Chem Soc 134:9796–9804

Fiebiger E, Hirsch C, Vyas JM, Gordon E, Ploegh HL, Tortorella D (2004) Dissection of the dislocation pathway for type I membrane proteins with a new small molecule inhibitor, eeyarestatin. Mol Biol Cell 15:1635–1646

Fitzgerald J, Lamande SR, Bateman JF (1999) Proteasomal degradation of unassembled mutant type I collagen pro-α1(I) chains. J Biol Chem 274:27392–27398

Forlino A, Marini JC (2016) Osteogenesis imperfecta. Lancet 387:1657–1671

Fu S et al (2011) Aberrant lipid metabolism disrupts calcium homeostasis causing liver endoplasmic reticulum stress in obesity. Nature 473:528–531

Fumagalli F et al (2016) Translocon component Sec62 acts in endoplasmic reticulum turnover during stress recovery. Nat Cell Biol 18:1173–1184

Gallagher CM, Walter P (2016) Ceapins inhibit ATF6α signaling by selectively preventing transport of ATF6α to the Golgi apparatus during ER stress. eLife 5:e11880

Gass JN, Gifford NM, Brewer JW (2002) Activation of an unfolded protein response during differentiation of antibody-secreting B cells. J Biol Chem 277:49047–49054

Genereux JC et al (2015) Unfolded protein response-induced ERdj3 secretion links ER stress to extracellular proteostasis. EMBO J 34:4–19

Gershenson A, Gierasch LM, Pastore A, Radford SE (2014) Energy landscapes of functional proteins are inherently risky. Nat Chem Biol 10:884–891

Ghosh R et al (2014) Allosteric inhibition of the IRE1α RNase preserves cell viability and function during endoplasmic reticulum stress. Cell 158:534–548

Hammarstrom P, Schneider F, Kelly JW (2001) Trans-suppression of misfolding in an amyloid disease. Science 293:2459–2462

Hammarstrom P, Wiseman RL, Powers ET, Kelly JW (2003) Prevention of transthyretin amyloid disease by changing protein misfolding energetics. Science 299:713–716

Han D et al (2009) IRE1α kinase activation modes control alternate endoribonuclease outputs to determine divergent cell fates. Cell 138:562–575

Han DY, Di XJ, Fu YL, Mu TW (2015) Combining valosin-containing protein (VCP) inhibition and suberanilohydroxamic acid (SAHA) treatment additively enhances the folding, trafficking, and function of epilepsy-associated γ-aminobutyric acid, type A $GABA_A$ receptors. J Biol Chem 290:325–337

Han D, Upton JP, Hagen A, Callahan J, Oakes SA, Papa FR (2008) A kinase inhibitor activates the IRE1α RNase to confer cytoprotection against ER stress. Biochem Biophys Res Commun 365:777–783

Harding HP, Novoa I, Zhang Y, Zeng H, Wek R, Schapira M, Ron D (2000) Regulated translation initiation controls stress-induced gene expression in mammalian cells. Mol Cell 6:1099–1108

Harding HP, Zhang Y, Ron D (1999) Protein translation and folding are coupled by an endoplasmic-reticulum-resident kinase. Nature 397:271–274

Hartl FU, Bracher A, Hayer-Hartl M (2011) Molecular chaperones in protein folding and proteostasis. Nature 475:324–332

Hassan IH, Zhang MS, Powers LS, Shao JQ, Baltrusaitis J, Rutkowski DT, Legge K, Monick MM (2012) Influenza A viral replication is blocked by inhibition of the inositol-requiring enzyme 1 (IRE1) stress pathway. J Biol Chem 287:4679–4689

Haze K, Yoshida H, Yanagi H, Yura T, Mori K (1999) Mammalian transcription factor ATF6 is synthesized as a transmembrane protein and activated by proteolysis in response to endoplasmic reticulum stress. Mol Biol Cell 10:3787–3799

Hempstead AD, Isberg RR (2015) Inhibition of host cell translation elongation by *Legionella pneumophila* blocks the host cell unfolded protein response. Proc Natl Acad Sci USA 112: E6790–E6797

Hollien J, Lin JH, Li H, Stevens N, Walter P, Weissman JS (2009) Regulated Ire1-dependent decay of messenger RNAs in mammalian cells. J Cell Biol 186:323–331

Iwamoto M, Björklund T, Lundberg C, Kirik D, Wandless TJ (2010) A general chemical method to regulate protein stability in the mammalian central nervous system. Chem Biol 17:981–988

Jamora C, Dennert G, Lee AS (1996) Inhibition of tumor progression by suppression of stress protein GRP78/BiP induction in fibrosarcoma B/C10ME. Proc Natl Acad Sci USA 93:7690–7694

Johnson SM, Connelly S, Fearns C, Powers ET, Kelly JW (2012) The transthyretin amyloidoses: from delineating the molecular mechanism of aggregation linked to pathology to a regulatory-agency-approved drug. J Mol Biol 421:185–203

Kampinga HH et al (2009) Guidelines for the nomenclature of the human heat shock proteins. Cell Stress Chaperones 14:105–111

Khaminets A et al (2015) Regulation of endoplasmic reticulum turnover by selective autophagy. Nature 522:354–358

Kisselev AF, van der Linden WA, Overkleeft HS (2012) Proteasome inhibitors: an expanding army attacking a unique target. Chem Biol 19:99–115

Kono N, Amin-Wetzel N, Ron D (2017) Generic membrane spanning features endow IRE1α with responsiveness to membrane aberrancy. Mol Biol Cell 28:2318–2332

Laguesse S et al (2015) A dynamic unfolded protein response contributes to the control of cortical neurogenesis. Dev Cell 35:553–567

Lee AH, Iwakoshi NN, Glimcher LH (2003) XBP-1 regulates a subset of endoplasmic reticulum resident chaperone genes in the unfolded protein response. Mol Cell Biol 23:7448–7459

Li X et al (2013) Analogues of the allosteric heat shock protein 70 (Hsp70) inhibitor, MKT-077, as anti-cancer agents. ACS Med Chem Lett 4:1042–1047

Li X et al (2015) Validation of the Hsp70-Bag3 protein-protein interaction as a potential therapeutic target in cancer. Mol Cancer Ther 14:642–648

Lin JH, Li H, Zhang YH, Ron D, Walter P (2009) Divergent effects of PERK and IRE1 signaling on cell viability. PLoS ONE 4:e0004170
Lindholm D, Wootz H, Korhonen L (2006) ER stress and neurodegenerative diseases. Cell Death Differ 13:385–392
Liu S, Street TO (2016) 5′-N-ethylcarboxamidoadenosine is not a paralog-specific Hsp90 inhibitor. Protein Sci 25:2209–2215
Lu PD et al (2004) Cytoprotection by pre-emptive conditional phosphorylation of translation initiation factor 2. EMBO J 23:169–179
Magnaghi P et al (2013) Covalent and allosteric inhibitors of the ATPase VCP/p97 induce cancer cell death. Nat Chem Biol 9:548–556
Maji B, Moore CL, Zetsche B, Volz SE, Zhang F, Shoulders MD, Choudhary A (2017) Multidimensional chemical control of CRISPR-Cas9. Nat Chem Biol 13:9–11
Malhotra JD, Kaufman RJ (2007) Endoplasmic reticulum stress and oxidative stress: a vicious cycle or a double-edged sword? Antiox Redox Signal 9:2277–2293
Margariti A et al (2013) XBP1 mRNA splicing triggers an autophagic response in endothelial cells through BECLIN-1 transcriptional activation. J Biol Chem 288:859–872
Martinon F, Chen X, Lee AH, Glimcher LH (2010) TLR activation of the transcription factor XBP1 regulates innate immune responses in macrophages. Nat Immunol 11:411–418
McCleese JK et al (2009) The novel HSP90 inhibitor STA1474 exhibits biologic activity against osteosarcoma cell lines. Int J Cancer 125:2792–2801
Mendez AS et al (2015) Endoplasmic reticulum stress-independent activation of unfolded protein response kinases by a small molecule ATP-mimic. eLife 4:e05434
Meyer H, Bug M, Bremer S (2012) Emerging functions of the VCP/p97 AAA-ATPase in the ubiquitin system. Nat Cell Biol 14:117–123
Mimura N et al (2012) Blockade of XBP1 splicing by inhibition of IRE1 α is a promising therapeutic option in multiple myeloma. Blood 119:5772–5781
Mirigian LS, Makareeva E, Mertz EL, Omari S, Roberts-Pilgrim AM, Oestreich AK, Phillips CL, Leikin S (2016) Osteoblast malfunction caused by cell stress response to procollagen misfolding in α2(I)-G610C mouse model of osteogenesis imperfecta. J Bone Miner Res 31:1608–1616
Moore CL, Dewal MB, Nekongo EE, Santiago S, Lu NB, Levine SS, Shoulders MD (2016) Transportable, chemical genetic methodology for the small molecule-mediated inhibition of heat shock factor 1. ACS Chem Biol 11:200–210
Moore K, Hollien J (2015) IRE1-mediated decay in mammalian cells relies on mRNA sequence, structure, and translational status. Mol Biol Cell 26:2873–2884
Mu TW, Fowler DM, Kelly JW (2008a) Partial restoration of mutant enzyme homeostasis in three distinct lysosomal storage disease cell lines by altering calcium homeostasis. PLoS Biol 6:e26
Mu TW, Ong DS, Wang YJ, Balch WE, Yates JR III, Segatori L, Kelly JW (2008b) Chemical and biological approaches synergize to ameliorate protein-folding diseases. Cell 134:769–781
Obeng EA, Carlson LM, Gutman DM, Harrington WJ Jr, Lee KP, Boise LH (2006) Proteasome inhibitors induce a terminal unfolded protein response in multiple myeloma cells. Blood 107:4907–4916
Okada T, Yoshida H, Akazawa R, Negishi M, Mori K (2002) Distinct roles of activating transcription factor 6 (ATF6) and double-stranded RNA-activated protein kinase-like endoplasmic reticulum kinase (PERK) in transcription during the mammalian unfolded protein response. Biochem J 366:585–594
Ong DS, Mu TW, Palmer AE, Kelly JW (2010) Endoplasmic reticulum Ca^{2+} increases enhance mutant glucocerebrosidase proteostasis. Nat Chem Biol 6:424–432
Ong DS, Wang YJ, Tan YL, Yates JR III, Mu TW, Kelly JW (2013) FKBP10 depletion enhances glucocerebrosidase proteostasis in Gaucher disease fibroblasts. Chem Biol 20:403–415
Papa FR, Zhang C, Shokat K, Walter P (2003) Bypassing a kinase activity with an ATP-competitive drug. Science 302:1533–1537
Patel PD et al (2013) Paralog-selective Hsp90 inhibitors define tumor-specific regulation of HER2. Nat Chem Biol 9:677–684

Pincus D, Chevalier MW, Aragon T, van Anken E, Vidal SE, El-Samad H, Walter P (2010) BiP binding to the ER-stress sensor Ire1 tunes the homeostatic behavior of the unfolded protein response. PLoS Biol 8:e1000415

Plate L et al (2016) Small molecule proteostasis regulators that reprogram the ER to reduce extracellular protein aggregation. eLife 5:e15550

Preissler S, Rato C, Perera LA, Saudek V, Ron D (2016) FICD acts bi-functionally to AMPylate and de-AMPylate the endoplasmic reticulum chaperone BiP. Nat Struct Mol Biol 24:23–29

Rabinovich E, Kerem A, Frohlich KU, Diamant N, Bar-Nun S (2002) AAA-ATPase p97/Cdc48p, a cytosolic chaperone required for endoplasmic reticulum-associated protein degradation. Mol Cell Biol 22:626–634

Raina K, Noblin DJ, Serebrenik YV, Adams A, Zhao C, Crews CM (2014) Targeted protein destabilization reveals an estrogen-mediated ER stress response. Nat Chem Biol 10:957–962

Reimold AM et al (2001) Plasma cell differentiation requires the transcription factor XBP-1. Nature 412:300–307

Richardson CE, Kooistra T, Kim DH (2010) An essential role for XBP-1 in host protection against immune activation in *C. elegans*. Nature 463:1092–1095

Ron I, Horowitz M (2005) ER retention and degradation as the molecular basis underlying Gaucher disease heterogeneity. Hum Mol Genet 14:2387–2398

Roy CR, Salcedo SP, Gorvel JP (2006) Pathogen-endoplasmic-reticulum interactions: in through the out door. Nat Rev Immunol 6:136–147

Ruiz-Canada C, Kelleher DJ, Gilmore R (2009) Cotranslational and posttranslational N-glycosylation of polypeptides by distinct mammalian OST isoforms. Cell 136:272–283

Rutkowski DT et al (2006) Adaptation to ER stress is mediated by differential stabilities of pro-survival and pro-apoptotic mRNAs and proteins. PLoS Biol 4:e374

Ryoo HD, Domingos PM, Kang MJ, Steller H (2007) Unfolded protein response in a Drosophila model for retinal degeneration. EMBO J 26:242–252

Schuck S, Prinz WA, Thorn KS, Voss C, Walter P (2009) Membrane expansion alleviates endoplasmic reticulum stress independently of the unfolded protein response. J Cell Biol 187:525–536

Schulte TW, Neckers LM (1998) The benzoquinone ansamycin 17-allylamino-17-demethoxygeldanamycin binds to HSP90 and shares important biologic activities with geldanamycin. Cancer Chemother Pharmocal 42:273–279

Shaffer AL et al (2004) XBP1, downstream of Blimp-1, expands the secretory apparatus and other organelles, and increases protein synthesis in plasma cell differentiation. Immunity 21:81–93

Shen X et al (2001) Complementary signaling pathways regulate the unfolded protein response and are required for *C. elegans* development. Cell 107:893–903

Shi J et al (2012) EC144 is a potent inhibitor of the heat shock protein 90. J Med Chem 55:7786–7795

Shoulders MD, Ryno LM, Cooley CB, Kelly JW, Wiseman RL (2013a) Broadly applicable methodology for the rapid and dosable small molecule-mediated regulation of transcription factors in human cells. J Am Chem Soc 135:8129–8132

Shoulders MD et al (2013b) Stress-independent activation of XBP1s and/or ATF6 reveals three functionally diverse ER proteostasis environments. Cell Rep 3:1279–1292

Sidrauski C, McGeachy AM, Ingolia NT, Walter P (2015) The small molecule ISRIB reverses the effects of eIF2α phosphorylation on translation and stress granule assembly. eLife 4:e05033

Smith MH, Ploegh HL, Weissman JS (2011a) Road to ruin: targeting proteins for degradation in the endoplasmic reticulum. Science 334:1086–1090

Smith SE, Granell S, Salcedo-Sicilia L, Baldini G, Egea G, Teckman JH, Baldini G (2011b) Activating transcription factor 6 limits intracellular accumulation of mutant α1-antitrypsin Z and mitochondrial damage in hepatoma cells. J Biol Chem 286:41563–41577

Soldano KL, Jivan A, Nicchitta CV, Gewirth DT (2003) Structure of the N-terminal domain of GRP94—basis for ligand specificity and regulation. J Biol Chem 279:48330–48338

Stein A, Ruggiano A, Carvalho P, Rapoport TA (2014) Key steps in ERAD of luminal ER proteins reconstituted with purified components. Cell 158:1375–1388

Tan YL, Genereux JC, Pankow S, Aerts JM, Yates JR 3rd, Kelly JW (2014) ERdj3 is an endoplasmic reticulum degradation factor for mutant glucocerebrosidase variants linked to Gaucher's disease. Chem Biol 21:967–976

Taylor RC, Dillin A (2013) XBP-1 is a cell-nonautonomous regulator of stress resistance and longevity. Cell 153:1435–1447

Timms RT et al (2016) Genetic dissection of mammalian ERAD through comparative haploid and CRISPR forward genetic screens. Nat Commun 7:11786

Tirasophon W, Welihinda AA, Kaufman RJ (1998) A stress response pathway from the endoplasmic reticulum to the nucleus requires a novel bifunctional protein kinase/endoribonuclease (Ire1p) in mammalian cells. Genes Dev 12:1812–1824

Treacy-Abarca S, Mukherjee S (2015) Legionella suppresses the host unfolded protein response via multiple mechanisms. Nat Commun 6:7887

Tsaytler P, Harding HP, Ron D, Bertolotti A (2011) Selective inhibition of a regulatory subunit of protein phosphatase 1 restores proteostasis. Science 332:91–94

Tu BP, Weissman JS (2004) Oxidative protein folding in eukaryotes: mechanisms and consequences. J Cell Biol 164:341–346

Vattem KM, Wek RC (2004) Reinitiation involving upstream ORFs regulates *ATF4* mRNA translation in mammalian cells. Proc Natl Acad Sci USA 101:11269–11274

Volkmann K et al (2011) Potent and selective inhibitors of the inositol-requiring enzyme 1 endoribonuclease. J Biol Chem 286:12743–12755

Volmer R, Ron D (2015) Lipid-dependent regulation of the unfolded protein response. Curr Opin Cell Biol 33:67–73

Walter P, Ron D (2011) The unfolded protein response: from stress pathway to homeostatic regulation. Science 334:1081–1086

Wang F, Song WS, Brancati G, Segatori L (2011) Inhibition of endoplasmic reticulum-associated degradation rescues native folding in loss of function protein misfolding diseases. J Biol Chem 286:43454–43464

Wang L et al (2012) Divergent allosteric control of the IRE1α endoribonuclease using kinase inhibitors. Nat Chem Biol 8:982–989

Wang QY, Li LY, Ye YH (2008) Inhibition of p97-dependent protein degradation by eeyarestatin I. J Biol Chem 283:7445–7454

Wang ZV et al (2014) Spliced X-box binding protein 1 couples the unfolded protein response to hexosamine biosynthetic pathway. Cell 156:1179–1192

Whitesell L, Mimnaugh EG, De Costa B, Myers CE, Neckers LM (1994) Inhibition of heat shock protein HSP90-pp60^{v-src} heteroprotein complex formation by benzoquinone ansamycins: essential role for stress proteins in oncogenic transformation. Proc Natl Acad Sci USA 91:8324–8328

Wu J et al (2007) ATF6α optimizes long-term endoplasmic reticulum function to protect cells from chronic stress. Dev Cell 13:351–364

Yoshida H, Matsui T, Yamamoto A, Okada T, Mori K (2001) XBP1 mRNA is induced by ATF6 and spliced by IRE1 in response to ER stress to produce a highly active transcription factor. Cell 107:881–891

Zhang P, Su C, Jiang Z, Zheng C (2017) Herpes simplex virus 1 UL41 protein suppresses the IRE1/XBP1 signal pathway of the unfolded protein response via its RNase activity. J Virol 91: e02056

Cell Non-autonomous UPR^{ER} Signaling

Soudabeh Imanikia, Ming Sheng and Rebecca C. Taylor

Abstract The UPR^{ER} is an important regulator of secretory pathway homeostasis, and plays roles in many physiological processes. Its broad range of targets and ability to modulate secretion and membrane trafficking make it perfectly positioned to influence intercellular communication, enabling the UPR^{ER} to coordinate physiological processes between cells and tissues. Recent evidence suggests that the activation of the UPR^{ER} can itself be communicated between cells. This cell non-autonomous route to UPR^{ER} activation occurs in multiple species, and enables organism-wide responses to stress that involve processes as diverse as immunity, metabolism, aging and reproduction. It may also play roles in disease progression, making the pathways that mediate cell non-autonomous UPR^{ER} signaling a potential source of novel future therapeutics.

Contents

1 Introduction	28
1.1 Signaling in the UPR^{ER}	28
1.2 Activation of the UPR^{ER}	29
2 *Caenorhabditis elegans* (*C. elegans*)	30
2.1 Innate Immunity	31
2.2 Aging and Reproduction	32
3 *Drosophila melanogaster* (*D. melanogaster*)	33
3.1 Barrier Epithelia	34
4 Mammalian Systems	35
4.1 Metabolic Regulation	35
4.2 Immune Responses	38
4.3 Tumorigenesis	39
5 Conclusions	40
References	41

S. Imanikia · M. Sheng · R.C. Taylor (✉)
MRC Laboratory of Molecular Biology, Cambridge CB2 0QH, UK
e-mail: rtaylor@mrc-lmb.cam.ac.uk

1 Introduction

Homeostatic regulation is critical to the functioning of living organisms. The ability to monitor biological variables and return them to physiologically acceptable levels upon perturbation allows complex and interconnected systems to operate successfully. One means by which organisms maintain homeostasis is through cellular stress responses, signaling systems that detect departures from homeostasis, or the presence of stressors that might cause disequilibrium, and enact downstream mechanisms that enable organisms to return to homeostasis, or to adapt to a new set of environmental norms.

Stress responses monitor specific subcellular compartments for homeostatic disturbances, and one key element of the cell's stress response arsenal is the endoplasmic reticulum (ER) unfolded protein response (UPR^{ER}), responsible for monitoring homeostasis within the secretory compartment. Given the extent of protein folding within the secretory pathway—as many as one-third of a cell's proteins fold within the ER—as well as the organelle's roles in lipid homeostasis, signaling, and its connections with other organelles, maintaining secretory pathway homeostasis is particularly critical to the cell.

1.1 Signaling in the UPR^{ER}

To maintain homeostasis in the secretory system, three major pathways monitor and respond to homeostatic perturbations within the ER, including protein homeostasis (proteostasis) and lipid disequilibrium (Walter and Ron 2011). Each pathway comprises an upstream molecule within the ER membrane that monitors ER homeostasis, and a downstream molecule or molecules that mediate transcriptional or translational changes to return ER homeostasis to its previous equilibrium, or to induce changes that allow the cell to adapt to new conditions.

One of these pathways utilizes the membrane-localized kinase/endoribonuclease inositol-requiring protein-1 (IRE1), which splices a specific intron from the mRNA of the transcription factor X-box binding protein-1 (XBP1) to allow the translation of the active transcription factor. XBP1 then induces a wide range of transcriptional targets, including genes involved in protein degradation, protein folding, and lipid metabolism (Calfon et al. 2002; Cox and Walter 1996; Sidrauski and Walter 1997). IRE1 is also able to degrade ER-localized mRNAs in a process called regulated IRE1-dependent decay (RIDD), reducing the load of proteins and therefore the protein folding burden entering the ER (Hollien et al. 2009). In addition, IRE1 is able to activate other signaling pathways, including the c-Jun N-terminal kinase (JNK) pathway, to mediate further transcriptional changes (Urano et al. 2000).

A second UPR^{ER} pathway relies upon the membrane-localized kinase protein kinase R (PKR)-like endoplasmic reticulum kinase (PERK), which phosphorylates the translational regulator elongation initiation factor-2α (eIF2α) (Harding et al. 1999).

eIF2α phosphorylation leads to a reduction in protein translation, while also allowing preferential translation of specific transcripts encoding proteins that alleviate stress, changing the ratio of newly synthesized proteins to homeostasis effectors in order to encourage productive protein folding (Harding et al. 2000a, b). Some of the proteins upregulated subsequent to preferential translation promote programmed cell death, allowing cells to undergo apoptosis in the face of unresolvable stress (Lin et al. 2007; Marciniak et al. 2004).

The third of the canonical UPR^{ER} pathways utilizes the transcription factor activating transcription factor-6 (ATF6) as both sensor and effector (Haze et al. 1999). ATF6 is positioned within the ER membrane; upon detection of misfolded proteins within the ER lumen, ATF6 is translocated to the Golgi apparatus, where it is cleaved by the Site 1 and Site 2 proteases. This releases the active transcription factor, which goes on to upregulate targets involved in proteostasis.

1.2 Activation of the UPR^{ER}

The activation of the different branches of the UPR^{ER} can occur as a cell-autonomous event, with cell-autonomous consequences for the maintenance of homeostasis—that is, the upstream sensor components are directly activated by the presence of disequilibrium within the ER of a cell, and genes are induced that enable that cell to reestablish homeostasis, or adjust to new environmental conditions (Ron and Walter 2007). This disequilibrium can be caused by the presence of misfolded proteins, changes in lipid levels, or alterations in cellular Ca^{2+} levels within the secretory pathway. Misfolded proteins accumulating to levels beyond the physiologically acceptable homeostatic range are detected by the luminal domains of IRE1, PERK and ATF6 through the chaperone binding-immunoglobulin protein (BiP): BiP, normally present in excess within the ER, binds and inhibits these domains under conditions of equilibrium. When misfolded proteins accumulate, free BiP is titrated away from the UPR^{ER} sensor molecules and is sequestered among these misfolded species, leaving IRE1, PERK and ATF6 free to oligomerize and become active (Bertolotti et al. 2000). This mechanism renders the UPR^{ER} extremely sensitive to the environmental conditions experienced by a cell.

However, the maintenance of homeostasis works most efficiently when regulated at multiple levels—within and between the tissues of an organism, as well as within individual cells. Given that the vast majority of molecules involved in intercellular communication traffic through the secretory pathway, to gain access to other cells through localization in the plasma membrane or secretion to the extracellular environment, the UPR is perfectly placed to influence intercellular signaling. Due to this position at the intersection between the sensing of stress and the modulation of secretory pathway trafficking, the UPR^{ER} has been co-opted to influence multiple organism-wide physiological processes, allowing it to connect the detection of changes in environmental conditions with the regulation of systems that must be coordinated throughout the organism.

UPR^{ER} activation can affect a variety of different cellular processes, due to the range and breadth of its downstream effector genes (Acosta-Alvear et al. 2007). This makes UPR^{ER} activation itself an attractive target for long-range regulation, in order to coordinate homeostasis between different cells and tissues. Correspondingly, in addition to its ability to regulate the activation of other signaling pathways, an accumulating body of evidence has suggested that the UPR^{ER} is also able to initiate its own activation in distal cells, which have not themselves been exposed to stress. This ability to cell non-autonomously activate the UPR^{ER}, coordinating stress response activation between cells and tissues through the release of intercellular signaling mediators, has only been appreciated relatively recently (Taylor et al. 2014). However, its importance is clear—these cell non-autonomous routes to stress response activation allow anticipatory adaptation, enabling cells to prepare for environmental changes without first undergoing damage, and also allow coordination of secretory pathway homeostasis between cells that must work together to function as tissue- or organism-wide systems.

The ability of the UPR^{ER} to coordinate organism-wide physiology and upregulate stress resistance cell non-autonomously has been demonstrated in multiple species, with implications for how these organisms regulate homeostasis and coordinate processes as diverse as immunity, glucose homeostasis, and aging. In this review, we will survey the roles that cell non-autonomous UPR^{ER} signaling plays in coordinating physiology and the mechanisms by which these are achieved, and finally discuss aspects and implications of these signaling processes that are yet to be understood.

2 Caenorhabditis elegans (C. elegans)

The nematode *C. elegans* organizes its 959 cells (in the hermaphrodite) into several discrete tissues. 302 of these cells comprise the animal's nervous system, enabling the worm to respond to its environment in a surprisingly diverse array of ways. As well as regulating behavior, the *C. elegans* nervous system can directly influence physiology and cell fate decisions in other tissues. Recent evidence has suggested that one means by which *C. elegans* connects its physiological state to environmental conditions is through the use of stress responses in the nervous system to control the activation of signaling pathways, including the stress responses themselves, in other tissues (Taylor et al. 2014). These mechanisms enable worms to upregulate homeostatic mechanisms and control elements of their metabolism and life cycle in response to environmental changes.

In *C. elegans*, as in other metazoans, three UPR^{ER}-regulating proteins are located at the ER membrane—IRE-1, the PERK homolog PEK-1, and ATF-6—and one key means by which the nervous system coordinates organism-wide processes is through the activation of the UPR^{ER} in neurons to regulate communication between neurons and other tissues. Studies in this organism are increasing our understanding of the ways in which UPR^{ER} communication is utilized to influence

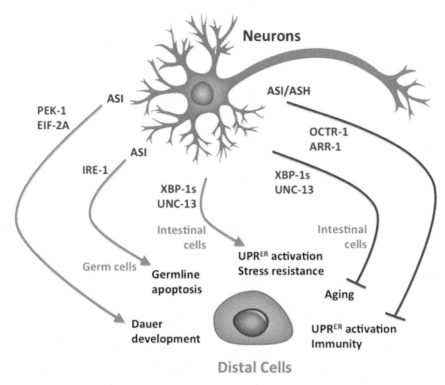

Fig. 1 Processes coordinated by neuronal UPR^{ER} activation in *C. elegans*. Activation of the PERK branch of the UPR^{ER} in ASI neurons leads to the coordination of dauer development throughout the organism. Active IRE-1 in ASI neurons can trigger apoptosis in germ cells. Splicing of XBP-1 in neurons causes the UNC-13-mediated release of secreted signals that activate the UPR^{ER} in intestinal cells, leading to increased stress resistance and slowed aging. OCTR-1 and ARR-1 function in ASI and ASH neurons to inhibit UPR^{ER} activation and immunity in other tissues

physiological decisions and responses that include innate immunity, dauer formation and longevity (Fig. 1).

2.1 Innate Immunity

Innate immunity in *C. elegans* requires UPR^{ER} activation in intestinal cells, to enable the secretion of anti-microbial effectors (Richardson et al. 2010). The detection of pathogens by sensory neurons may regulate the activation of the UPR^{ER} throughout the organism, promoting preemptive ER stress response upregulation. This neuronal coordination of the immune response depends, at least in part, upon the neuronal G protein-coupled catecholamine receptor (GPCR) octopamine receptor-1 (OCTR-1) and the adapter protein arrestin-1 (ARR-1) (Singh and Aballay 2012; Sun et al. 2011). OCTR-1 functions in sensory neurons to actively suppress innate immune responses in other tissues, downregulating the expression of both canonical IRE-1/XBP-1-regulated and non-canonical UPR^{ER} gene targets in non-neuronal cells (Sun et al. 2011, 2012). Upon mutation of *octr-1*, these genes are constitutively upregulated, and animals survive for longer when exposed to pathogens. The OCTR-1-expressing ASH and ASI sensory neurons play a critical role in the process, suggesting a neuronal system to suppress UPR^{ER} activation until environmental circumstances demand it, tuning UPR^{ER} activation to external conditions that are sensed by specific neurons.

2.2 Aging and Reproduction

Activation of the UPR^{ER} within the nervous system can also directly induce UPR^{ER} activation in other tissues. Neuronally derived XBP-1s is sufficient to activate the UPR^{ER} in the *C. elegans* intestine, a tissue in which proteostasis is critical to health and longevity, through a cell non-autonomous mechanism that involves neuronal secretion via the conserved synaptic secretion regulator uncoordinated-13 (UNC-13) (Taylor and Dillin 2013). This communication rescues the loss of ER stress resistance associated with age, and increases lifespan, suggesting that the nervous system utilizes the UPR^{ER} to connect the perception of stress with organism-wide decisions regarding resource allocation. Encouraging long-term survival over rapid reproduction when adverse conditions are detected may allow the organism to survive long enough to breed once conditions have improved and the survival of offspring is more likely.

In keeping with this idea, activation of the UPR^{ER} in neurons can also influence the health of the germline, connecting perception of the environment with reproduction. Activation of IRE-1 in sensory neurons—specifically the ASI neurons—can cell non-autonomously regulate germ cell apoptosis, dependent on the same molecular machinery that mediates germ cell death after DNA damage (Levi-Ferber et al. 2014). This indicates that neurons have the ability to communicate with germ cells to promote their death in response to stress in the ER of neuronal cells. The downstream IRE-1 target XBP-1 is not required for this process, nor is the downstream c-Jun N-terminal kinase (JNK) signaling pathway, and neuronal system dysfunction does not lead to changes in germ cell apoptosis. This likely indicates a role for a novel form of IRE-1-mediated communication. Again, these findings suggest that the sensing of stress may trigger UPR^{ER} signaling that leads to a reallocation of resources towards somatic cell maintenance and away from reproduction, prolonging longevity, and delaying reproduction until conditions improve.

Another environment-dependent life cycle decision made by *C. elegans* is whether to continue through larval development into reproductive adulthood, or whether instead to enter the dauer diapause. Dauer is a stress-resistant and long-lived alternative developmental stage, triggered in response to adverse environmental conditions that include starvation and overcrowding. Entry into the dauer stage leads to organism-wide developmental and metabolic changes that increase longevity and stress resistance until environmental conditions improve sufficiently to allow development into adulthood. Again, ER stress in the ASI sensory neurons appears to play a role in coordinating this decision, as UPR^{ER} activation in these neurons promotes dauer formation (Kulalert and Kim 2013). This communication is dependent upon PERK-mediated phosphorylation of eIF2α, and the expression of a phospho-eIF2α mimetic in the ASI neurons is sufficient to trigger entry into dauer. This suggests that branches of the UPR^{ER} other than that controlled by IRE-1 are also involved in cell non-autonomous communication, and confirms the role of the UPR^{ER} as a systemic regulator of organism-wide physiological decisions concerning aging and reproduction in *C. elegans*.

3 *Drosophila melanogaster (D. melanogaster)*

In comparison to *C. elegans*, less is known about the roles of the UPR^{ER} in intercellular communication in another widely studied invertebrate model organism, the fruit fly *D. melanogaster*. However, the UPR^{ER} is now emerging as a key

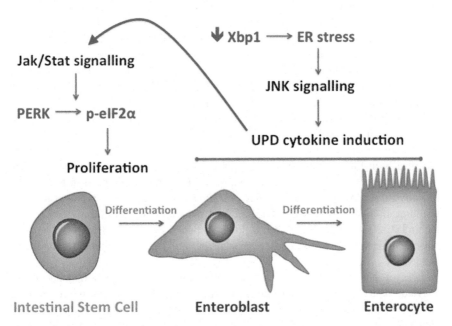

Fig. 2 Cell non-autonomous UPR^{ER} activation in the intestinal epithelium of *Drosophila melanogaster*. ER stress in enteroblasts and enterocytes leads to the activation of JNK signaling and the production of UPD cytokines, which trigger the downstream activation of Jak/Stat signaling in intestinal stem cells (ISCs). This in turn leads to the phosphorylation of eIF2α by PERK, and subsequent proliferation of ISCs

regulator of tissue homeostasis in barrier epithelia in this organism, with cell non-autonomous regulation of UPR^{ER} activation playing a central role (Fig. 2).

3.1 Barrier Epithelia

For barrier epithelia to function correctly, precise control of cell proliferation is required. This must also be balanced with responses to stress and inflammation. With age, over-proliferation of the intestinal epithelium becomes widespread; reducing (but not totally inhibiting) this proliferation can increase lifespan (Biteau et al. 2008, 2010). The transcription factor nuclear factor (erythroid-derived 2)-like 2 (Nrf2), a master regulator of redox state, specifically controls the proliferation of intestinal stem cells (ISCs) (Hochmuth et al. 2011). The UPR^{ER} also plays a central role in the regulation of regeneration in the intestinal epithelium, through the control of intestinal stem cell proliferation: increased ER stress across the intestinal epithelium is associated with age-related dysplasia mediated by redox signaling through JNK (Heazlewood et al. 2008; Wang et al. 2014). Indeed, improving ER homeostasis in aging ISCs is sufficient to limit age-associated epithelial dysplasia and extend lifespan. The ER-stress responsive transcription factor Xbp1s and the ER-associated degradation pathway component ER-associated degradation (ERAD)-associated E3 ubiquitin-protein ligase Hrd1 act in ISCs to limit their proliferative activity. However, the activation of PERK and downstream phosphorylation of eIF2a plays an opposing role, encouraging the proliferation of these cells, and this is believed to play a major role in age-related dysplasia. Knocking down PERK specifically in ISCs improves homeostasis and barrier function, and extends lifespan by reducing pro-mitotic signals to ISCs (Wang et al. 2015).

Recent studies have shown that epithelial regeneration by *Drosophila* intestinal stem cells can be regulated by PERK cell non-autonomously (Wang et al. 2015). Within the intestinal epithelium, PERK is activated and eIF2α is phosphorylated specifically in ISCs, even when ER stress is induced in entirely different cells, the enteroblasts (EBs) or enterocytes (ECs). When ER stress is induced in the EBs or ECs, by knockdown of Xbp1, PERK is activated in ISCs cell non-autonomously, resulting in phosphorylation of eIF2α and activation of proliferative processes within ISCs. This effect is not seen following knockdown of Xbp1 in the fat body or muscle, indicating that this is an event occurring only between cells of the intestinal epithelium.

Loss of Xbp1 in ECs activates the Janus kinase (Jak)/signal transducer and activator of transcription (Stat) signaling pathway in ISCs, which is required for eIF2α phosphorylation and proliferation in these cells: knockdown of Jak/Stat pathway components in ISCs prevents cell non-autonomously-induced eIF2α phosphorylation (Wang et al. 2015). Knockdown of Xbp1 in EBs also requires the *Drosophila* JNK Basket (Bsk) and the cytokine Unpaired-3 (Upd3) for its non-autonomous effects on ISCs. Therefore, ER stress-induced JNK-mediated induction of Upds—inflammatory cytokines—in EBs leads to the activation of

PERK via JAK/Stat signaling in ISCs, triggering ISC proliferation. This therefore indicates a possible mechanism for cell non-autonomous UPR^{ER} communication in *Drosophila*.

These results demonstrate the importance of intercellular transmission of UPR^{ER} activation in the maintenance of homeostasis within a tissue. Again, this illustrates the emerging theme that UPR^{ER} communication can play a major role in the coordination of tissue-wide processes, and in the determination of age-associated health and lifespan.

4 Mammalian Systems

In mammals, UPR^{ER} activation enables organism-wide coordination of several physiological systems, including glucose and lipid metabolism, and immunity (Fig. 3a). However, this ability to regulate intercellular communication may also be co-opted by cancer cells in ways that increase their survival, making the UPR^{ER} a potential target for cancer therapies, as well as treatments for metabolic and immune diseases (Fig. 3b).

4.1 Metabolic Regulation

The UPR^{ER} modulates many facets of mammalian metabolism. It is of major importance in the regulation of systemic glucose homeostasis, through its roles in the differentiation of pancreatic β cells and in secretory adaptation to allow insulin production (Volchuk and Ron 2010). However, these requirements for the UPR^{ER} in insulin signaling mean that its stress-induced activation can affect global metabolism, leading to systemic metabolic disease. High cellular lipid levels associated with obesity, for example, can lead to the onset of local insulin resistance, a disruption of systemic glucose metabolism, and the development of type II diabetes (Hotamisligil 2010; Ozcan et al. 2004; Volchuk and Ron 2010). Stress-induced UPR^{ER} activation can also disrupt other metabolic hormone signaling systems—mice fed on high fat diets activate the UPR^{ER} in the hypothalamus, which interferes with leptin signaling, leading to leptin resistance in peripheral tissues, and increased obesity (Ozcan et al. 2009). The liver, a major metabolic organ, seems particularly susceptible to UPR^{ER} activation in models of metabolic disease: BiP is present at significantly higher levels in the liver of obese and diabetic mice compared to control animals, while JNK is activated and both PERK and eIF2α are highly phosphorylated in the hepatic tissue of obese mice (Nakatani et al. 2005; Ozcan et al. 2004).

Relevantly, recent evidence suggests that activation of the UPR^{ER} in the hypothalamus can trigger cell non-autonomous induction of the UPR^{ER} in the liver. Expression of the constitutively active form of Xbp1, Xbp1s, in pro-opiomelanocortin

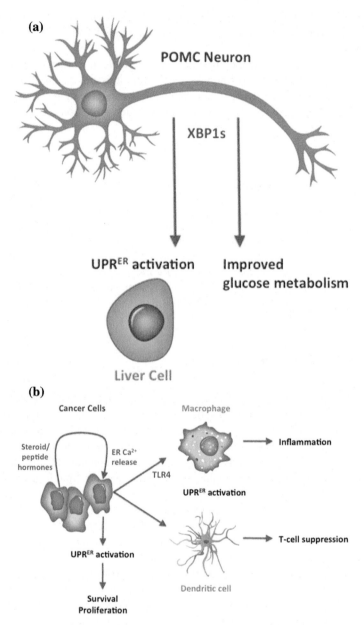

Fig. 3 Intercellular UPR^{ER} activation in mammals. **a** Expression of spliced XBP1s in POMC neurons of the murine hypothalamus leads to the activation of the UPR^{ER} in liver cells, and to improved systemic glucose metabolism. **b** UPR^{ER} activation in cancer cells can lead to the anticipatory upregulation of the UPR^{ER} in neighboring cancer cells, through the release of steroid or peptide hormones that cause PLCγ-induced release of Ca^{2+} from the ER to the cytoplasm in receiving cells. This triggers UPR^{ER} activation, leading to increased survival and proliferation. In addition, cancer cells can trigger UPR^{ER} activation in macrophages and dendritic cells through the release of secreted signals, received in macrophages by TLR4, which lead to pro-tumorigenic inflammatory and T-cell-suppressive responses

(POMC) neurons of the murine hypothalamus, results in an increase in the levels of Xbp1s targets in hepatic cells (Williams et al. 2014)(Fig. 3a). This communication of UPR^{ER} activation between neurons and the liver leads to higher energy expenditure and resistance to diet-induced obesity, regardless of changes in food intake. In addition, glucose circulation is decreased, leading to better glucose regulation. Confirming the capacity for cell non-autonomous UPR^{ER} activation in liver cells, injection of conditioned medium from ER-stressed tumor cells into wild-type mice leads to a general UPR^{ER} induction in the liver (Mahadevan et al. 2011). This also implies that UPR^{ER}-activating signals can be released directly from stressed cells into the extracellular environment, to induce UPR^{ER} activation in distal cells. This communication of UPR^{ER} activation may play important roles in the global regulation of metabolism, and also in the development of metabolic disease following UPR^{ER} activation in the brain.

No signaling mediator for this cell non-autonomous UPR^{ER} activation has yet been identified. One candidate may be the fibroblast growth factor (FGF) family member, FGF21. FGF21 is a regulator of metabolism that has broad effects on multiple metabolic pathways (Fisher and Maratos-Flier 2016). It has a hormone-like cytokine role in several tissues that enable it to regulate carbohydrate and lipid metabolism, and it is known to stimulate glucose uptake independently of insulin, as well as facilitating cells in dealing with hyperglycemia and dyslipidemia. Furthermore, FGF21 has roles in protecting and preserving pancreatic β cell function and inducing insulin sensitivity, through hepatic and peripheral mechanisms, as well as in resisting diet-induced obesity and promoting fatty acid oxidation (Arner et al. 2008; Kharitonenkov et al. 2005; Wente et al. 2006; Xu et al. 2009).

FGF21 seems to connect cellular stress to systemic metabolic effects—in both human and animal studies, significantly higher levels of circulating FGF21 have been reported when stress responses are activated, including in individuals suffering metabolic stress (such as obesity or type 2 diabetes). ER stress can directly regulate FGF21 expression, facilitating its global metabolic effects (Schaap et al. 2013; Wan et al. 2014). Upregulation of FGF21 is mediated by activating transcription factor-4 (ATF4), an ER stress transducer that functions downstream of PERK activation. C/EBP homologous protein (CHOP), a transcription factor that is upregulated by ATF4, induces FGF21 expression by activating transcription through promoter elements and enhancing mRNA stability under ER stress (Wan et al. 2014). ATF4-mediated FGF21 expression has beneficial effects in models of stress—in mice with a deficiency in autophagy-related protein-7 (ATG7) fed on a high fat diet, elevated FGF21 levels improve glucose metabolism and reduce obesity (Kim et al. 2013). Could FGF21 therefore be a key mediator of systemic UPR^{ER} activation, helping to explain the importance of the FGF21 signaling pathway in metabolic diseases?

4.2 Immune Responses

The immune response depends heavily upon intercellular signaling. It operates by recognizing danger or damage to cells, and enabling the body to produce an appropriate inflammatory response. The UPR^{ER} plays an important and complex role in promoting inflammation through its ability to trigger the release of a variety of proinflammatory molecules from cells undergoing ER stress, including tumor necrosis factor (TNF), granulocyte macrophage colony-stimulating factor (GM-CSF), and interleukins (ILs). These inflammatory mediators are induced by the UPR^{ER}-mediated activation of transcription factors that include nuclear factor-κB (NF-κB) and activator protein 1 (AP-1)—processes that have been extensively reviewed elsewhere (Garg et al. 2012a). The release of these proinflammatory factors then contributes to the activation of cells of the immune system, and the mounting of an inflammatory response, which may contribute to the resolution of infection, or to pathogenesis, depending upon context.

The UPR^{ER} is also a central regulator of cascades that produce immunogenic cell death (ICD) signals—molecular indicators produced by dying cells to alert the immune system that a defensive response is needed. UPR^{ER} activation promotes ICD, and ER stress leads to the release of both surface calreticulin (CRT) and secreted adenosine triphosphate (ATP), signals that activate receptors on the surface of immune cells to trigger an inflammatory response (Garg et al. 2012b). The UPR-dependent release of these signals requires secretory pathway function and phosphoinositol-3 kinase (PI3K)-dependent plasma membrane/extracellular trafficking. PERK activation appears to be involved in all cases of ICD. However, different pathways have been proposed downstream of PERK activation. In some cases, surface exposure of CRT requires phosphorylation of eIF2α, and leads to a downstream signaling cascade involving the apoptotic protease caspase-8, cleavage of the ER protein B-cell receptor-associated protein-31 (BAP31), opening of ER membrane channels, release of Ca^{2+} into the cytosol, and exocytosis. In others, an eIF2α- and caspase-independent downstream pathway may be involved (Garg et al. 2012b; Panaretakis et al. 2009). This suggests that, regardless of the origin of the stimulus, ER stress responses may trigger the emission of ICD-promoting signals, enabling the UPR^{ER} to coordinate an immune response.

UPR^{ER} activation can also contribute to inflammation in another important way: cells undergoing ER stress can cell non-autonomously initiate UPR^{ER} activation in cells of the immune system, leading to their activation and the mounting of an immune response (Fig. 3b). When cancer cells are subjected to physiological or pharmacological ER stress, they are able to release signals that activate the UPR^{ER} in macrophages and dendritic cells (DCs), leading to pro-tumor phenotypes (Mahadevan et al. 2011). *In vitro* cultures of macrophages treated with conditioned medium from ER-stressed tumor cells show ER stress response activation, resulting in the expression of the downstream genes BiP, growth arrest and DNA damage-inducible protein-34 (Gadd34), CHOP and XBP-1s. This leads to a proinflammatory response: macrophages release proinflammatory cytokines, aiding tumorigenesis. The transmissible factor(s) released by ER-stressed tumor cells may be detected by toll-like receptor-4 (TLR4), as transmission of UPR^{ER} activation is reduced in TLR4-knockout macrophages, and these signals can be augmented by concomitant TLR4 signaling through lipopolysaccharide (LPS). This suggests that endogenous TLR4 ligands or infection by gram-negative bacteria might act as cofactors in tumorigenesis. Cell non-autonomous transmission of UPR^{ER} activation to DCs also promotes the release of proinflammatory cytokines from these cells, encouraging a T cell-suppressive phenotype through the downregulation of cross-presentation of high-affinity antigens, and a failure to efficiently cross-prime CD8+ T cells. This leads to T cell activation without proliferation, facilitating tumor growth (Mahadevan et al. 2012). In this way, the pathway to cell non-autonomous UPR^{ER} activation can be co-opted by cancer cells, to aid in tumorigenesis.

4.3 Tumorigenesis

As well as allowing cancer cells to influence the immune response to tumors, UPR^{ER} activation plays other important roles in cancer development. Cancer cells activate the UPR^{ER} due to stresses that result from tumor growth, including hypoxia, protein misfolding that results from mutation, and nutrient deprivation. UPR^{ER} activation enables cancer cells to survive these conditions, and is often required for further tumor development; it also decreases sensitivity towards chemotherapeutic compounds (Ma and Hendershot 2004). While the UPR^{ER} is activated cell-autonomously in cancer cells, through environmental conditions, UPR^{ER} activation may also be communicated cell non-autonomously—to influence the inflammatory profiles and differentiation states of cells of the immune system, and also to activate the response in neighboring tumor cells, enabling them to preemptively increase their protein folding capacity (Shapiro et al. 2016).

Activation of the UPR^{ER} in neighboring tumor cells, termed "anticipatory activation", may occur as a result of the secretion of steroid and peptide hormones. These include the mitogenic hormones estrogen [17b-estradiol] (E$_2$), androgen [dihydrotestosterone] (DHT), and ecdysone (Ec), and the peptide hormones epidermal growth factor (EGF) and vascular endothelial growth factor (VEGF) (Andruska et al. 2015a, b; Karali et al. 2014; Yu et al. 2016). Any of these signals can elicit UPR^{ER} activation in other tumor cells, in the absence of ER stress. One mechanism by which this may occur involves the release of calcium from the ER, where high concentrations of Ca^{2+} are stored. Anticipatory activators of the UPR^{ER} induce the opening of inositol triphosphate receptor (IP$_3$R) calcium channels in the ER membrane, which open upon binding of IP$_3$, generated by the activity of phosphorylated phospholipase Cγ (PLCγ). Binding of E$_2$ or EGF to their receptors increases the intracellular concentration of inositol triphosphate (IP$_3$), and leads to the activation of PLCγ (Andruska et al. 2015b; Yu et al. 2016) Inhibition or knockdown of PLCγ prevents the anticipatory activation of the UPR^{ER}, while inhibition or knockdown of the IP_3R prevents increases in cytosolic calcium and UPR^{ER} activation, confirming the role of this mechanism. VEGF, however, may activate an alternative mechanism involving PLCγ and mTORC, suggesting that cells may have co-opted a variety of molecular pathways to ensure cell non-autonomous UPR^{ER} activation (Karali et al. 2014).

All of these activating hormones converge upon the PLCγ-dependent activation of the ATF6 and IRE1 branches of the UPR^{ER}, resulting in the splicing of XBP1 and upregulation of BiP. Each pathway also activates PERK, but this activation is mild and transient. Anticipatory UPR^{ER} activation therefore promotes pro-survival branches of the UPR^{ER}, without interfering with the synthesis of proteins required for tumorigenesis. Cell proliferation is triggered, which can be inhibited by knockdown of PLCγ, and which may depend upon the proliferative properties of increased intracellular calcium; UPR^{ER} activation also promotes VEGF-induced blood vessel formation within tumors, and increases resistance to chemotherapeutics. In ERα-positive breast cancer patients, for example, UPR^{ER} activation is tightly correlated with resistance to tamoxifen (Andruska et al. 2015a). However, one recent line of study has utilized strong induction of the anticipatory UPR^{ER} pathway by the noncompetitive ERα biomodulator BHPI, with concomitant high levels of PERK activation, to induce apoptosis in tumor cells, suggesting that this pathway may be of interest as a target for cancer therapies (Andruska et al. 2015b). Much remains to be learned, however, including one key question: does the same anticipatory pathway operate in other, healthy tissues?

Cell non-autonomous UPR^{ER} activation can also be induced by at least one cancer-causing agent—the bacteria *Helicobacter pylori (H. pylori)*, a key element in the development of gastric cancer. Upon infection by *H. pylori*, the gastric epithelium interacts with the bacteria and chronic inflammation is followed by oxynitric atrophy, which leads to mucous metaplasia, intestinal metaplasia and spasmolytic polypeptide expressing metaplasia (SPEM); intestinal metaplasia and SPEM are essential for the progression of gastric carcinogenesis (Correa and Houghton 2007; Weis and Goldenring 2009). In mouse models of gastric cancer, the UPR^{ER} is cell non-autonomously activated to induce helicobacter-initiated metaplasia and dysplasia, an effect dependent upon the long-term ER stress initiated by chronic *H. pylori* infection (Baird et al. 2013). CHOP expression then causes cell death, leading to the development of oxyntic atrophy, while proliferation is supported by the upregulation of molecular chaperones. This promotes ongoing tumor formation, driving the metaplasia-dysplasia pathway, and emphasizing the important role of the UPR^{ER} in disease progression upon *H. pylori* infection.

5 Conclusions

Cell non-autonomous UPR^{ER} signaling is a phenomenon that occurs across phyla, leading to the ability to coordinate stress resistance, physiology, and cell fate decisions between cells and tissues. As well as adding to our fundamental understanding of how cells respond to stress, these discoveries have the potential to transform our understanding of the development of multiple diseases, and how they might be treated. However, to make these opportunities a reality, there are several key questions that need to be addressed.

First, in most cases the signals that transmit UPR^{ER} activation between cells have not been identified, and this knowledge is critical to the development of drugs that harness these pathways. In the case of the anticipatory UPR^{ER} activation mechanism identified in cancer cells, it will be interesting to determine whether the same pathway is utilized in healthy tissues, and in other physiological contexts. Second, how UPR^{ER}-activating signals are generated is largely still a mystery. Does the source of ER stress determine whether a signal is sent, and if so, which are the stresses that specifically induce cell–cell communication? And does UPR^{ER} activation trigger signal production, or release? Third, how does signal reception at the cell membrane lead to the activation of UPR^{ER} regulators that are located at the ER membrane—does activation require signal transmission through the ER lumen, or do novel, cytosolic activation sites exist in these molecules? Finally, are the targets of the cell non-autonomously activated UPR^{ER} the same as the targets of the autonomous response, or are different target genes, or different subsets of known UPR^{ER} targets, triggered? The use of model organisms to address these questions, and decipher the mechanisms and signaling pathways involved in UPR^{ER} communication, will likely continue to be invaluable. Once these issues are better understood, the possibility that these pathways may be harnessed to combat immunological, metabolic or neurodegenerative diseases, or cancer, will become an exciting prospect.

Acknowledgements SI, MS and RCT are supported by the Medical Research Council.

References

Acosta-Alvear D et al (2007) XBP1 controls diverse cell type- and condition-specific transcriptional regulatory networks. Mol Cell 27:53–66

Andruska N et al (2015a) Anticipatory estrogen activation of the unfolded protein response is linked to cell proliferation and poor survival in estrogen receptor alpha-positive breast cancer. Oncogene 34:3760–3769

Andruska ND et al (2015b) Estrogen receptor alpha inhibitor activates the unfolded protein response, blocks protein synthesis, and induces tumor regression. Proc Natl Acad Sci U S A. 112:4737–4742

Arner P et al (2008) FGF21 attenuates lipolysis in human adipocytes—a possible link to improved insulin sensitivity. FEBS Lett 582:1725–1730

Baird M et al (2013) The unfolded protein response is activated in Helicobacter-induced gastric carcinogenesis in a non-cell autonomous manner. Lab Invest 93:112–122

Bertolotti A et al (2000) Dynamic interaction of BiP and ER stress transducers in the unfolded-protein response. Nat Cell Biol 2:326–332

Biteau B, Hochmuth CE, Jasper H (2008) JNK activity in somatic stem cells causes loss of tissue homeostasis in the aging Drosophila gut. Cell Stem Cell 3:442–455

Biteau B et al (2010) Lifespan extension by preserving proliferative homeostasis in Drosophila. PLoS Genet 6:e1001159

Calfon M et al (2002) IRE1 couples endoplasmic reticulum load to secretory capacity by processing the XBP-1 mRNA. Nature 415:92–96

Correa P, Houghton J (2007) Carcinogenesis of Helicobacter pylori. Gastroenterology 133: 659–672

Cox JS, Walter P (1996) A novel mechanism for regulating activity of a transcription factor that controls the unfolded protein response. Cell 87:391–404

Fisher FM, Maratos-Flier E (2016) Understanding the Physiology of FGF21. Annu Rev Physiol 78:223–241

Garg AD et al (2012a) ER stress-induced inflammation: does it aid or impede disease progression? Trends Mol Med. 18:589–598

Garg AD et al (2012b) A novel pathway combining calreticulin exposure and ATP secretion in immunogenic cancer cell death. EMBO J 31:1062–1079

Harding HP, Zhang Y, Ron D (1999) Protein translation and folding are coupled by an endoplasmic-reticulum-resident kinase. Nature 397:271–274

Harding HP et al (2000a) Regulated translation initiation controls stress-induced gene expression in mammalian cells. Mol Cell 6:1099–1108

Harding HP et al (2000b) Perk is essential for translational regulation and cell survival during the unfolded protein response. Mol Cell 5:897–904

Haze K et al (1999) Mammalian transcription factor ATF6 is synthesized as a transmembrane protein and activated by proteolysis in response to endoplasmic reticulum stress. Mol Biol Cell 10:3787–3799

Heazlewood CK et al (2008) Aberrant mucin assembly in mice causes endoplasmic reticulum stress and spontaneous inflammation resembling ulcerative colitis. Plos Medicine. 5:440–460

Hochmuth CE et al (2011) Redox regulation by Keap1 and Nrf2 controls intestinal stem cell proliferation in Drosophila. Cell Stem Cell 8:188–199

Hollien J et al (2009) Regulated Ire1-dependent decay of messenger RNAs in mammalian cells. J Cell Biol 186:323–331

Hotamisligil GS (2010) Endoplasmic reticulum stress and the inflammatory basis of metabolic disease. Cell 140:900–917

Karali E et al (2014) VEGF Signals through ATF6 and PERK to promote endothelial cell survival and angiogenesis in the absence of ER stress. Mol Cell 54:559–572

Kharitonenkov A et al (2005) FGF-21 as a novel metabolic regulator. J Clin Invest. 115:1627–1635

Kim KH et al (2013) Autophagy deficiency leads to protection from obesity and insulin resistance by inducing Fgf21 as a mitokine. Nat Med 19:83–92

Kulalert W, Kim DH (2013) The unfolded protein response in a pair of sensory neurons promotes entry of C. elegans into dauer diapause. Curr Biol 23:2540–2545

Levi-Ferber M et al (2014) It's all in your mind: determining germ cell fate by neuronal IRE-1 in C-elegans. Plos Genetics 10

Lin JH et al (2007) IRE1 signaling affects cell fate during the unfolded protein response. Science 318:944–949

Ma Y, Hendershot LM (2004) The role of the unfolded protein response in tumour development: friend or foe? Nat Rev Cancer 4:966–977

Mahadevan NR et al (2011) Transmission of endoplasmic reticulum stress and pro-inflammation from tumor cells to myeloid cells. Proc Natl Acad Sci U S A. 108:6561–6566

Mahadevan NR et al (2012) Cell-extrinsic effects of tumor ER stress imprint myeloid dendritic cells and impair CD8(+) T cell priming. PLoS ONE 7:e51845

Marciniak SJ et al (2004) CHOP induces death by promoting protein synthesis and oxidation in the stressed endoplasmic reticulum. Genes Dev 18:3066–3077

Nakatani Y et al (2005) Involvement of endoplasmic reticulum stress in insulin resistance and diabetes. J Biol Chem 280:847–851

Ozcan L et al (2009) Endoplasmic reticulum stress plays a central role in development of leptin resistance. Cell Metab 9:35–51

Ozcan U et al (2004) Endoplasmic reticulum stress links obesity, insulin action, and type 2 diabetes. Science 306:457–461

Panaretakis T et al (2009) Mechanisms of pre-apoptotic calreticulin exposure in immunogenic cell death. EMBO J 28:578–590

Richardson CE, Kooistra T, Kim DH (2010) An essential role for XBP-1 in host protection against immune activation in C. elegans. Nature 463:1092–1095

Ron D, Walter P (2007) Signal integration in the endoplasmic reticulum unfolded protein response. Nat Rev Mol Cell Biol 8:519–529

Schaap FG et al (2013) Fibroblast growth factor 21 is induced by endoplasmic reticulum stress. Biochimie 95:692–699

Shapiro DJ et al (2016) Anticipatory UPR activation: a protective pathway and target in cancer. Trends Endocrinol Metab 27:731–741

Sidrauski C, Walter P (1997) The transmembrane kinase Ire1p is a site-specific endonuclease that initiates mRNA splicing in the unfolded protein response. Cell 90:1031–1039

Singh V, Aballay A (2012) Endoplasmic reticulum stress pathway required for immune homeostasis is neurally controlled by arrestin-1. J Biol Chem 287:33191–33197

Sun J et al (2011) Neuronal GPCR controls innate immunity by regulating noncanonical unfolded protein response genes. Science 332:729–732

Sun JR, Liu YY, Aballay A (2012) Organismal regulation of XBP-1-mediated unfolded protein response during development and immune activation. EMBO Rep 13:855–860

Taylor RC, Dillin A (2013) XBP-1 is a cell-nonautonomous regulator of stress resistance and longevity. Cell 153:1435–1447

Taylor RC, Berendzen KM, Dillin A (2014) Systemic stress signalling: understanding the cell non-autonomous control of proteostasis. Nat Rev Mol Cell Biol 15:211–217

Urano F et al (2000) Coupling of stress in the ER to activation of JNK protein kinases by transmembrane protein kinase IRE1. Science 287:664–666

Volchuk A, Ron D (2010) The endoplasmic reticulum stress response in the pancreatic beta-cell. Diabetes Obes Metab 12(Suppl 2):48–57

Walter P, Ron D (2011) The unfolded protein response: from stress pathway to homeostatic regulation. Science 334:1081–1086

Wan XS et al (2014) ATF4- and CHOP-dependent induction of FGF21 through endoplasmic reticulum stress. Biomed Res Int 2014:807874

Wang LF et al (2014) Integration of UPRER and oxidative stress signaling in the control of intestinal stem cell proliferation. Plos Genetics 10

Wang LF et al (2015) PERK limits drosophila lifespan by promoting intestinal stem cell proliferation in response to ER stress. Plos Genetics 11

Weis VG, Goldenring JR (2009) Current understanding of SPEM and its standing in the preneoplastic process. Gastric Cancer 12:189–197

Wente W et al (2006) Fibroblast growth factor-21 improves pancreatic beta-cell function and survival by activation of extracellular signal-regulated kinase 1/2 and Akt signaling pathways. Diabetes 55:2470–2478

Williams KW et al (2014) Xbp1s in Pomc neurons connects ER stress with energy balance and glucose homeostasis. Cell Metab 20:471–482

Xu J et al (2009) Fibroblast growth factor 21 reverses hepatic steatosis, increases energy expenditure, and improves insulin sensitivity in diet-induced obese mice. Diabetes 58:250–259

Yu L et al (2016) Anticipatory activation of the unfolded protein response by epidermal growth factor is required for immediate early gene expression and cell proliferation. Mol Cell Endocrinol 422:31–41

The Unfolded Protein Response in the Immune Cell Development: Putting the Caretaker in the Driving Seat

Simon J. Tavernier, Bart N. Lambrecht and Sophie Janssens

Abstract The endoplasmic reticulum (ER) is the primary site for the folding of proteins destined for the membranous compartment and the extracellular space. This elaborate function is coordinated by the unfolded protein response (UPR), a stress-activated cellular program that governs proteostasis. In multicellular organisms, cells have adopted specialized functions, which required functional adaptations of the ER and its UPR. Recently, it has become clear that in immune cells, the UPR has acquired functions that stretch far beyond its original scope. In this review, we will discuss the role of the UPR in the immune system and highlight the plasticity of this signaling cascade throughout immune cell development.

Contents

1	Introduction..	46
2	Coordinating an Unfolded Protein Response..	47
3	The UPR: Instrumental in Immune Cell Development..................................	52
	3.1 The UPR and Stem Cells	52

S.J. Tavernier · B.N. Lambrecht (✉) · S. Janssens (✉)
Laboratory of Immunoregulation and Mucosal Immunology,
VIB Inflammation Research Center, Ghent, Belgium
e-mail: bart.lambrecht@irc.vib-ugent.be

S. Janssens
e-mail: sophie.janssens@irc.vib-ugent.be

S.J. Tavernier · B.N. Lambrecht · S. Janssens
GROUP-ID Consortium, Ghent University and University Hospital, Ghent, Belgium

S.J. Tavernier · B.N. Lambrecht · S. Janssens
Department of Internal Medicine, Ghent University, Ghent, Belgium

B.N. Lambrecht · S. Janssens
VIB Inflammation Research Center, Ghent University, Technologiepark 927,
B-9052 Zwijnaarde, Belgium

3.2 The UPR in Lymphopoiesis.. 56
3.3 Granulocytes and UPR.. 58
3.4 The UPR in the Mononuclear Phagocytic System....................................... 59
3.5 UPR Regulation at the Mucosal Barrier... 61
4 Discussion... 63
References.. 64

1 Introduction

Up to one-third of the whole proteome is destined for the extracellular space or the membranous compartment. Prior to secretion, these proteins undergo maturation in the endoplasmic reticulum (ER) through a complex process involving glycosylation, folding, and/or disulfide formation. These posttranslational modifications are managed by ER-residing glycolases, chaperones, and protein disulfide isomerases (PDIs) (Hetz et al. 2015). Given that a cell produces up to 4×10^6 proteins every minute, control of proteostasis (protein-folding homeostasis) in the ER is essential (Princiotta et al. 2003). Indeed, a disruption of proteostasis has been observed in a number of neurological (e.g., disease of Parkinson), endocrinological (e.g., type 2 diabetes), and inflammatory (e.g., inflammatory bowel disease) disorders and has spurred research to elucidate its contribution to these diseases (Wang and Kaufman 2016). A large set of physiological (e.g., hypoglycemia, hypoxia, and pattern recognition receptor ligation) and pathophysiological circumstances (viral infection, bacterial toxins, or uncontrolled cell replication) impair the equilibrium between protein load and protein processing, resulting into protein misfolding and a condition called ER stress. To restore homeostasis, cells activate an ER stress-responsive signaling cascade called the unfolded protein response (UPR). Importantly, additional metabolic processes operate at the ER (e.g., lipogenesis, glucogenesis, and autophagosome formation) and are also coordinated by the UPR, stretching its functionalities far beyond the regulation of the secretory apparatus (Rutkowski and Hegde 2010).

Recently, gene candidate and genome-wide association studies in inflammatory diseases, such as asthma and inflammatory bowel disease, have identified single-nucleotide polymorphisms (SNPs) in genes controlling key functions of the UPR (Wang and Kaufman 2016; Grootjans et al. 2016). These studies underpin the deregulation of the UPR as an inciting event in inflammatory diseases, and uncovered an emerging role of the UPR in regulating immune responses (recently reviewed in Janssens et al. 2014; Grootjans et al. 2016). In addition, the UPR has been increasingly implicated in the development of immune cells (Iwakoshi et al. 2007; Asada et al. 2012; van Galen et al. 2014; Bettigole et al. 2015). In this current review, we questioned how the UPR coordinates the development of the immune system and highlight how this can contribute to disease pathogenesis.

2 Coordinating an Unfolded Protein Response

With slight alterations in protein function having potential detrimental effects, proteins need to be carefully folded in the ER. Polypeptide strands entering the ER can be co-translationally *N*-glycosylated and subsequently bound to the folding chaperones calnexin and calreticulin, to coordinate protein folding and prevent aggregation. A stringent quality control will sense proteins that are not properly folded and prevents their subsequent egress from the ER to the Golgi (Tannous et al. 2015). Upon a number of folding iterations, chronically misfolded proteins are translocated back to the cytosol for proteasomal degradation (a process called endoplasmic reticulum-associated protein degradation or ERAD). Alternatively, specific autophagy of the ER (ER-phagy) may also degrade chronically stressed ER (Smith et al. 2011).

In metazoa, accumulation of these misfolded proteins in the ER is detected by three UPR sensors residing in the ER membrane, called activating transcription factor (ATF)-6, inositol-requiring enzyme (IRE)-1, and protein kinase R-like endoplasmic reticulum kinase (PERK). Initially, the UPR aims to restore homeostasis by reducing protein synthesis, enhancing folding capacity, and bolstering ERAD machinery. If unresolved, chronic pathological ER stress drives a pro-apoptotic UPR, although the molecular mechanisms are incompletely understood (Tabas and Ron 2011; Walter and Ron 2011).

The mode of activation of the UPR sensors is still debated. The UPR must be carefully tuned to quickly respond to physiological fluctuations in proteostasis, while avoiding unnecessary induction of cell death. Similar to HSP90 inhibiting HSF1 by direct binding in the cytosol, BiP (also named GRP78), the main chaperone in the ER, is able to limit the activation of the UPR sensors by occupying protein domains located in the ER lumen (Bertolotti et al. 2000; Pincus et al. 2010; Anckar and Sistonen 2011). BiP has a high affinity for the hydrophobic regions of misfolded proteins, which results into a sequestration of BiP during ER stress and release of the UPR sensors (Bertolotti et al. 2000). In the case of IRE-1 and PERK, this induces dimerization, oligomerization, and subsequent activation. Although BiP release was originally believed to be sufficient to activate the UPR, there is biochemical and structural data suggesting that the regulation may be more complex (Korennykh and Walter 2012). Particularly in *Saccharomyces cerevisiae*, where Ire1 (yeast homologue of IRE-1) is the sole protein managing all functions of the UPR, the (genetically engineered) release of BiP was insufficient to induce a response of the UPR machinery (Kimata et al. 2004). Furthermore, crystallization of an Ire1 dimer revealed the presence of a groove predicted to bind misfolded peptide strands (Credle et al. 2005). The ability of a model unfolded protein to bind to and cluster Ire1 into higher-order protein complexes has led to a unifying concept. In this model, BiP restrains UPR sensor activation by keeping these into inactive monomers while unfolded proteins both sequester BiP and act as direct UPR sensor ligands, inducing oligomerization and activation (Kimata et al. 2007; Gardner and Walter 2011). At present, crystal structures of mammalian IRE-1 fail to reveal a similar peptide-binding groove. Although this could represent an alternative conformational state, the direct binding of peptides to mammalian IRE-1 remains a matter of debate (reviewed in Gardner et al. 2013). In the case of IRE-1 and PERK, studies have illustrated that these sensors also detect alterations in the lipid content of the ER membrane and that the flavonol quercetin can bind and activate IRE-1 through its cytosolic domain (Wiseman et al. 2010; Volmer et al. 2013; Kitai et al. 2013). These observations indicate that the UPR is responsive to a broader set of cues aside of misfolded proteins.

IRE-1 is an ancient branch of the UPR, preserved among most species examined. It is a type I membranous protein with a dual function, bearing both a kinase and an endonuclease in its C-terminal domain (Sidrauski and Walter 1997). In mammalia, two paralogues of IRE-1 have been discovered: IRE-1α ubiquitously expressed, and IRE-1β which is restricted to the epithelia of lung and gut (Bertolotti et al. 2001). During ER stress, IRE-1 molecules cluster into oligomers, facilitating *trans*-autophosphorylation, conformational shift, and activation of its endonuclease domain (Aragón et al. 2009; Korennykh et al. 2009). The endonuclease recognizes and cleaves a specific mRNA substrate with a dual stem loop, called XBP1 (Calfon et al. 2002). The severed XBP1 mRNA strands are joined by the RTCB tRNA ligase, ultimately removing 26 basepairs (Lu et al. 2014; Kosmaczewski et al. 2014). This unconventional splicing activity induces a frame shift resulting in a novel C-terminus and the translation of a potent cAMP response element binding (CREB)/ATF basic leucine zipper (bZIP)-containing transcription factor (TF) XBP1s (s stands for spliced) (Yoshida et al. 2001 and Fig. 1a). In ER-stressed cells, XBP1s activates the expression of genes encoding chaperones (e.g., BiP) and PDIs, the ERAD machinery (e.g., EDEM), and enzymes coordinating phospholipid biosynthesis and ER expansion (Yoshida et al. 2003; Lee et al. 2003b; Sriburi et al. 2007). More recently, a number of studies have indicated that XBP1s has a role beyond the classical UPR. XBP1s was found to be essential for driving cell differentiation, tuning cytokine responses, fueling the HIF1 pathway, and regulating

Fig. 1 The multipronged UPR. Unfolded proteins trigger the three sensors of UPR by ▶ sequestering the BiP chaperone. Activation of the UPR initiates both transcriptional and translational processes, resulting in enhanced protein folding machinery, degradation of unfolded proteins, and reduced protein translation. **a** Release of the inactive monomers by BiP results in IRE-1 autophosphorylation and subsequent formation of oligomers, stabilized by the binding of unfolded proteins. A conformational shift activates the RNase domain and removes a short intron in *XBP1u* mRNA. Subsequent religation by the RNA ligase RtcB results into *XBP1s* that is translated into the transcription factor XBP1s, controlling the expression of various UPR genes. Through ill-defined mechanisms, the RNase can switch to RIDD; the cleavage and degradation of ER-localized RNAs. **b** Upon activation, PERK phosphorylates its downstream targets NRF2 and eIF2α. Phosphorylated eIF2α in turn reduces global translation but enhances the selective translation of mRNAs with short uORFs such as ATF4. This activates the integrated stress response, including CHOP and Gadd34, the latter a regulatory subunit of the phosphatase PP1. The PP1-Gadd34 complex can dephosphorylate P-eIF2α and reinitiate translation. **c** Release of ATF6 results in the translocation to the Golgi complex. Subsequent cleavage by S1P and S2P proteases releases the cytosolic domain, resulting in a potent transcription factor

The Unfolded Protein Response in the Immune Cell Development ...

a. IRE-1 signaling

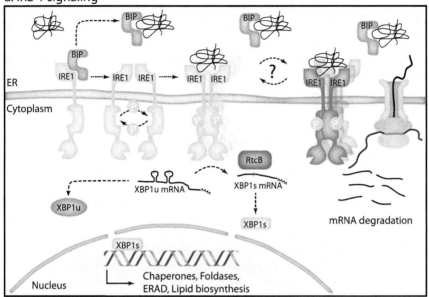

b. PERK signaling

c. ATF6 signaling

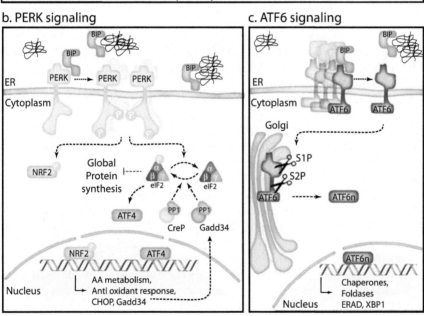

the hexosamine biosynthetic pathway (Acosta-Alvear et al. 2007; Martinon et al. 2010; Wang et al. 2014; Chen et al. 2014).

In 2006, Hollien and colleagues revealed a second activity of IRE-1 endonuclease in *Drosophilla melanogaster*. In this organism, chemically induced ER stress resulted into the degradation of ER-localized RNAs in an ER signal sequence-dependent manner. This process was termed regulated IRE-1-dependent decay of RNA or RIDD (Fig. 1a). It is suggested that RIDD reduces ER stress in a refined way, alleviating protein translation in a spatially restricted manner and liberating ER-localized ribosomes to synthesize protein-folding machinery (Hollien and Weissman 2006; Hollien et al. 2009). This alternate function of IRE-1 has been described in other species and appears relevant to cell differentiation, fate regulation, metabolism, and immune responses (Coelho and Domingos 2014; Osorio et al. 2014; Maurel et al. 2014). Although the molecular basis that allows the IRE-1 endonuclease to alternate between the unconventional splicing of XBP1 and the degradation of a broader set of RNAs is still unclear (Ghosh et al. 2014; Tam et al. 2014), from data obtained in *Schizosaccharomyces pombe*, it appears that RIDD is the most ancient function of IRE-1 (Kimmig et al. 2012).

Adding an extra layer of complexity, the metazoan IRE-1 kinase also signals independently of its endonuclease activity. Through multiprotein complexes containing adaptor proteins such as TNF receptor-associated factor (TRAF) 2 and apoptosis signal-regulating kinase (ASK), the IRE-1 kinase is able to activate both nuclear factor (NF)-κB and mitogen-activated protein kinase (MAPK) pathways such as p38, Jun N-terminal kinase (JNK), and extracellular signal-regulated kinase (ERK). These signaling events connect the UPR to inflammatory and pro-apoptotic pathways (Hetz et al. 2015).

Similar to IRE-1, PERK is a type I transmembrane protein with a BiP-binding ER luminal domain and a kinase activity at the cytosolic side. Indeed, exchange of the luminal domain of IRE-1 and PERK results in similar activation upon ER stress (Liu et al. 2000). Recent structural data revealed that upon ligand binding, PERK forms dimers and tetramers, inducing its *trans*-autophosphorylation and kinase activation, phosphorylating its downstream target eIF2α (Carrara et al. 2015). eIF2α is part of the ternary complex, consisting of the eIF2 proteins, methionyl tRNA, and GTP. Together with the 40S ribosomal subunit, the ternary complex binds to the 5′-cap of mRNA and initiates protein translation upon recognition of AUG initiation codon. During this process, GTP gets hydrolyzed and eIF2a. GDP needs to be recharged with GTP before it can initiate a new round of translation (Hershey et al. 2012). Phosphorylation of eIF2α inhibits the GTP exchange factor eIF2B, depletes GTP-loaded ternary complexes, and reduces overall translation initiation (Scheuner et al. 2001 and Fig. 1b). Protein translation inhibition is crucial for ER homeostasis, underscored by the widespread pancreatic β-cell failure in PERK$^{-/-}$ mice and the identification of two loss of function mutations in the *EIF2AK3* gene (PERK) inducing the Wolcott–Rallison syndrome, a severe form of neonatal TI diabetes (Delépine et al. 2000; Harding et al. 2001).

Paradoxically, reduced cap-dependent protein synthesis during ER stress enhances the translation of a number of mRNAs with short upstream open reading frames (uORFs) (Sidrauski et al. 2015). Among these proteins, the bZIP-activated transcription factor 4 ATF-4 regulates many aspects of the integrated stress response (ISR). The ISR is an evolutionary conserved pathway downstream of phosphorylated eIF2α that regulates protein translation, amino acid metabolism, antioxidant defenses, and autophagy. Aside of ER stress, also amino acid deprivation, heme shortage or viral infection can activate the ISR through three other eIF2α kinases: general control nonderepressible 2 (GCN2), heme-regulated inhibitor kinase (HRI), and protein kinase R (PKR) (Walter and Ron 2011). ATF-4 also enhances the expression of the transcriptional regulator CCAAT-/enhancer-binding protein homologous protein (CHOP), which further tunes the ATF-4 response (Walter and Ron 2011). Both ATF-4 and CHOP mediate the expression of GADD34, a regulatory subunit of the protein phosphatase 1 (PP1). The PP1-GADD34 complex will dephosphorylate eIF2α and reinitiate protein translation (Novoa et al. 2001; Kojima et al. 2003 and Fig. 1b). Although restoration of protein translation is crucial for the induction of the UPR machinery and cellular survival, engaging the translational machinery without elimination of the initial ER stress agent can be deleterious, inducing oxidative stress and cell death (Novoa et al. 2003; Han et al. 2013).

Aside from eIF2α, nuclear factor erythroid-related factor 2 (NRF2) has been proposed as a second target of PERK (Cullinan et al. 2003 and Fig. 1b). Its phosphorylation releases NRF2 from Kelch-like ECH-associated protein 1 (KEAP1), allowing NRF2 to migrate to the nucleus and mount antioxidant defenses. The activity of both ATF-4 and NRF2 has proven to be important for cell survival under conditions of ER stress (Harding et al. 2003; Cullinan et al. 2003).

The third branch of the UPR is controlled by ATF-6α, a type II transmembrane protein and member of the large CREB-like transcription factor family (Bailey and O'Hare 2007). In contrast to IRE-1 and PERK, ATF-6α is present in oligomeric structures during homeostatic conditions (Nadanaka et al. 2007). Release of BiP and changes in ATF6 redox status disassemble the oligomer and expose an export signal (Shen et al. 2002; Nadanaka et al. 2007). ATF-6α is subsequently shuttled to the Golgi where it is cleaved by site-1 and site-2 proteases in a process called regulated intramembrane proteolysis (Ye et al. 2000). Once released, the N-terminal domain of ATF6α translocates to the nucleus and activates genes encoding chaperones, ERAD machinery, and notably the transcription factor XBP1 (Wu et al. 2007 and Fig. 1c). In contrast to IRE-1 and PERK, whose deletion results in early embryonic or neonatal lethality, respectively, the absence of ATF-6α does not cause gross pathology. Subsequent studies confirmed a secondary role for ATF-6α in optimizing the UPR during ER stress (Wu et al. 2007). Strikingly, a double knockout mouse lacking both ATF-6α and ATF-6β displays embryonic lethality suggesting that ATF-6 members share important redundancy (Yamamoto et al. 2007).

Meanwhile, several additional sensors coordinating the ER stress response have been revealed, putting pressure on the model of a three-pronged UPR. For instance, a number of ATF-6α-related proteins (e.g., LUMAN, OASIS, and CREB-H) respond to ER stress and are able to mount UPR responses (Chin et al. 2005).

Epithelial cells of the gut and lung contain two IRE-1 paralogues, IRE-1α and IRE-1β with diverging affinity for XBP1 and RIDD targets (Imagawa et al. 2008). Also, two eIF2α kinases, GCN2 and PKR, have been identified to be activated upon ER stress and coordinate the ISR (Hamanaka et al. 2005; Nakamura et al. 2010). Given the tissue- and cell-specific distribution of these different UPR sensors, it is likely that the increasing specialization of cells in multicellular organisms throughout evolution required the need for a wide array of UPR responses to accommodate its broad physiological functions.

3 The UPR: Instrumental in Immune Cell Development

The immune system has evolved to efficiently recognize and eliminate threats to the organism. This myriad of threats (e.g., viruses, parasites, bacteria but also cancerous cells) requires tailored responses, coordinated by distinct cells of the innate and adaptive immune system, as well as epithelial cells. Some of these cells (e.g., plasma cells and epithelial goblet cells) depend on a strongly developed and highly structured ER to facilitate their secretory function. The development of genetically engineered mice lacking core components of the UPR confirmed a major role for the UPR in the development of these professional secretory cells. Surprisingly, these mice also harbored widespread defects in their immune system and subsequent analysis revealed that (specific) UPR sensors are essential for proper development and functioning of non-professional secretory cells. These insights expanded the role of the UPR beyond the realm of protein folding and refined our view on its physiological activation. In the following section, we will give an overview of the most recent insights on the role of the UPR in the development of the immune system.

3.1 The UPR and Stem Cells

3.1.1 Hematopoietic Stem Cells

The hematopoietic lineage relies on stem cells for the replenishment of aging cells. These stem cells are characterized by their multilineage potential, self-renewal capacity, and longevity. Recently, it has become clear that these properties critically depend on their strict regulation of proteostasis to avoid cellular stresses caused by accumulation of reactive oxygen species (ROS) or ER stress (Rouault-Pierre et al. 2013; Ito and Suda 2014; Buszczak et al. 2014).

Compared to their faster cycling progeny, hematopoietic stem cells (HSC) in the bone marrow produce fewer proteins and genetic manipulation of their translation rates significantly impairs HSC function (Reavie et al. 2010; Signer et al. 2014). Introduction of a mutation in the ribosomal subunit RPL24 (RPL24^{Bst} mutant allele) in HSCs decreased their protein synthesis capacity and crippled their ability to engraft and reconstitute irradiated hosts. On the flip side, loss of the phosphatase and tensin homologue (PTEN) in HSCs activated AKT signaling and induced a steep increase in protein synthesis. This resulted in HSC loss and promoted tumor formation (Signer et al. 2014). Interestingly, reducing the protein synthesis in PTEN-deficient animals by the introduction of the RPL24^{Bst} allele rescued this phenotype, implicating protein synthesis as an important regulator of self-renewal and differentiation capacity in HSCs (Signer et al. 2014). In general, HSCs display specific activation of the PERK signaling branch, whereas the IRE-1α/XBP1 branch is hardly activated (van Galen et al. 2014). Furthermore, human cord blood HSCs were shown to be particularly sensitive to genetically or chemically induced proteotoxic stress in a PERK-dependent manner. Overexpression of the chaperone Erdj4 in HSCs increased their folding capacity, protected them from ER stress-induced cytotoxicity, and augmented their engraftment rate (van Galen et al. 2014). These data raise the concept that the stem cell pool maintains clonal integrity by removing stressed and damaged cells, ultimately preventing tumorigenesis (Fig. 2).

In contrast, direct descendants of HSCs, such as multipotent progenitors, activate the proto-oncogen c-MYC, rally the 'anabolic' IRE-1/XBP1 pathway, and increase their protein synthesis rate to underpin the increased proliferative capacity (Zhang et al. 2005; Reavie et al. 2010; Signer et al. 2014; van Galen et al. 2014). The mechanism of this differential UPR regulation remains largely enigmatic. One study suggested that developmental pluripotency associated 5 (dppa5), highly expressed in long-term stem cells, governs HSC reconstitution capacity through regulation of UPR, mitochondrial respiration, and glycolysis (Miharada et al. 2014). Although all aforementioned studies focused on the cell differentiation to granulocytes, translational control and UPR regulation seem to be a conserved feature across hematopoietic lineages as also the macrophage and dendritic cell (DC) progenitors display higher protein synthesis rates and increased IRE-1 activity compared to HSCs (Tavernier SJ et al., under review, Cannoodt R et al., under review). Indeed, overexpression of XBP1 in hematopoietic cells enhances DC development (Iwakoshi et al. 2007).

During fetal development, the expansion of the hematopoietic system relies on fetal liver residing HSCs (FL-HSCs). In contrast to dormant adult BM-HSCs, FL-HSCs actively proliferate and similarly increase their protein synthesis rates (Morrison and Spradling 2008; Ito and Suda 2014; Sigurdsson et al. 2016). Strikingly, rather than increasing the folding capacity of the ER, FL-HSCs similarly display low XBP1 splicing. Instead, FL-HSCs rely on the chaperone functions of fetal and maternal liver-derived taurine-conjugated bile acids to prevent protein aggregates and PERK activation (Sigurdsson et al. 2016 and Fig. 2). The importance of these non-cell autonomous chaperones is further highlighted by in vitro culture studies of HSCs. In culture, HSCs quickly lose their self-renewal capacity and undergo apoptosis, a process that is associated with increased protein synthesis and activation of the UPR (Miharada et al. 2014). This HSC exhaustion can be prevented by the coculture of HSCs with rapamycin, a potent inhibitor of mTOR

Regulation of the UPR during immune cell differentiation

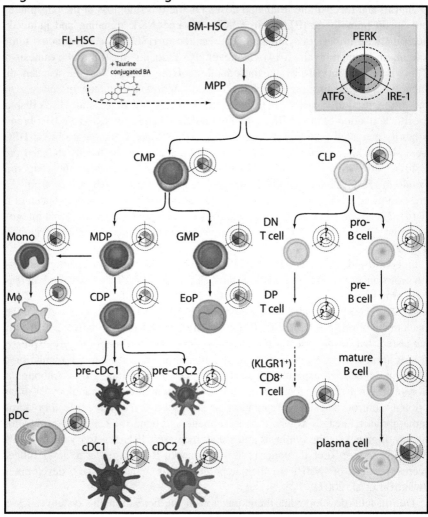

and protein synthesis, or by coculture in the presence of bile acids and derivatives (Huang et al. 2012; Miharada et al. 2014).

3.1.2 Intestinal Stem Cells

Although the data on UPR regulation in intestinal stem cells (ISCs) are scarce, some interesting parallels and differences can be noted. In homeostatic conditions, ISCs are located at the bottom of the crypts, where they generate the rapidly dividing

◄**Fig. 2 UPR regulation during immune cell differentiation.** Activation of the three UPR branches throughout immune cell differentiation (PERK *blue*, IRE-1 *green*, and ATF6 *purple*). Quiescent stem cells attain longevity and self-renewal through reduced protein synthesis, low ROS levels, and enhanced PERK signaling. Both fetus and maternal bile acids (BA) sustain protein folding in fetal liver stem cells, precluding deleterious PERK activation. Stem cell progeny increasingly activate IRE-1/XBP1 signaling underpinning the increased proliferative capacity. In the mononuclear phagocytic branch, CSF-1 ligation induces macrophage differentiation and results in a broad activation of the UPR. DC development reveals subset-specific changes in the UPR, and cDC1s profoundly activate the IRE1/XBP1 axis in the pre-DC stage whereas pDCs display a balanced UPR. Among granulocytes, eosinophils selectively require XBP1 during differentiation. Correspondingly, eosinophil maturation is associated with enhanced XBP1 splicing. DP-T cells and pro-B cells, lymphoid cell precursors, display an IRE-1/XBP1 signature. Plasma cells induce both IRE-1 and ATF6 while actively suppressing PERK. During infection, IL-2 and TCR signaling activate a full UPR in KLRG1$^+$ CD8$^+$ T cells. (*BM-HSC* bone marrow hematopoietic stem cell, *FL-HSC* fetal liver hematopoietic stem cell, *MPP* multipotent progenitor, *CLP* common lymphoid progenitor, *CMP* common myeloid progenitor, *MDP* macrophage dendritic cell progenitor, *GMP* granulocyte macrophage progenitor, *CDP* common dendritic cell progenitor, and *pDC* plasmacytoid dendritic cell)

transit-amplifying cells (TA cells). These cells have lost self-renewal capacity and after a number of cycles give rise to the different subtypes of intestinal epithelial cells (absorptive cells, Paneth cells, neuroendocrine cells, and goblet cells). In marked contrast with the TA cells and intestinal epithelial cells (IECs), ISCs only weakly stain for BiP, XBP1s, and P-eIF2α, key members of the UPR (Schwitalla et al. 2013; Heijmans et al. 2013). Indeed, gene expression arrays revealed an inverse correlation between markers of stemness and UPR genes (Muñoz et al. 2012). Similar to HSCs, both chemically and genetically generated proteotoxic stress quickly resulted in the loss of the Lgr5$^+$, Olfm4$^+$ stem cells, both in vitro and in vivo (Heijmans et al. 2013; Rosekrans et al. 2015). Rather than enhanced apoptosis, as observed in HSCs, ER stress quickly induced the differentiation of ISCs into TA cells and IECs. Selective activation of PERK-eIF2α was both sufficient and necessary for the ISC depletion and IEC differentiation, presumably through the translational control of cell fate proteins such as c-MYC (Heijmans et al. 2013).

Aside of this cell autonomous UPR activation, damaged or stressed IECs can alert the stem cell pool. The absence of XBP1 in the intestinal epithelium induces expansion of the ISC pool and predisposes for tumor formation in an IRE-1α-dependent manner. Furthermore, XBP1-deficient IECs enhance TA cell proliferation through a NF-κB/IL-6/Stat3 paracrine loop (Niederreiter et al. 2013). In Drosophila, a similar pathway (Bsk/Upd/Stat) originating from ER-stressed IECs induces stem cell proliferation in a PERK-dependent but IRE-1- and ATF-6-independent manner. In this model, excessive or chronic PERK activation induced epithelial dysplasia and repression of PERK in ISCs improved intestinal homeostasis and extended life span (Wang et al. 2015).

These reports suggest that a tight UPR control is essential for stem cell function. This is considered as a safeguard against loss of function or tumorigenesis, due to ROS-mediated DNA damage (van Galen et al. 2014). HSCs and ISCs are both exquisitely sensitive to ER stress but the ensuing response—apoptosis versus differentiation—is fundamentally different. In contrast to HSCs residing in the bone marrow, mucosal stem cells are continuously exposed to a hostile environment with various microbial and inflammatory stresses. The integration of these signals into the UPR pathway might be a possible explanation of these different outcomes. Further research on how the UPR determines these cell responses in these different settings will be instrumental for our understanding how the UPR gives rise to diverging cell fates.

3.2 The UPR in Lymphopoiesis

3.2.1 B-cell Development

Proteostasis seems to be essential in early lymphocyte development. Deletion of BiP in hematopoietic cells induces widespread ER stress and reduces common lymphocyte progenitors in the bone marrow, thymus atrophy, and lymphopenia (Wey et al. 2012a). During lymphocyte development, both pro-B cells in the bone marrow and $CD4^+CD8^+$ T cells in the thymus strongly activate IRE-1 (Brunsing et al. 2008 and Fig. 2). In the case of B cells, this is associated with high levels of XBP1 splicing at the level of pro-B cells and enhanced expression of IRE-1, XBP1, and BiP at the pre-B-cell stage (Kharabi Masouleh et al. 2014). This activation of the IRE-1/XBP1 pathway coincides with the development of the critical pre-B- and pre-T-cell receptors and V(D)J antigen receptor gene arrangements (Zhang et al. 2005). Although these results predict that loss of IRE-1 signaling would block B-cell development beyond the pro-B-cell stage or T-cell differentiation in the thymus, mice lacking XBP1 in the hematopoietic lineage do not show obvious signs of lymphopenia (Bettigole et al. 2015), suggesting that the IRE-1 activity is independent of XBP1. Indeed, the cytoplasmic domain of IRE-1 but not its enzymatic activity appears required for the rearrangement of the B-cell receptor (BCR) and B-cell differentiation, in an as-yet poorly understood mechanism (Zhang et al. 2005).

The terminal differentiation of B cells into immunoglobulin (Ig)-secreting plasma cells (PC) requires a fundamental adjustment of the cellular architecture. In particular, the ER needs to expand to cope with the enormous increase in protein secretion. These changes occur in an anticipatory fashion, i.e., metabolic changes and ER folding capacity precede Ig production. This ensures that the cell is well adapted for its new secretory role (Gass et al. 2002; van Anken et al. 2003; Hu et al. 2009). The expansion of the ER is associated with a selective activation of the UPR, with increased nuclear expression of XBP1s and ATF-6, whereas PERK is deactivated during PC differentiation (Gass et al. 2002; Zhang et al. 2005; Ma et al. 2010; Goldfinger et al. 2011 and Fig. 2). In a seminal study using $XBP1/Rag2^{-/-}$ chimeric mice, XBP1 was found to be essential for PC generation, Ig production, and control of virus infections such as polyomavirus (Reimold et al. 2001).

The specific XBP1 dependency of PC is underscored by the otherwise normal maturation of B cells and isotype switching in the absence of XBP1 (Hu et al. 2009; Todd et al. 2009). The advent of more elegant tools such as B-cell-specific XBP1-deficient mice allowed more refined observations: Although XBP1 guides metabolic adaptations in differentiating B cells and sustains Ig production, PCs can still develop in the absence of XBP1 (Hu et al. 2009; Todd et al. 2009; Taubenheim et al. 2012). In a series of publications, the molecular basis for this XBP1 dependency was elucidated. Activation of B cells by IL-4 or LPS induces the expression of IRF4 and BLIMP1, a transcriptional repressor essential in PC differentiation (Reimold et al. 2001; Gass et al. 2002; Iwakoshi et al. 2003). Both the BLIMP1-orchestrated repression of PAX5 and activation of ATF-6 induce *XBP1u* mRNA and activate the IRE-1 endonuclease, generating XBP1s protein (Reimold et al. 1996; Lin et al. 2002; Iwakoshi et al. 2003; Shaffer et al. 2004; Tellier et al. 2016). Together with BLIMP1, XBP1 sets the stage for terminal differentiation of B cells, including cell growth, enhanced protein synthesis, and expansion of cell organelles such as mitochondria, lysosomes, and the ER (Shaffer et al. 2004; Tellier et al. 2016). Furthermore, XBP1 specifically promotes IgM synthesis and translation (Tirosh et al. 2005). The ensuing UPR activation acts as a feedforward mechanism, generating more BLIMP1 and XBP1 protein and bolstering IgM antibody production (Iwakoshi et al. 2003; Gunn et al. 2004; Doody et al. 2006).

The importance of XBP1 in PCs is underscored by the finding that the mere overexpression of XBP1 in B cells is sufficient to drive the development of frank multiple myeloma (MM), a cancerous transformation of terminally differentiated B cells (Carrasco et al. 2007). It was consequently revealed that the current state-of-the-art therapeutics for MM, proteasome inhibitors (PIs), target XBP1, activate the UPR, and induce apoptosis specifically in MM cells (Lee et al. 2003a; Obeng et al. 2006; Mimura et al. 2007). The therapeutic efficacy of PIs could be enhanced by the combination therapy with compounds blocking the IRE-1 endonuclease activity (Papandreou et al. 2011; Mimura et al. 2012; Tang et al. 2014). On the other hand, XBP1-negative tumor cells emerge during PI treatment and are a hallmark of relapse and PI resistance (Leung-Hagesteijn et al. 2013), limiting the use of combination therapies. Future studies should address whether and how IRE-1 inhibitors can be harnessed as potential therapeutics, not only for MM but also for other B-cell malignancies and antibody-driven autoimmune diseases (Neubert et al. 2008; Todd et al. 2009; Kharabi Masouleh et al. 2014; Tang et al. 2014).

Although these data collectively give rise to an elegant model of UPR activation in PCs, some outstanding questions remain unanswered. How does the absence of IRE-1, independently of XBP1, block early B-cell development (Zhang et al. 2005)? Several publications have indicated that ER chaperone induction precedes UPR activation, rising assumptions about a non-UPR driven ER expansion (Gass et al. 2002; van Anken et al. 2003; Iwakoshi et al. 2003). What drives this early expansion of the folding machinery? Furthermore, PCs generate up to 2ng of antibodies per day (Brinkmann and Heusser 1993). How can long-lived PCs combine longevity in the bone marrow and this laborious and potentially toxic workload? Indicative of an essential role of autophagy in controlling ER homeostasis and PC maintenance, Atg5-deficient PCs contain an enlarged ER, display overt UPR activation, and fail to generate memory responses (Pengo et al. 2013).

3.2.2 T-cell Development

In stark contrast to B cells, the UPR is understudied in T-cell development. Similar to B cells, naïve T cells increase ER chaperone expression, transcription of XBP1u mRNA, and XBP1 splicing upon antigen recognition by the T-cell receptor (TCR) and IL-2 stimulation (Kamimura and Bevan 2008; Pino et al. 2008). This suggests that the anticipatory activation of the UPR by an extracellular signal may be a general mechanism of multiple cell types (Pino et al. 2008). In T cells, the TCR-mediated UPR activation depends on protein kinase C and occurs regardless of the subsequent differentiation of the naïve T cells (Pino et al. 2008). The tight regulation of key metabolic switches such as mTOR or the UPR is crucial in T-cell development as inappropriate activation of these pathways in quiescent thymocytes results in partial activation, proliferation defects, and enhanced apoptosis (Ozcan et al. 2008; Omar et al. 2016). In the spleen, a subset of KLRG1$^+$ CD8 T cells activates XBP1 which contributes to their terminal differentiation (Brunsing et al. 2008; Kamimura and Bevan 2008 and Fig. 2). Using a transgenic mouse reporting on XBP1 splicing, one study found a similar activation of the IRE-1/XBP1 signaling axis in natural killer cells, revealing an intriguing parallel between these two immune cell types with cytotoxic properties (Osorio et al. 2014).

3.3 *Granulocytes and UPR*

Compared to the lymphoid compartment, a similar theme emerges in the development of granulocytes, although with a couple of surprising twists. During granulocyte maturation, cytokines and mitogens trigger the transcription of *XBP1u* mRNA and selectively activate the IRE-1 but not the PERK branch of the UPR. The ensuing transcriptional program modifies cell cycle and sustains granulocyte differentiation. Failure to properly acquire this checkpoint results in enhanced/premature apoptosis (Kurata et al. 2011; Bettigole et al. 2015).

Granulocyte progenitors display *XBP1* splicing in vivo, paralleling the enhanced protein synthesis to underpin granulocyte production (Signer et al. 2014; Bettigole et al. 2015). In vitro, IL-3 stimulation of the granulocyte cell line 32Dcl3 results in STAT5- and PI3K-dependent XBP1 splicing. Upon overexpression of XBP1s, this cell line exhibits features of granulocyte maturation (Kurata et al. 2011). Remarkably, the absence of XBP1 in vivo cripples eosinophil differentiation without noteworthy defects in the neutrophil and basophil lineages (Bettigole et al. 2015). In this report, the authors found that the absence of XBP1 resulted into enhanced apoptosis of eosinophil progenitors and a developmental block with reduced formation of eosinophilic granules. These granules contain the ER-derived proteoglycan 2 (PRG2) and eosinophil peroxidase (EPX), whose mRNA is remarkably reduced in the absence of XBP1. The authors speculate that the absence of these lineage-specific granule proteins halts the developmental program by reducing GATA1, in an attempt to allow the UPR to restore homeostasis. The prolonged PERK activation might then drive the subsequent demise of the eosinophil-committed cells (Bettigole et al. 2015 and Fig. 2).

Since loss of XBP1 activates the IRE-1 endonuclease activity in a number of cells (Hur et al. 2012; Benhamron et al. 2013; Osorio et al. 2014), the relative contribution of RIDD in the pathogenesis remains to be investigated. Furthermore, activation of both neutrophils and basophils during inflammation induces the secretion of a large amount of inflammatory agents, putting a heavy burden on the ER. The relative contribution of XBP1 during these conditions to the functional properties of granulocytes remains to be elucidated.

3.4 The UPR in the Mononuclear Phagocytic System

Considered as a separate branch in the immune system, the mononuclear phagocytic system (MPS) is comprised of macrophages, monocytes, and DCs (Yona and Gordon 2015). As a distinct lineage, these cells originate from early committed progenitors and have specialized but pleiotropic functionalities (Naik et al. 2013). Whereas monocytes are considered as a quiescent pool of BM residing and circulating cells that can be quickly recruited in the advent of inflammation, macrophages and DCs are tissue-resident cells (Guilliams et al. 2014). DCs are short-lived and continuously replenished from a pool of BM-derived precursors called pre-DCs. Furthermore, DCs are paramount in the priming and skewing of T cells: Residing in the tissue, these cells integrate antigenic and environmental cues, migrate to the T-cell compartment of lymphoid organs, and relay the antigen-specific information to mount an appropriate T-cell response (Merad et al. 2013). In contrast, most macrophages develop from yolk sac-derived embryonic precursors and acquire functionalities beyond the classical immune functions, including tissue homeostasis regulation (Ginhoux and Jung 2014).

DCs constitutively activate the IRE-1/XBP1 axis in the absence of ER stress (Iwakoshi et al. 2007; Osorio et al. 2014). The first studies on the role of the UPR in DC development were conducted in chimeric mice produced by the injection of XBP1$^{-/-}$ embryonic stem cells into C57BL/6-RAG-2$^{-/-}$ blastocysts (Iwakoshi et al. 2007). In these mice, DC subsets, but particularly the plasmacytoid DCs (pDC), were dramatically reduced in number and the remaining cells expressed high levels of the surface molecules CD86 and MHCII, generally associated with DC activation. Furthermore, these cells did not survive ex vivo culture and displayed enhanced cleavage of caspases, a hallmark of apoptosis. Indicative of a developmental block, cell cultures of the XBP1/RAG2$^{-/-}$ chimeric BM cells failed to generate DCs (Iwakoshi et al. 2007). These findings were particularly peculiar since DCs are known to be mainly derived from RAG2-independent ancestors (Naik et al. 2013). In a more recent study using a mouse model with widespread loss of XBP1 in hematopoietic cells, the authors reaffirmed the reduction of the CD11c^{high} compartment in the absence of XBP1, although the relative contributions of compensatory mechanisms, such as UPR activation and RIDD, are still to be assessed (Bettigole et al. 2015). Indeed, it was found that splenic CD8$^+$ conventional DCs (cDC) critically depend on a tightly regulated IRE-1/XBP1 axis for cross-presentation of particulate antigen. The conditional deletion of XBP1 in DCs activates RIDD in CD8$^+$ but not CD11b$^+$ cDCs and results in the decay of a set of RNAs coding for antigen processing and presentation machinery (Osorio et al. 2014 and Fig. 2). Despite reduced CD11c expression in cDC1s, a consequence of the degradation of *Itgb2* mRNA, terminally differentiated splenic DCs are present in appropriate numbers and phenotypically normal. Since the promotor of the CRE-recombinase used in this study (Itgax-cre) is only activated at the pre-DC level, these results indicate that XBP1 is not crucial for the terminal development of DCs (Osorio et al. 2014).

Reminiscent of differentiation of B cells into PCs, monocytes increase overall cell size, number of organelles, and protein synthesis upon differentiation into macrophages. This is accompanied by the structural and functional reorganization of the ER and a 'physiological' UPR activation (Dickhout et al. 2011 and Fig. 2). Macrophages are plastic cells, and it has been suggested that UPR activation polarizes macrophages toward a M2 phenotype, promoting cholesterol uptake and development into foam cells upon exposition to cholesterol (Oh et al. 2012; Riek et al. 2012). Indeed, monocytes infiltrating in early atherosclerotic lesions display M2 polarization and signs of UPR activation (Zhou and Tabas 2013). In mouse models of atherosclerosis, the uptake of oxidized cholesterol by macrophages elicits ER stress, subverting the physiological UPR into a chronic and toxic UPR, resulting in CHOP-dependent apoptosis. This UPR-mediated death of foam cells is associated with plaque instability and disease progression (Thorp et al. 2009).

In particular inflammatory settings, monocytes can also acquire most of the phenotypical and functional characteristics of DCs (Plantinga et al. 2013; Tamoutounour et al. 2013). These monocyte-derived DCs (moDCs) can outnumber cDCs, locally sustain antigen-specific T-cell responses, and produce large quantities of chemokines and inflammatory cytokines (Plantinga et al. 2013). Both in patients and mouse models of cancer, CD11b$^+$ DCs and moDCs become densely laden with triglycerides, which profoundly cripple their ability to present tumor-associated antigens (Herber et al. 2010). CD11c$^+$ phagocytes acquire large quantities of lipid products present in the tumor environment. The subsequent intracellular lipid oxidation by ROS generates reactive byproducts shown to generate ER stress. The ensuing XBP1 activation upregulates triglyceride biosynthetic genes (e.g., *Agpat6*, *Fasn*, *Scd2*, and *Lpar1*), induces the formation of lipid bodies, and inhibits T-cell activation. Indeed, silencing of XBP1 in tumor phagocytes enhances protective T-cell immune responses and host survival (Cubillos-Ruiz et al. 2015).

The contribution of the UPR in tissue-resident macrophage development remains largely unchartered territory, mediated in part by the embryonic origin of these cells. This research area is ripe for investigation and is bound to reveal unexpected roles in macrophage development and function.

3.5 UPR Regulation at the Mucosal Barrier

Mucosal cells are our foremost barrier, separating tissues such as lung and intestinal parenchym from the luminal content and commensal flora. In the intestine, up to 6 different intestinal epithelial cells (IECs) have been identified: absorptive enterocytes, Paneth cells, enteroendocrine cells, goblet cells, microfold villus cells, and tuft cells. Among these, Paneth cells and goblet cells are considered professional secretory cells and are an integral part of the innate immune system (Hosomi et al. 2015; Cao 2016).

A thick layer of mucus, which retains antimicrobial products in its framework and forms a physical barrier against environmental threats, lines the intestinal mucosa. Mucus is composed of mucins, large O-glycosylated proteins that are secreted by the goblet cells as polymers (Allen et al. 1998). The posttranslational processing of these proteins requires an elaborate secretory system to handle this workload, and a physiological UPR activation is crucial for the optimal functioning of goblet cells. For example, goblet cells of mice carrying missense mutations in the *Muc2* gene (the *Winnie* and *Eeyore* mice) retain intracellular MUC2 precursor aggregates and suffer from ER stress and enhanced apoptosis (Heazlewood et al. 2008; Eri et al. 2011). Deletion of anterior gradient protein 2 homolog (AGR2), a PDI facilitating the correct folding of MUC2 and MUC5AC, results in UPR activation and deficient mucin production in intestinal and airway goblet cells (Park et al. 2009; Zhao et al. 2010; Schroeder et al. 2012). As such, it is of no surprise that the UPR has been implicated in the correct development of the goblet cell. OASIS is an ER membrane-residing TF highly expressed in the colon and belongs to the same CREB/ATF family as ATF-6 (Asada et al. 2012). Similarly, OASIS is activated by regulated intramembranous proteolysis during the terminal differentiation of the goblet cell, which is associated with UPR activation. In the absence of OASIS, goblet cells retain an immature phenotype, produce less MUC2, harbor an abnormally expanded ER, and are reduced in numbers (Asada et al. 2012). IRE-1β, a paralog of IRE-1α, is specifically expressed in the gastrointestinal and respiratory tracts, and studies have implicated IRE-1β in the maintenance of goblet cell homeostasis (Bertolotti et al. 2001; Martino et al. 2013). In allergic airways, IRE-1β but not IRE-1α induces both *Agr2* mRNA through *XBP1* splicing and *Muc5ac* mRNA, increasing the production and secretion of mucins (Martino et al. 2013). Currently, the exact molecular mechanisms downstream of IRE-1β remain elusive, as the physiological role of the IRE-1β to splice *XBP1* mRNA has been questioned (Imagawa et al. 2008). Alternatively, both suppression of global protein synthesis by cleavage of 28S ribosomal RNA and optimized MUC2 synthesis through specific degradation of *Muc2* mRNA have been proposed as candidate functions of the IRE-1β endonuclease (Iwawaki et al. 2001; Tsuru et al. 2013). Mice lacking IRE-1β exhibits a distended ER and signs of UPR activation early in goblet cell differentiation, accompanied by accumulation of mucin precursor aggregates in the ER (Schroeder et al. 2012; Tsuru et al. 2013). Surprisingly, goblet cells were normal in transgenic mice bearing the non-phosphorylatable eIF2α-Ser51Ala mutant allele (Cao et al. 2014). These findings indicate that the PERK and other eIF2α kinases are redundant for goblet cell differentiation and, more importantly, are not required for mucin protein translation regulation.

Paneth cells are essential to maintain mucosal integrity and regulate the microbial populations by the production of large amounts of antimicrobial products such as defensins, phospholipase A2, and lysozyme. Differentiation, proliferation, antimicrobial peptide secretion, and survival of Paneth cells rely on various components of the UPR. In the absence of AGR2, the Paneth cell population expands beyond the base of the crypts into the villi of the small intestine (Zhao et al. 2010; Muñoz et al. 2012). This is associated with ER stress activation measured by elevated levels of *Hspa5*, *Pdia3*, and *Ddit3*. It remains to be determined whether the direct loss of AGR2 or the indirect effects of UPR activation are accountable for the enhanced Paneth cell proliferation (Zhao et al. 2010). Loss of XBP1 dramatically reduces the number of Paneth cells with the few remaining cells demonstrating a disorganized ER, hypomorphic granule formation, and displaying signs of ER stress, autophagy induction, and apoptosis (Kaser et al. 2008). In contrast to goblet cells, ATF-6-deficient Paneth cells are normal and rather rely on PERK- and PKR-mediated phosphorylation of eIF2α to sustain their function (Cao et al. 2012, 2013, 2014). The absence of phosphorylatable eIF2α in Paneth cells results in a reduced number of lysozyme-containing granules, a fragmented ER, damaged mitochondria, and a defective UPR response despite phosphorylation of both IRE-1 and PERK (Cao et al. 2012, 2014).

Except for OASIS, all of the above mutants acquire spontaneous intestinal inflammation or are more susceptible to gastrointestinal infections and experimental models of inflammatory bowel disease (IBD). Compared to these mutant mice, mice devoid of Paneth cells or lacking MUC2 in goblet cells display a reduced penetrance of spontaneous inflammatory disease. This indicates that the ensuing ER stress response in these genetically manipulated mice, rather than the sole loss of function of the IECs, contributes to disease induction and progression (Garabedian et al. 1997; Van der Sluis et al. 2006). Most notably, the absence of XBP1 in Paneth cells results in ER stress, IRE-1α-dependent proinflammatory NF-κB signaling, and increased sensitivity to TNFα and flagellin, ultimately culminating in spontaneous, microbiota-dependent enteritis (Adolph et al. 2013). These observations give rise to a conceptual model with a central role for IECs in IBD pathogenesis. In this model, IECs integrate a number of cues derived from the intestinal lumen (e.g., colitogenic microflora), cell-intrinsic stress signals (ER stress and autophagy), and pattern recognition and cytokine signaling (STAT and NF-κB signaling), which endow these cells with the capability to maintain quiescence or drive the activation of the immune system (Hosomi et al. 2015). Indeed, human genetic research has identified mutations in a number of genes involved in UPR (*XBP1, Agr2, and Ormdl3*), autophagy (*Atg16l1* and *Irgm*), and immune signaling (e.g., *Nod2, Stat3, and Il10R*) that are associated with IBD (Hasnain et al. 2013; Hosomi et al. 2015; Cao 2016). Biopsies of healthy and diseased tissue of IBD patients display features of ER stress, and treatment with folding chaperones such as TUDCA and BPA alleviates inflammation in various experimental models of IBD (Shkoda et al. 2007; Cao et al. 2013). In the *Winnie* mice, IL-10 and corticosteroids limit intestinal inflammation and the anti-inflammatory properties of these molecules partly rely on their capacity to enhance ERAD machinery and optimize MUC2 production (Hasnain et al. 2013; Das et al. 2013). Altogether, these findings indicate that chemical UPR modulation might be harnessed as therapeutics and supplement current IBD therapy.

4 Discussion

The ER is interconnected with the other cellular organelles and has adopted many functions beyond protein folding. As such, the UPR has evolved from an IRE-1p/Hac1 transcriptional program to a flexible multipronged signaling node interconnected with many other regulatory pathways in the cell. In this sense, even the classical IRE-1-XBP1/PERK-ATF-4/ATF-6 paradigm is probably insufficient to accommodate the wealth of functional specializations of the ER in metazoan cells. Moreover, it has become clear that during physiological processes, cells co-opt specific branches of the UPR while actively suppressing others. At present, we have only caught a glimpse of this physiological regulation and our current models of UPR regulation fail to offer a satisfying explanation (excellently reviewed in Rutkowski and Hegde 2010). This is beautifully demonstrated during the development and activation of the immune system. Although loss of chaperones such as BiP and p58IPK or accumulation of misfolded proteins results into classical UPR activation and perturbs differentiation and function of immune cells, it is clear that developing immune cells do not operate in this way but rather use a specific and preemptive activation of the UPR.

In the earliest steps of development, long-term stem cells rely on PERK to prevent proteotoxic stress and preserve a healthy niche. Interestingly, the output of (enhanced) PERK activation seems to be highly tissue-specific, and HSCs undergo apoptosis whereas PERK induces differentiation of stem cells in the crypts of the intestine, presumably to preserve continuity of the mucosal barrier (Heijmans et al. 2013; van Galen et al. 2014). The molecular basis of these distinct outputs is ill-defined. In contrast, more differentiated progenitor cells activate the adaptive IRE-1/XBP1 branch of the UPR to sustain proliferation. This seems to be conserved for the lymphoid, granulocytic, and mononuclear phagocytic lineage but also for intestinal cells (Zhang et al. 2005; Niederreiter et al. 2013; Bettigole et al. 2015). Deletion of BiP or IRE-1 results into lymphopenia and loss of B-cell progenitors, respectively, and we speculate that a functional UPR could be a metabolic checkpoint, similar to positive and negative selection of lymphocyte development (Zhang et al. 2005; Wey et al. 2012a).

The wide range of immune cell functions sets the stage for ER specialization and cell-specific UPR regulation. Professional secretory cells such as PCs, goblet cells, and Paneth cells rely on the IRE-1/XBP1 (and in some cells, ATF-6 paralogues) to accommodate the high rate of protein production. The role of PERK is intriguing, whereas PERK seems obsolete for PCs and goblet cells; the absence of phosphorylatable eIF2α is detrimental to Paneth cells (Cao et al. 2014). Rather, PCs require autophagy for sustainable immunoglobulin production and goblet cells seem to trim mucin production through IRE-1β-regulated mRNA decay, avoiding the activation of PERK to ensure continuous production of these proteins (Tsuru et al. 2013; Pengo et al. 2013).

In the case of the MPS, DCs and macrophages utilize the ERAD compartment for antigen presentation (Osorio et al. 2014). Recognition of microbial products results into the specific activation of UPR as a signal amplification cascade and optimizes cytokine production (Martinon et al. 2010; Clavarino et al. 2012). The integration of the innate immune response with the UPR has been proposed as a separate stress response pathway called the microbial stress response (for review see Cláudio et al. 2013).

Activation of the UPR in immune cells has been reported in a wide array of diseases (reviewed in Hasnain et al. 2012; Osorio et al. 2013; Janssens et al. 2014; Wang and Kaufman 2016). Although the exact contribution of the UPR in disease as a primary inducer or secondary propagator is still debated, it has become clear that modulating ER stress can be beneficial. Indeed, preclinical trials have identified compounds such as chemical chaperones (TUDCA, BPA) or UPR branch-specific small molecules (BI-09, salubrinal) with potential therapeutic opportunities for a number of diseases. Given the crucial nature of the UPR in (immune) cell development, careful use of these therapeutics will be mandatory and future research should aim to dissect the detrimental proinflammatory signals from the physiological branches.

References

Acosta-Alvear D, Zhou Y, Blais A et al (2007) XBP1 controls diverse cell type- and condition-specific transcriptional regulatory networks. Mol Cell 27:53–66. doi:10.1016/j.molcel.2007.06.011

Adolph TE, Tomczak MF, Niederreiter L et al (2013) Paneth cells as a site of origin for intestinal inflammation. Nature 503:272–276. doi:10.1038/nature12599

Allen A, Hutton DA, Pearson JP (1998) The MUC2 gene product: a human intestinal mucin. Int J Biochem Cell Biol 30:797–801

Anckar J, Sistonen L (2011) Regulation of H SF1 function in the heat stress response: implications in aging and disease. Annu Rev Biochem 80:1089–1115. doi:10.1146/annurev-biochem-060809-095203

Aragón T, van Anken E, Pincus D et al (2009) Messenger RNA targeting to endoplasmic reticulum stress signalling sites. Nature 457:736–740. doi:10.1038/nature07641

Asada R, Saito A, Kawasaki N et al (2012) The endoplasmic reticulum stress transducer OASIS is involved in the terminal differentiation of goblet cells in the large intestine. J Biol Chem 287:8144–8153. doi:10.1074/jbc.M111.332593

Bailey D, O'Hare P (2007) Transmembrane bZIP transcription factors in ER stress signaling and the unfolded protein response. Antioxid Redox Signal 9:2305–2321. doi:10.1089/ars.2007.1796

Benhamron S, Hadar R, Iwawaky T et al (2013) Regulated IRE-1 dependent decay participates in curtailing immunoglobulin secretion from plasma cells. Eur J Immunol. doi:10.1002/eji.201343953

Bertolotti A, Wang X, Novoa I et al (2001) Increased sensitivity to dextran sodium sulfate colitis in IRE1beta-deficient mice. J Clin Invest 107:585–593. doi:10.1172/JCI11476

Bertolotti A, Zhang Y, Hendershot LM et al (2000) Dynamic interaction of BiP and ER stress transducers in the unfolded-protein response. Nat Cell Biol 2:326–332. doi:10.1038/35014014

Bettigole SE, Lis R, Adoro S et al (2015) The transcription factor XBP1 is selectively required for eosinophil differentiation. Nat Immunol 16:829–837. doi:10.1038/ni.3225

Brinkmann V, Heusser CH (1993) T cell-dependent differentiation of human B cells into IgM, IgG, IgA, or IgE plasma cells: high rate of antibody production by IgE plasma cells, but limited clonal expansion of IgE precursors. Cell Immunol 152:323–332. doi:10.1006/cimm.1993.1294

Brunsing R, Omori SA, Weber F et al (2008) B- and T-cell development both involve activity of the unfolded protein response pathway. J Biol Chem 283:17954–17961. doi:10.1074/jbc.M801395200

Buszczak M, Signer RAJ, Morrison SJ (2014) Cellular differences in protein synthesis regulate tissue homeostasis. Cell 159:242–251. doi:10.1016/j.cell.2014.09.016

Calfon M, Zeng H, Urano F et al (2002) IRE1 couples endoplasmic reticulum load to secretory capacity by processing the XBP-1 mRNA. Nature 415:92–96. doi:10.1038/415092a

Cao SS (2016) Epithelial ER stress in Crohn's disease and ulcerative colitis. Inflamm Bowel Dis 22:984–993. doi:10.1097/MIB.0000000000000660

Cao SS, Song B, Kaufman RJ (2012) PKR protects colonic epithelium against colitis through the unfolded protein response and prosurvival signaling. Inflamm Bowel Dis 18:1735–1742. doi:10.1002/ibd.22878

Cao SS, Wang M, Harrington JC et al (2014) Phosphorylation of eIF2α is dispensable for differentiation but required at a posttranscriptional level for paneth cell function and intestinal homeostasis in mice. Inflamm Bowel Dis 20:712–722. doi:10.1097/MIB.0000000000000010

Cao SS, Zimmermann EM, Chuang B-M et al (2013) The unfolded protein response and chemical chaperones reduce protein misfolding and colitis in mice. Gastroenterology 144(989–1000):e6. doi:10.1053/j.gastro.2013.01.023

Carrara M, Prischi F, Nowak PR, Ali MM (2015) Crystal structures reveal transient PERK luminal domain tetramerization in endoplasmic reticulum stress signaling. EMBO J 34:1589–1600. doi:10.15252/embj.201489183

Carrasco DR, Sukhdeo K, Protopopova M et al (2007) The differentiation and stress response factor XBP-1 drives multiple myeloma pathogenesis. Cancer Cell 11:349–360. doi:10.1016/j.ccr.2007.02.015

Chen X, Iliopoulos D, Zhang Q et al (2014) XBP1 promotes triple-negative breast cancer by controlling the HIF1α pathway. Nature. doi:10.1038/nature13119

Chin K-T, Zhou H-J, Wong C-M et al (2005) The liver-enriched transcription factor CREB-H is a growth suppressor protein underexpressed in hepatocellular carcinoma. Nucleic Acids Res 33:1859–1873. doi:10.1093/nar/gki332

Clavarino G, Cláudio N, Couderc T et al (2012) Induction of GADD34 is necessary for dsRNA-dependent interferon-β production and participates in the control of Chikungunya virus infection. PLoS Pathog 8:e1002708. doi:10.1371/journal.ppat.1002708

Cláudio N, Dalet A, Gatti E, Pierre P (2013) Mapping the crossroads of immune activation and cellular stress response pathways. EMBO J 32:1214–1224. doi:10.1038/emboj.2013.80

Coelho DS, Domingos PM (2014) Physiological roles of regulated Ire1 dependent decay. Front Genet 5:76. doi:10.3389/fgene.2014.00076

Credle JJ, Finer-Moore JS, Papa FR et al (2005) On the mechanism of sensing unfolded protein in the endoplasmic reticulum. Proc Natl Acad Sci U S A 102:18773–18784. doi:10.1073/pnas.0509487102

Cubillos-Ruiz JR, Silberman PC, Rutkowski MR, et al (2015) ER stress sensor XBP1 controls anti-tumor immunity by disrupting dendritic cell homeostasis. Cell 1–13. doi:10.1016/j.cell.2015.05.025

Cullinan SB, Zhang D, Hannink M et al (2003) Nrf2 is a direct PERK substrate and effector of PERK-dependent cell survival. Mol Cell Biol 23:7198–7209

Das I, Png CW, Oancea I et al (2013) Glucocorticoids alleviate intestinal ER stress by enhancing protein folding and degradation of misfolded proteins. J Exp Med. doi:10.1084/jem.20121268

Delépine M, Nicolino M, Barrett T et al (2000) EIF2AK3, encoding translation initiation factor 2-alpha kinase 3, is mutated in patients with Wolcott-Rallison syndrome. Nat Genet 25:406–409. doi:10.1038/78085

Dickhout JG, Lhoták Š, Hilditch BA et al (2011) Induction of the unfolded protein response after monocyte to macrophage differentiation augments cell survival in early atherosclerotic lesions. FASEB J 25:576–589. doi:10.1096/fj.10-159319

Doody GM, Stephenson S, Tooze RM (2006) BLIMP-1 is a target of cellular stress and downstream of the unfolded protein response. Eur J Immunol 36:1572–1582. doi:10.1002/eji.200535646

Eri RD, Adams RJ, Tran TV et al (2011) An intestinal epithelial defect conferring ER stress results in inflammation involving both innate and adaptive immunity. Mucosal Immunol 4:354–364. doi:10.1038/mi.2010.74

Garabedian EM, Roberts LJ, McNevin MS, Gordon JI (1997) Examining the role of Paneth cells in the small intestine by lineage ablation in transgenic mice. J Biol Chem 272:23729–23740

Gardner BM, Pincus D, Gotthardt K et al (2013) Endoplasmic reticulum stress sensing in the unfolded protein response. Cold Spring Harbor Perspect Biol 5:a013169. doi:10.1101/cshperspect.a013169

Gardner BM, Walter P (2011) Unfolded proteins are Ire1-activating ligands that directly induce the unfolded protein response. Science 333:1891–1894. doi:10.1126/science.1209126

Gass JN, Gifford NM, Brewer JW (2002) Activation of an unfolded protein response during differentiation of antibody-secreting B cells. J Biol Chem 277:49047–49054. doi:10.1074/jbc.M205011200

Ghosh R, Wang L, Wang ES et al (2014) Allosteric inhibition of the IRE1a RNase preserves cell viability and function during endoplasmic reticulum stress. Cell 158:534–548. doi:10.1016/j.cell.2014.07.002

Ginhoux F, Jung S (2014) Monocytes and macrophages: developmental pathways and tissue homeostasis. Nat Publishing Group 14:392–404. doi:10.1038/nri3671

Goldfinger M, Shmuel M, Benhamron S, Tirosh B (2011) Protein synthesis in plasma cells is regulated by crosstalk between endoplasmic reticulum stress and mTOR signaling. Eur J Immunol 41:491–502. doi:10.1002/eji.201040677

Grootjans J, Kaser A, Kaufman RJ, Blumberg RS (2016) The unfolded protein response in immunity and inflammation. Nat Publishing Group. doi:10.1038/nri.2016.62

Guilliams M, Ginhoux F, Jakubzick C et al (2014) Dendritic cells, monocytes and macrophages: a unified nomenclature based on ontogeny. Nat Rev Immunol 14:571–578. doi:10.1038/nri3712

Gunn KE, Gifford NM, Mori K, Brewer JW (2004) A role for the unfolded protein response in optimizing antibody secretion. Mol Immunol 41:919–927. doi:10.1016/j.molimm.2004.04.023

Hamanaka RB, Bennett BS, Cullinan SB, Diehl JA (2005) PERK and GCN2 contribute to eIF2alpha phosphorylation and cell cycle arrest after activation of the unfolded protein response pathway. Mol Biol Cell 16:5493–5501. doi:10.1091/mbc.E05-03-0268

Han J, Back SH, Hur J et al (2013) ER-stress-induced transcriptional regulation increases protein synthesis leading to cell death. Nat Cell Biol 15:481–490. doi:10.1038/ncb2738

Harding HP, Zeng H, Zhang Y et al (2001) Diabetes mellitus and exocrine pancreatic dysfunction in perk −/− mice reveals a role for translational control in secretory cell survival. Mol Cell 7:1153–1163

Harding HP, Zhang Y, Zeng H et al (2003) An integrated stress response regulates amino acid metabolism and resistance to oxidative stress. Mol Cell 11:619–633

Hasnain SZ, Lourie R, Das I et al (2012) The interplay between endoplasmic reticulum stress and inflammation. Immunol Cell Biol 90:260–270. doi:10.1038/icb.2011.112

Hasnain SZ, Tauro S, Das I et al (2013) IL-10 promotes production of intestinal mucus by suppressing protein misfolding and endoplasmic reticulum stress in goblet cells. Gastroenterology 144(357–368):e9. doi:10.1053/j.gastro.2012.10.043

Heazlewood CK, Cook MC, Eri R et al (2008) Aberrant mucin assembly in mice causes endoplasmic reticulum stress and spontaneous inflammation resembling ulcerative colitis. PLoS Med 5:e54. doi:10.1371/journal.pmed.0050054

Heijmans J, van Lidth de Jeude JF, Koo B-K et al (2013) ER stress causes rapid loss of intestinal epithelial stemness through activation of the unfolded protein response. Cell Rep 3:1128–1139. doi:10.1016/j.celrep.2013.02.031

Herber DL, Cao W, Nefedova Y et al (2010) Lipid accumulation and dendritic cell dysfunction in cancer. Nat Med 16:880–886. doi:10.1038/nm.2172

Hershey JWB, Sonenberg N, Mathews MB (2012) Principles of translational control: an overview. Cold Spring Harbor Perspect Biol. doi:10.1101/cshperspect.a011528

Hetz C, Chevet E, Oakes SA (2015) Proteostasis control by the unfolded protein response. Nat Cell Biol 17:829–838. doi:10.1038/ncb3184

Hollien J, Lin JH, Li H et al (2009) Regulated Ire1-dependent decay of messenger RNAs in mammalian cells. J Cell Biol 186:323–331. doi:10.1083/jcb.200903014

Hollien J, Weissman JS (2006) Decay of endoplasmic reticulum-localized mRNAs during the unfolded protein response. Science 313:104–107. doi:10.1126/science.1129631

Hosomi S, Kaser A, Blumberg RS (2015) Role of endoplasmic reticulum stress and autophagy as interlinking pathways in the pathogenesis of inflammatory bowel disease. Curr Opin Gastroenterol 31:81–88. doi:10.1097/MOG.0000000000000144

Hu C-CA, Dougan SK, McGehee AM et al (2009) XBP-1 regulates signal transduction, transcription factors and bone marrow colonization in B cells. EMBO J 28:1624–1636. doi:10.1038/emboj.2009.117

Huang J, Nguyen-McCarty M, Hexner EO et al (2012) Maintenance of hematopoietic stem cells through regulation of Wnt and mTOR pathways. Nat Med 18:1778–1785. doi:10.1038/nm.2984

Hur KY, So J-S, Ruda V et al (2012) IRE1α activation protects mice against acetaminophen-induced hepatotoxicity. J Exp Med 209:307–318. doi:10.1084/jem.20111298

Imagawa Y, Hosoda A, Sasaka S-I et al (2008) RNase domains determine the functional difference between IRE1alpha and IRE1beta. FEBS Lett 582:656–660. doi:10.1016/j.febslet.2008.01.038

Ito K, Suda T (2014) Metabolic requirements for the maintenance of self-renewing stem cells. Nat Rev Mol Cell Biol 15:243–256. doi:10.1038/nrm3772

Iwakoshi NN, Lee A-H, Vallabhajosyula P et al (2003) Plasma cell differentiation and the unfolded protein response intersect at the transcription factor XBP-1. Nat Immunol 4:321–329. doi:10.1038/ni907

Iwakoshi NN, Pypaert M, Glimcher LH (2007) The transcription factor XBP-1 is essential for the development and survival of dendritic cells. J Exp Med 204:2267–2275. doi:10.1084/jem.20070525

Iwawaki T, Hosoda A, Okuda T et al (2001) Translational control by the ER transmembrane kinase/ribonuclease IRE1 under ER stress. Nat Cell Biol 3:158–164. doi:10.1038/35055065

Janssens S, Pulendran B, Lambrecht BN (2014) Emerging functions of the unfolded protein response in immunity. Nat Immunol 15:910–919. doi:10.1038/ni.2991

Kamimura D, Bevan MJ (2008) Endoplasmic reticulum stress regulator XBP-1 contributes to effector CD8 + T cell differentiation during acute infection. J Immunol 181:5433–5441

Kaser A, Lee A-H, Franke A et al (2008) XBP1 links ER stress to intestinal inflammation and confers genetic risk for human inflammatory bowel disease. Cell 134:743–756. doi:10.1016/j.cell.2008.07.021

Kharabi Masouleh B, Geng H, Hurtz C et al (2014) Mechanistic rationale for targeting the unfolded protein response in pre-B acute lymphoblastic leukemia. Proc Nat Acad Sci 111: E2219–E2228. doi:10.1073/pnas.1400958111

Kimata Y, Ishiwata-Kimata Y, Ito T et al (2007) Two regulatory steps of ER-stress sensor Ire1 involving its cluster formation and interaction with unfolded proteins. J Cell Biol 179:75–86. doi:10.1083/jcb.200704166

Kimata Y, Oikawa D, Shimizu Y et al (2004) A role for BiP as an adjustor for the endoplasmic reticulum stress-sensing protein Ire1. J Cell Biol 167:445–456. doi:10.1083/jcb.200405153

Kimmig P, Diaz M, Zheng J et al (2012) The unfolded protein response in fission yeast modulates stability of select mRNAs to maintain protein homeostasis. Elife 1:e00048. doi:10.7554/eLife.00048

Kitai Y, Ariyama H, Kono N et al (2013) Membrane lipid saturation activates IRE1α without inducing clustering. Genes Cells 18:798–809. doi:10.1111/gtc.12074

Kojima E, Takeuchi A, Haneda M et al (2003) The function of GADD34 is a recovery from a shutoff of protein synthesis induced by ER stress: elucidation by GADD34-deficient mice. FASEB J 17:1573–1575. doi:10.1096/fj.02-1184fje

Korennykh A, Walter P (2012) Structural basis of the unfolded protein response. Annu Rev Cell Dev Biol 28:251–277. doi:10.1146/annurev-cellbio-101011-155826

Korennykh AV, Egea PF, Korostelev AA et al (2009) The unfolded protein response signals through high-order assembly of Ire1. Nature 457:687–693. doi:10.1038/nature07661

Kosmaczewski SG, Edwards TJ, Han SM et al (2014) The RtcB RNA ligase is an essential component of the metazoan unfolded protein response. EMBO Rep. doi:10.15252/embr.201439531

Kurata M, Yamazaki Y, Kanno Y et al (2011) Anti-apoptotic function of Xbp1 as an IL-3 signaling molecule in hematopoietic cells. Cell Death Dis 2:e118. doi:10.1038/cddis.2011.1

Lee A-H, Iwakoshi NN, Anderson KC, Glimcher LH (2003a) Proteasome inhibitors disrupt the unfolded protein response in myeloma cells. Proc Natl Acad Sci U S A 100:9946–9951. doi:10.1073/pnas.1334037100

Lee A-H, Iwakoshi NN, Glimcher LH (2003b) XBP-1 regulates a subset of endoplasmic reticulum resident chaperone genes in the unfolded protein response. Mol Cell Biol 23:7448–7459

Leung-Hagesteijn C, Erdmann N, Cheung G et al (2013) Xbp1s-negative tumor B cells and pre-plasmablasts mediate therapeutic proteasome inhibitor resistance in multiple myeloma. Cancer Cell 24:289–304. doi:10.1016/j.ccr.2013.08.009

Lin K-I, Angelin-Duclos C, Kuo TC, Calame K (2002) Blimp-1-dependent repression of Pax-5 is required for differentiation of B cells to immunoglobulin M-secreting plasma cells. Mol Cell Biol 22:4771–4780

Liu CY, Schröder M, Kaufman RJ (2000) Ligand-independent dimerization activates the stress response kinases IRE1 and PERK in the lumen of the endoplasmic reticulum. J Biol Chem 275:24881–24885. doi:10.1074/jbc.M004454200

Lu Y, Liang F-X, Wang X (2014) A synthetic biology approach identifies the mammalian UPR RNA ligase RtcB. Mol Cell 55:758–770. doi:10.1016/j.molcel.2014.06.032

Ma Y, Shimizu Y, Mann MJ et al (2010) Plasma cell differentiation initiates a limited ER stress response by specifically suppressing the PERK-dependent branch of the unfolded protein response. Cell Stress Chaperones 15:281–293. doi:10.1007/s12192-009-0142-9

Martino MB, Jones L, Brighton B et al (2013) The ER stress transducer IRE1β is required for airway epithelial mucin production. Mucosal Immunol 6:639–654. doi:10.1038/mi.2012.105

Martinon F, Chen X, Lee A-H, Glimcher LH (2010) TLR activation of the transcription factor XBP1 regulates innate immune responses in macrophages. Nat Immunol 11:411–418. doi:10.1038/ni.1857

Maurel M, Chevet E, Tavernier J, Gerlo S (2014) Getting RIDD of RNA: IRE1 in cell fate regulation. Trends Biochem Sci. doi:10.1016/j.tibs.2014.02.008

Merad M, Sathe P, Helft J et al (2013) The dendritic cell lineage: ontogeny and function of dendritic cells and their subsets in the steady state and the inflamed setting. Annu Rev Immunol 31:563–604. doi:10.1146/annurev-immunol-020711-074950

Miharada K, Sigurdsson V, Karlsson S (2014) Dppa5 improves hematopoietic stem cell activity by reducing endoplasmic reticulum stress. Cell Rep 7:1381–1392. doi:10.1016/j.celrep.2014.04.056

Mimura N, Fulciniti M, Gorgun G et al (2012) Blockade of XBP1 splicing by inhibition of IRE1α is a promising therapeutic option in multiple myeloma. Blood 119:5772–5781. doi:10.1182/blood-2011-07-366633

Mimura N, Hamada H, Kashio M et al (2007) Aberrant quality control in the endoplasmic reticulum impairs the biosynthesis of pulmonary surfactant in mice expressing mutant BiP. Cell Death Differ 14:1475–1485. doi:10.1038/sj.cdd.4402151

Morrison SJ, Spradling AC (2008) Stem cells and niches: mechanisms that promote stem cell maintenance throughout life. Cell 132:598–611. doi:10.1016/j.cell.2008.01.038

Muñoz J, Stange DE, Schepers AG et al (2012) The Lgr5 intestinal stem cell signature: robust expression of proposed quiescent "+4" cell markers. EMBO J 31:3079–3091. doi:10.1038/emboj.2012.166

Nadanaka S, Okada T, Yoshida H, Mori K (2007) Role of disulfide bridges formed in the luminal domain of ATF6 in sensing endoplasmic reticulum stress. Mol Cell Biol 27:1027–1043. doi:10.1128/MCB.00408-06

Naik SH, Perié L, Swart E et al (2013) Diverse and heritable lineage imprinting of early haematopoietic progenitors. Nature 496:229–232. doi:10.1038/nature12013

Nakamura T, Furuhashi M, Li P et al (2010) Double-stranded RNA-dependent protein kinase links pathogen sensing with stress and metabolic homeostasis. Cell 140:338–348. doi:10.1016/j.cell.2010.01.001

Neubert K, Meister S, Moser K et al (2008) The proteasome inhibitor bortezomib depletes plasma cells and protects mice with lupus-like disease from nephritis. Nat Med 14:748–755. doi:10.1038/nm1763

Niederreiter L, Fritz TMJ, Adolph TE et al (2013) ER stress transcription factor Xbp1 suppresses intestinal tumorigenesis and directs intestinal stem cells. J Exp Med 210:2041–2056. doi:10.1084/jem.20122341

Novoa I, Zeng H, Harding HP, Ron D (2001) Feedback inhibition of the unfolded protein response by GADD34-mediated dephosphorylation of eIF2alpha. J Cell Biol 153:1011–1022

Novoa I, Zhang Y, Zeng H et al (2003) Stress-induced gene expression requires programmed recovery from translational repression. EMBO J 22:1180–1187. doi:10.1093/emboj/cdg112

Obeng EA, Carlson LM, Gutman DM et al (2006) Proteasome inhibitors induce a terminal unfolded protein response in multiple myeloma cells. Blood 107:4907–4916. doi:10.1182/blood-2005-08-3531

Oh J, Riek AE, Weng S et al (2012) Endoplasmic reticulum stress controls M2 macrophage differentiation and foam cell formation. J Biol Chem 287:11629–11641. doi:10.1074/jbc.M111.338673

Omar I, Lapenna A, Cohen-Daniel L et al (2016) Schlafen2 mutation unravels a role for chronic ER stress in the loss of T cell quiescence. Oncotarget. doi:10.18632/oncotarget.9818

Osorio F, Lambrecht B, Janssens S (2013) The UPR and lung disease. Semin Immunopathol. doi:10.1007/s00281-013-0368-6

Osorio F, Tavernier SJ, Hoffmann E et al (2014) The unfolded-protein-response sensor IRE-1α regulates the function of CD8α(+) dendritic cells. Nat Immunol. doi:10.1038/ni.2808

Ozcan U, Ozcan L, Yilmaz E et al (2008) Loss of the tuberous sclerosis complex tumor suppressors triggers the unfolded protein response to regulate insulin signaling and apoptosis. Mol Cell 29:541–551. doi:10.1016/j.molcel.2007.12.023

Papandreou I, Denko NC, Olson M et al (2011) Identification of an Ire1alpha endonuclease specific inhibitor with cytotoxic activity against human multiple myeloma. Blood 117:1311–1314. doi:10.1182/blood-2010-08-303099

Park S-W, Zhen G, Verhaeghe C et al (2009) The protein disulfide isomerase AGR2 is essential for production of intestinal mucus. Proc Nat Acad Sci 106:6950–6955. doi:10.1073/pnas.0808722106

Pengo N, Scolari M, Oliva L et al (2013) Plasma cells require autophagy for sustainable immunoglobulin production. Nat Immunol 14:298–305. doi:10.1038/ni.2524

Pincus D, Chevalier MW, Aragón T et al (2010) BiP binding to the ER-stress sensor Ire1 tunes the homeostatic behavior of the unfolded protein response. PLoS Biol 8:e1000415. doi:10.1371/journal.pbio.1000415

Pino SC, O'Sullivan-Murphy B, Lidstone EA et al (2008) Protein kinase C signaling during T cell activation induces the endoplasmic reticulum stress response. Cell Stress Chaperones 13:421–434. doi:10.1007/s12192-008-0038-0

Plantinga M, Guilliams M, Vanheerswynghels M et al (2013) Conventional and monocyte-derived CD11b(+) dendritic cells initiate and maintain T helper 2 cell-mediated immunity to house dust mite allergen. Immunity 38:322–335. doi:10.1016/j.immuni.2012.10.016

Princiotta MF, Finzi D, Qian S-B et al (2003) Quantitating protein synthesis, degradation, and endogenous antigen processing. Immunity 18:343–354

Reavie L, Gatta Della G, Crusio K et al (2010) Regulation of hematopoietic stem cell differentiation by a single ubiquitin ligase-substrate complex. Nat Immunol 11:207–215. doi:10.1038/ni.1839

Reimold AM, Iwakoshi NN, Manis J et al (2001) Plasma cell differentiation requires the transcription factor XBP-1. Nature 412:300–307. doi:10.1038/35085509

Reimold AM, Ponath PD, Li YS et al (1996) Transcription factor B cell lineage-specific activator protein regulates the gene for human X-box binding protein 1. J Exp Med 183:393–401

Riek AE, Oh J, Sprague JE et al (2012) Vitamin D suppression of endoplasmic reticulum stress promotes an antiatherogenic monocyte/macrophage phenotype in type 2 diabetic patients. J Biol Chem 287:38482–38494. doi:10.1074/jbc.M112.386912

Rosekrans SL, Heijmans J, Büller NVJA et al (2015) ER stress induces epithelial differentiation in the mouse oesophagus. Gut 64:195–202. doi:10.1136/gutjnl-2013-306347

Rouault-Pierre K, Lopez-Onieva L, Foster K et al (2013) HIF-2α protects human hematopoietic stem/progenitors and acute myeloid leukemic cells from apoptosis induced by endoplasmic reticulum stress. Cell Stem Cell 13:549–563. doi:10.1016/j.stem.2013.08.011

Rutkowski DT, Hegde RS (2010) Regulation of basal cellular physiology by the homeostatic unfolded protein response. J Cell Biol 189:783–794. doi:10.1083/jcb.201003138

Scheuner D, Song B, McEwen E et al (2001) Translational control is required for the unfolded protein response and in vivo glucose homeostasis. Mol Cell 7:1165–1176

Schroeder BW, Verhaeghe C, Park S-W et al (2012) AGR2 is induced in asthma and promotes allergen-induced mucin overproduction. Am J Respir Cell Mol Biol 47:178–185. doi:10.1165/rcmb.2011-0421OC

Schwitalla S, Fingerle AA, Cammareri P et al (2013) Intestinal tumorigenesis initiated by dedifferentiation and acquisition of stem-cell-like properties. Cell 152:25–38. doi:10.1016/j.cell.2012.12.012

Shaffer AL, Shapiro-Shelef M, Iwakoshi NN et al (2004) XBP1, downstream of Blimp-1, expands the secretory apparatus and other organelles, and increases protein synthesis in plasma cell differentiation. Immunity 21:81–93. doi:10.1016/j.immuni.2004.06.010

Shen J, Chen X, Hendershot L, Prywes R (2002) ER stress regulation of ATF6 localization by dissociation of BiP/GRP78 binding and unmasking of Golgi localization signals. Dev Cell 3:99–111

Shkoda A, Ruiz PA, Daniel H et al (2007) Interleukin-10 blocked endoplasmic reticulum stress in intestinal epithelial cells: impact on chronic inflammation. YGAST 132:190–207. doi:10.1053/j.gastro.2006.10.030

Sidrauski C, McGeachy AM, Ingolia NT, Walter P (2015) The small molecule ISRIB reverses the effects of eIF2α phosphorylation on translation and stress granule assembly. Elife. doi:10.7554/eLife.05033

Sidrauski C, Walter P (1997) The transmembrane kinase Ire1p is a site-specific endonuclease that initiates mRNA splicing in the unfolded protein response. Cell 90:1031–1039

Signer RAJ, Magee JA, Salic A, Morrison SJ (2014) Haematopoietic stem cells require a highly regulated protein synthesis rate. Nature 509:49–54. doi:10.1038/nature13035

Sigurdsson V, Takei H, Soboleva S et al (2016) Bile acids protect expanding hematopoietic stem cells from unfolded protein stress in fetal liver. Stem Cell 18:522–532. doi:10.1016/j.stem. 2016.01.002

Smith MH, Ploegh HL, Weissman JS (2011) Road to ruin: targeting proteins for degradation in the endoplasmic reticulum. Science 334:1086–1090. doi:10.1126/science.1209235

Sriburi R, Bommiasamy H, Buldak GL et al (2007) Coordinate regulation of phospholipid biosynthesis and secretory pathway gene expression in XBP-1(S)-induced endoplasmic reticulum biogenesis. J Biol Chem 282:7024–7034. doi:10.1074/jbc.M609490200

Tabas I, Ron D (2011) Integrating the mechanisms of apoptosis induced by endoplasmic reticulum stress. Nat Cell Biol 13:184–190. doi:10.1038/ncb0311-184

Tam AB, Koong AC, Niwa M (2014) Ire1 has distinct catalytic mechanisms for XBP1/HAC1 splicing and RIDD. Cell Rep 9:850–858. doi:10.1016/j.celrep.2014.09.016

Tamoutounour S, Guilliams M, Montanana Sanchis F et al (2013) Origins and functional specialization of macrophages and of conventional and monocyte-derived dendritic cells in mouse skin. Immunity 39:925–938. doi:10.1016/j.immuni.2013.10.004

Tang C-HA, Ranatunga S, Kriss CL et al (2014) Inhibition of ER stress-associated IRE-1/XBP-1 pathway reduces leukemic cell survival. J Clin Invest 124:2585–2598. doi:10.1172/JCI73448

Tannous A, Pisoni GB, Hebert DN, Molinari M (2015) N-linked sugar-regulated protein folding and quality control in the ER. Semin Cell Dev Biol 41:79–89. doi:10.1016/j.semcdb.2014.12. 001

Taubenheim N, Tarlinton DM, Crawford S et al (2012) High rate of antibody secretion is not integral to plasma cell differentiation as revealed by XBP-1 deficiency. J Immunol 189:3328–3338. doi:10.4049/jimmunol.1201042

Tellier J, Shi W, Minnich M et al (2016) Blimp-1 controls plasma cell function through the regulation of immunoglobulin secretion and the unfolded protein response. Nat Immunol 17:323–330. doi:10.1038/ni.3348

Thorp E, Li G, Seimon TA et al (2009) Reduced apoptosis and plaque necrosis in advanced atherosclerotic lesions of Apoe −/− and Ldlr −/− mice lacking CHOP. Cell Metab 9:474–481. doi:10.1016/j.cmet.2009.03.003

Tirosh B, Iwakoshi NN, Glimcher LH, Ploegh HL (2005) XBP-1 specifically promotes IgM synthesis and secretion, but is dispensable for degradation of glycoproteins in primary B cells. J Exp Med 202:505–516. doi:10.1084/jem.20050575

Todd DJ, McHeyzer-Williams LJ, Kowal C et al (2009) XBP1 governs late events in plasma cell differentiation and is not required for antigen-specific memory B cell development. J Exp Med 206:2151–2159. doi:10.1084/jem.20090738

Tsuru A, Fujimoto N, Takahashi S et al (2013) Negative feedback by IRE1β optimizes mucin production in goblet cells. Proc Nat Acad Sci 110:2864–2869. doi:10.1073/pnas.1212484110

van Anken E, Romijn EP, Maggioni C et al (2003) Sequential waves of functionally related proteins are expressed when B cells prepare for antibody secretion. Immunity 18:243–253

Van der Sluis M, De Koning BAE, De Bruijn ACJM et al (2006) Muc2-deficient mice spontaneously develop colitis, indicating that MUC2 is critical for colonic protection. YGAST 131:117–129. doi:10.1053/j.gastro.2006.04.020

van Galen P, Kreso A, Mbong N et al (2014) The unfolded protein response governs integrity of the haematopoietic stem-cell pool during stress. Nature 510:268–272. doi:10.1038/nature13228

Volmer R, van der Ploeg K, Ron D (2013) Membrane lipid saturation activates endoplasmic reticulum unfolded protein response transducers through their transmembrane domains. Proc Nat Acad Sci 110:4628–4633. doi:10.1073/pnas.1217611110

Walter P, Ron D (2011) The unfolded protein response: from stress pathway to homeostatic regulation. Science 334:1081–1086. doi:10.1126/science.1209038

Wang L, Ryoo HD, Qi Y, Jasper H (2015) PERK Limits drosophila lifespan by promoting intestinal stem cell proliferation in response to ER stress. PLoS Genet 11:e1005220. doi:10.1371/journal.pgen.1005220

Wang M, Kaufman RJ (2016) Protein misfolding in the endoplasmic reticulum as a conduit to human disease. Nature 529:326–335. doi:10.1038/nature17041

Wang ZV, Deng Y, Gao N et al (2014) Spliced X-box binding protein 1 couples the unfolded protein response to hexosamine biosynthetic pathway. Cell 156:1179–1192. doi:10.1016/j.cell.2014.01.014

Wey S, Luo B, Lee AS (2012) Acute inducible ablation of GRP78 reveals its role in hematopoietic stem cell survival, lymphogenesis and regulation of stress signaling. PLoS ONE 7:e39047. doi:10.1371/journal.pone.0039047

Wiseman RL, Zhang Y, Lee KPK et al (2010) Flavonol activation defines an unanticipated ligand-binding site in the kinase-RNase domain of IRE1. Mol Cell 38:291–304. doi:10.1016/j.molcel.2010.04.001

Wu J, Rutkowski DT, Dubois M et al (2007) ATF6alpha optimizes long-term endoplasmic reticulum function to protect cells from chronic stress. Dev Cell 13:351–364. doi:10.1016/j.devcel.2007.07.005

Yamamoto K, Sato T, Matsui T et al (2007) Transcriptional induction of mammalian ER quality control proteins is mediated by single or combined action of ATF6alpha and XBP1. Dev Cell 13:365–376. doi:10.1016/j.devcel.2007.07.018

Ye J, Rawson RB, Komuro R et al (2000) ER stress induces cleavage of membrane-bound ATF6 by the same proteases that process SREBPs. Mol Cell 6:1355–1364

Yona S, Gordon S (2015) From the reticuloendothelial to mononuclear phagocyte system—the unaccounted years. Front Immunol 6:328. doi:10.3389/fimmu.2015.00328

Yoshida H, Matsui T, Hosokawa N et al (2003) A time-dependent phase shift in the mammalian unfolded protein response. Dev Cell 4:265–271

Yoshida H, Matsui T, Yamamoto A et al (2001) XBP1 mRNA is induced by ATF6 and spliced by IRE1 in response to ER stress to produce a highly active transcription factor. Cell 107:881–891

Zhang K, Wong HN, Song B et al (2005) The unfolded protein response sensor IRE1alpha is required at 2 distinct steps in B cell lymphopoiesis. J Clin Invest 115:268–281. doi:10.1172/JCI21848

Zhao F, Edwards R, Dizon D et al (2010) Disruption of Paneth and goblet cell homeostasis and increased endoplasmic reticulum stress in Agr2 −/− mice. Dev Biol 338:270–279. doi:10.1016/j.ydbio.2009.12.008

Zhou AX, Tabas I (2013) The UPR in atherosclerosis. Semin Immunopathol 35:321–332. doi:10.1007/s00281-013-0372-x

Mitochondria-Associated Membranes and ER Stress

Alexander R. van Vliet and Patrizia Agostinis

Abstract The endoplasmic reticulum (ER) is a crucial organelle for coordinating cellular Ca^{2+} signaling and protein synthesis and folding. Moreover, the dynamic and complex membranous structures constituting the ER allow the formation of contact sites with other organelles and structures, including among others the mitochondria and the plasma membrane (PM). The contact sites that the ER form with mitochondria is a hot topic in research, and the nature of the so-called mitochondria-associated membranes (MAMs) is continuously evolving. The MAMs consist of a proteinaceous tether that physically connects the ER with mitochondria. The MAMs harness the main functions of both organelles to form a specialized subcompartment at the interface of the ER and mitochondria. Under homeostatic conditions, MAMs are crucial for the efficient transfer of Ca^{2+} from the ER to mitochondria, and for proper mitochondria bioenergetics and lipid synthesis. MAMs are also believed to be the master regulators of mitochondrial shape and motility, and to form a crucial site for autophagosome assembly. Not surprisingly, MAMs have been shown to be a hot spot for the transfer of stress signals from the ER to mitochondria, most notably under the conditions of loss of ER proteostasis, by engaging the unfolded protein response (UPR). In this chapter after an introduction on ER biology and ER stress, we will review the emerging and key signaling roles of the MAMs, which have a root in cellular processes and signaling cascades coordinated by the ER.

Abbreviations

ACAT	acyl-CoA:cholesterol acyltransferase
ATF4	Activating transcription factor 4
ATF6	Activating transcription factor 6

A.R. van Vliet
Laboratory of Cell Death Research and Therapy, Department of Cellular and Molecular Medicine, University of Leuven, KU Leuven, 3000 Leuven, Belgium

P. Agostinis (✉)
Laboratory of Cell Death Research & Therapy, University of Leuven, Campus Gasthuisberg O&N1, Herestraat 49, box 802, 3000 Leuven, Belgium
e-mail: Patrizia.Agostinis@kuleuven.be

ATG	Autophagy-related gene
BiP	Immunoglobulin heavy chain-binding protein
Ca^{2+}	Calcium
Cav-1	Caveolin-1
cER	Cortical endoplasmic reticulum
CHOP	C/EBP homologous protein
DAG	Diacylglycerol
Drp1	Dynamin-1-like protein
eIF2α	Eukaryotic initiation factor 2 alpha
ER	Endoplasmic reticulum
ERAD	ER-associated protein degradation
ERMES	ER–mitochondria encounter structure
Ero1α	ER oxidoreductin 1 alpha
Fis1	Mitochondrial fission protein 1
GADD34	Growth arrest and DNA damage-inducible gene 34
GED	GTPase effector domain
Grp78	Glucose-regulated protein 78
Grp75	Glucose-regulated protein 75
GST	Glutathione-S-tranferase
GTP	Guanosine tri-phosphate
HO-1	Heme-oxygenase
HSP	Heat-shock protein
IMM	Inner mitochondrial membrane
IP_3R	Inositol 1, 4, 5-trisphosphate receptor
IP_3	Inositol trisphosphate
IRE1	Inositol-requiring enzyme 1
JNK	c-Jun N-terminal kinase
MAMs	Mitochondria-associated membranes
MCU	Mitochondrial Ca^{2+} uniporter
MCUR1	Mitochondrial Ca^{2+} uniporter regulator
Mff	Mitochondrial fission factor
MFN	Mitofusin
MOMP	Mitochondrial outer membrane permeabilization
NLRP3	Nucleotide-binding oligomerization domain (NOD)-like receptor family 3
NRF2	Nuclear factor (erythroid-derived 2)-like 2
OMM	Outer mitochondrial membrane
Orai1	Calcium release-activated calcium channel protein 1
ORD	OSBP-related domain
ORP	Oxysterol-binding protein (OSBP)-related protein
PACS2	Phosphofurin acidic pluster porting protein 2
PC	Phosphatidylcholine
PDI	Protein disulfide isomerase
PE	Phosphatidylethanolamine

PERK	Double-stranded RNA-activated protein kinase (PKR)-like ER kinase
PH	Pleckstrin homology
PI4P	Phosphatidylinositol 4-Phosphate
PKA	Protein kinase A
PM	Plasma membrane
PML	Promyelocytic leukemia
PP1	Protein phosphatase 1
PS	Phosphatidylserine
PTEN	Phosphatase and tensin homolog deleted on chromosome 10
PTPIP51	Protein tyrosine phosphatase-interacting protein 51
RER	Rough ER
RHOT	Mitochondrial Rho GTPase
RIDD	Regulated IRE1-dependent decay of mRNA
ROS	Reactive oxygen species
RyR	Ryanodine receptor
S1T	Truncated form of SERCA
S1R	Sigma R1 receptor
SER	Smooth ER
SERCA	Sarco/endoplasmic reticulum Ca^{2+} ATPase
SOCE	Store-operated calcium entry
STIM1	Stromal interaction molecule 1
Stx17	Syntaxin 17
Tespa-1	Thymocyte-expressed, positive selection-associated gene 1
TXNIP	Thioredoxin-interacting protein
TRAF2	TNFR-associated factor 2
UPR	Unfolded protein response
VAP	VAMP-associated proteins
VAPB	Vesicle-associated membrane protein-associated protein B and C
VDAC1	Voltage-dependent anion channel 1
XBP1	X-box binding protein 1

Contents

1	The Endoplasmic Reticulum: From Basal Homeostasis to Protein Folding Stress	76
	1.1 Basics of ER Morphology: Rough Sheets and Smooth Tubules	76
	1.2 The ER as a Central Organelle in the Cell	77
	1.3 When Protein Folding Goes Wrong: ER Stress	79
2	The ER Connection: ER–Mitochondria Contact Sites	84
	2.1 A Brief Introduction on MAMs Discovery and Characterization	84
	2.2 MAMs—a Dynamic Connection at the ER–Mitochondria Interface Recruiting a Multitude of Proteins	85
	2.3 ER–Mitochondria Contact Sites: Bridging Ca^{2+} Signals to the Cell's Fate	86
	2.4 ER–Mitochondria Contact Sites; a Greasy Connection	89
	2.5 Regulation of the Mitochondrial Network Through the MAMs	91

3	UPR Signaling at the ER–Mitochondria Interface	93
	3.1 Modulation of ER Stress Signaling at the MAMs	93
	3.2 ER Stress, MAMs, and Autophagosome Formation	95
4	Conclusions	96
References		96

1 The Endoplasmic Reticulum: From Basal Homeostasis to Protein Folding Stress

The endoplasmic reticulum (ER) is a crucial organelle with specific signaling and biosynthetic functions. The ER represents, along with mitochondria, the major intracellular Ca^{2+} store and has a fundamental role in the synthesis of cholesterol, steroids, and numerous lipid constituents of cellular membrane structures (Park and Blackstone 2010; Penno et al. 2013; Vance 1990; Berridge 2002; Gorlach et al. 2006). However, importantly, the ER is the main subcellular organelle dedicated to protein folding and secretion, and constitutes the location where the synthesis of at least a third of the total proteome takes place. The ER is a plastic and dynamic organelle at the center of a tight communication network with other subcellular organelles and compartments. In the following subchapters, the main biological and morphological characteristics of this organelle will be discussed.

1.1 Basics of ER Morphology: Rough Sheets and Smooth Tubules

At the morphological level, the ER can be divided into two distinct but interconnected entities, namely the nuclear ER and the peripheral ER. Broadly speaking, these two categories are also covered by the terms "rough" and "smooth" ER (RER and SER), respectively (Park and Blackstone 2010; Voeltz et al. 2002). At the morphological level, ribosomes stud the outer RER membrane, and its compartments tend to organize in a series of flattened sheets. Proteins moving into the secretory pathway or destined to be incorporated in the plasma membrane (PM) will be translated by these ribosomes and co-translationally inserted into the ER lumen. The SER, on the other hand, is devoid of ribosomes and is the main site for fatty acid and steroid biosynthesis, Ca^{2+} storage, and drug detoxification. The SER can easily be distinguished by its vast tubular network spanning the length and breadth of the cell (Voeltz et al. 2002; Park and Blackstone 2010). Despite these distinct differences, the entire ER is connected and forms one continuous network, allowing biomolecules (like proteins) to travel between different compartments.

A characteristic of ER tubules is their high membrane curvatures, a requirement to form tubules. Distinct sets of proteins are required to induce and stabilize these atypical structures—the reticulons and DP1/Yop1p (Voeltz et al. 2006). These proteins are essential for the formation of ER tubules, since their deletion abolishes ER tubule formation. In contrast, overexpression of these proteins leads to long, unbranched tubules.

The formation of ER sheets relies to some extent on the reticulons and DP1/Yop1p, but only to stabilize the edges of the sheets, while the coiled coil protein climp63 serves as a luminal ER spacer. In line, overexpression of climp63 leads to the formation of ER sheets (Shibata et al. 2010). The ratio between amount of SER (tubules) and RER (sheets) is, however, cell-dependent. Specialized secretory cells like β-cells of the pancreas or antibody producing B-cells have a large RER since one of their main functions is protein folding and secretion. Cells that produce steroid hormones, such as adrenal cortex cells, or cells that mainly detoxify xenobiotics (like liver cells), will have substantially more SER. As it will be discussed later (Sect. 3), the tubular ER network is also largely engaged in the establishment of contact sites with various organelles and the PM. Moreover, we have to keep in mind that these ER subcompartments defined at the structural and functional levels are highly interconnected and that ER morphology dynamically changes during physiological and stress conditions.

1.2 The ER as a Central Organelle in the Cell

Besides forming morphologically distinct structures, as described briefly above, the ER also coordinates a variety of biosynthetic and signaling functions. These are primarily dedicated to the synthesis, folding, and posttranslational modifications of cell membrane-associated or extracellularly secreted proteins, lipid synthesis, and last but not least, Ca^{2+} storage and signaling (Vance 1990; Park and Blackstone 2010; Berridge 2002; Penno et al. 2013).

1.2.1 Protein Folding in the ER: The Basics

Proper protein folding is not only a crucial process to maintain homeostasis, but also a highly complex process, requiring the assistance of many folding enzymes and a specialized oxidizing environment (Dinner et al. 2000). The ER has thus evolved over time to provide an ideal location for proteins to fold properly, housing numerous chaperones, including the glucose-regulated protein 78 (Grp78, also known as BiP), the lectins, calreticulin, and calnexin, and enzymes dedicated to posttranslational modifications including disulfide bridge formation and redox control, like PDIs, and oxidoreductases like ER oxidoreductase 1 alpha (Ero1α) (Stevens and Argon 1999; Malhotra and Kaufman 2007; Ron and Walter 2007). Protein folding is an energetically demanding process where ATP is required for chaperone function as well as for maintaining ER redox and Ca^{2+} homeostasis, both crucial determinants of proper protein folding. When newly translated polypeptides enter the ER, many processes will take hold to bring them to their final proper conformation. Important processes include N-linked glycosylation, where N-linked glycans are attached to Asn residues on the polypeptide by oligosaccharyltransferases. Terminal failure in protein folding or in oligomer assembly results in ER-associated protein degradation (ERAD), in which non-native conformers are retrotranslocated to the cytosol and degraded by the 26S proteasome (McCracken and Brodsky 2003).

1.2.2 The ER as a Regulator of Ca^{2+} Signaling

Under resting conditions in most cells, steady-state ER Ca^{2+} ranges between 0.1 and 1 mM (Verkhratsky and Toescu 2003), although luminal ER Ca^{2+} levels are subjected to dynamic variations in response to various metabolic stimuli, cellular processes, and a diverse set of endogenous and exogenous insults. This relative high amount of intraluminal Ca^{2+} is required for the proper functioning of various chaperones (such as BiP, which is the most abundant ER chaperone) (Hendershot 2004). The maintenance of luminal ER Ca^{2+} steady-state levels is thus crucial for homeostasis and is regulated by the sarco-endoplasmic reticulum Ca^{2+} ATPase (SERCA) pumps, which couple ATP hydrolysis with active transport of Ca^{2+} from the cytosol back into the ER lumen. This is counteracted by ER "leak" channels, such as the inositol 1, 4, 5 triphosphate receptor (IP_3R) and the Ryanodine receptor (RyR) that can selectively discharge Ca^{2+} following extracellular cues, such as ATP or histamine, signaling via the formation of IP_3, or in the case of excitable cells via neurotransmitter binding (among others), respectively (Gorlach et al. 2006).

1.2.3 Lipid Synthesis in the ER

In mammalian cells, glycerophospholipids constitute the largest class of lipids. The ER plays a central role in the synthesis of glycerophospholipids through the presence of multiple enzymes that catalyze the production of the most important lipid species. The production of phosphatidylcholine (PC), produced by the transfer of a phosphocholine moiety from CDP-choline to diacylglycerol (DAG), is mediated by choline/ethanolamine phosphotransferase, localized in the ER membrane (Penno et al. 2013). Phosphatidylethanolamine (PE), the second most abundant phospholipid, is produced either at the ER, or at the inner mitochondrial membrane through two distinct pathways. In the ER, the reaction is mediated by choline/ethanolamine phosphotransferase, which transfers phosphoethanolamine from CDP-ethanolamine to diacylglycerol (Penno et al. 2013). The third major phospholipid produced in the ER is phosphatidylserine (PS), which is produced by exchanging the polar head group of PE or PC for serine. The enzymes responsible for this conversion, PS synthase 1 and 2, are present in the subcompartment of the ER, the mitochondria-associated membranes (MAMs), which will be discussed in detail later (Vance 1990).

1.3 When Protein Folding Goes Wrong: ER Stress

The ER is thus a balanced organelle that requires a tight homeostasis to perform its functions, of which protein folding is one of the most important. Perturbation in this homeostasis will lead to a condition termed ER stress, in which the ER protein folding machinery can no longer cope with the influx of newly synthesized proteins, leading to their accumulation (Malhotra and Kaufman 2007; Ron and Walter 2007). ER stress is set off by various intracellular and extracellular insults, which alter the protein folding capacity of the ER. These include among many others, agents causing ER Ca^{2+} depletion, oxidative stress, and glucose deprivation, which interferes with N-linked glycosylation and affects ATP levels and redox balance. Perturbation of ER homeostasis activates an evolutionarily conserved stress response, collectively called the unfolded protein response or UPR. The UPR is primarily a pro-survival response initiated to restore ER proteostasis. The following is a brief description of the major signaling arms of the UPR and the emerging role this adaptive response plays in certain pathological conditions. A more comprehensive discussion of the UPR can be found in recent reviews (Hetz 2012; Wang and Kaufman 2016).

1.3.1 The Unfolded Protein Response

The UPR aims to induce three major signaling outcomes, through an intricate system of distinct pathways: (1) short-term attenuation of translation in order to limit the protein folding load on the ER; (2) heightened clearance of unfolded or terminally misfolded proteins through the upregulation of ERAD; and (3) transcriptional activation of specific genes encoding ER-resident chaperones, in order to expand the ER folding capacity (see also Fig. 1). When these mechanisms are inadequate to restore homeostasis in the ER, the UPR switches to promote a cell death program, which in higher vertebrates is most likely activated in order to protect the host organism from an accumulation of dysfunctional cells.

At the molecular level, the UPR is characterized by the activation of three ER transmembrane proteins: the pancreatic ER kinase (PKR)-like ER kinase (PERK), the inositol-requiring enzyme 1 (IRE1), and the activating transcription factor 6 (ATF6). These transmembrane proteins contain a luminal domain, which functions as a sensor of the ER folding capacity, and a cytosolic effector domain that provides a signaling bridge connecting the ER to other cellular compartments. In unstressed cells, the luminal domain of the transmembrane receptors is bound to BiP. BiP belongs to the heat-shock protein (HSP) 70 class of molecular chaperones and can form complexes with heterologous proteins that are processed through the ER. The activity of BiP is regulated by its AMPylation on threonine 518 by FICD (FIC domain containing). AMPylated BiP loses its ability to bind to nascent unfolded proteins, in a rapid process that matches fluctuating protein load with ER folding capacity (Preissler et al. 2015). Furthermore, BiP cycles between oligomeric and

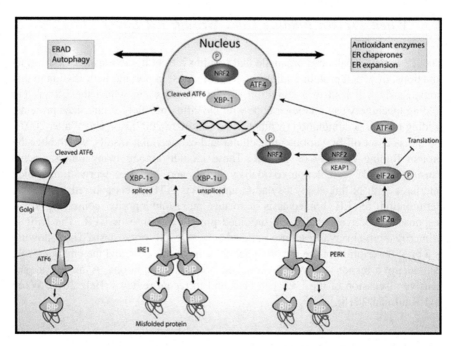

Fig. 1 Unfolded protein response in cellular survival. Simplified schematic overview of the unfolded protein response (UPR) in mammalian cells. The three arms of the UPR are shown, ATF6, IRE1 and PERK, and their most significant downstream effects. Upon ER stress, and the accumulation of unfolded proteins, BiP releases its inhibitory binding of the three ER stress proteins and activates them. Activated ATF6 will migrate toward the Golgi, to be cleaved, and will become a transcription factor. IRE1, when activated, will dimerize, which will activate its mRNase activity. Through this action, IRE1 will splice XBP-1 mRNA, enabling its translation. XBP-1 will then act as a transcription factor. Activated PERK will, like IRE1, also dimerize, inducing its autophosphorylation. Autophosphorylated PERK is then able to phosphorylate its two known substrates, eIF2α and NRF2. The phosphorylation of eIF2α will acutely inhibit the translation of new proteins, reducing the influx of newly synthesized proteins that require folding. Paradoxically, this translation stop will upregulate the transcription factor ATF4. Phosphorylation of NRF2 frees NRF2 from the inhibitory binding of KEAP1, allowing it to translocate to the nucleus to act as a transcription factor. The result of these three branches activating is initially a pro-survival response, leading to the upregulation of chaperones (like BiP) and anti-oxidant enzymes (like Ero1α) in order to increase the protein folding machinery to cope with the stress. This upregulation of chaperones goes hand in hand with the induction of ERAD and autophagy pathways, in order to clear (terminally) misfolded proteins. Furthermore, lipid synthesis will be increased in order for the ER to physically expand. Combined, these responses are geared to increase cell survival and re-establish ER proteostasis

monomeric states, the latter of which is thought to bind preferentially to unfolded proteins. The association of the ER transmembrane receptors with BiP locks them into an inactive, monomeric conformation that prevents their oligomerization or trafficking to other compartments. On accumulation of unfolded proteins, which promotes a dramatic increase in the luminal pool of monomeric BiP, BiP is competitively titrated away from the luminal domains of the three receptors. BiP dissociation triggers the activation of these proximal ER stress sensors and initiates the first molecular event of the complex transcriptional and translational program defining the UPR.

Moreover, the UPR works in conjunction with major proteolytic systems, such as the proteasome, autophagy, and the endosomal/lysosomal system to maintain the integrity of the proteome and attenuate proteotoxicity. Typically, under conditions of loss of ER proteostasis misfolded/abnormally folded proteins are re-translocated to the cytoplasm and subsequently degraded by the 26S proteasome through the ERAD pathway [reviewed in (McCracken and Brodsky 2003)]. Recently, ER protein overload has been shown to be alleviated by the secretory pathway, through a newly described mechanism, whereby unfolded glycosylphosphatidylinositol (GPI)-anchored proteins are exported via the Golgi apparatus to the PM for subsequent endocytosis and lysosome degradation (Noack and Molinari 2014). Other lysosomal degradative pathways contribute to the cytosolic removal of misfolded proteins. In line with this, autophagy has been shown to play an important cytoprotective role in restoring ER proteostasis under conditions of ER stress (Ogata et al. 2006).

a. The PERK branch of the UPR

PERK is a type I transmembrane S/T protein kinase. Dissociation of BiP from PERK drives PERK homo-oligomerization and self-activation by trans-autophosphorylation. PERK mediates the specific phosphorylation of nuclear factor (erythroid-derived 2)-like 2 (NRF2) and the eukaryotic initiation factor-2 α (eIF2α) at S51 (Harding et al. 1999; Cullinan et al. 2003). Phosphorylation of eIF2α interrupts the translation of most mRNAs, thereby reducing the load of newly synthesized proteins in the ER. However, while global translation is transiently repressed, the PERK-eIF2α pathway results in the preferential translation of a subset of mRNAs containing regulatory sequences in their 5' untranslated regions that allow them to bypass the eIF2α translational block. One of these mRNAs encodes ATF4, a member of the bZIP family of transcription factors. ATF4 promotes cell survival through the induction of several genes involved in restoring ER homeostasis. Under conditions of sustained ER stress, ATF4 prompts the expression of the transcription factor C/EBP homologous protein (CHOP) and of one of its target genes, the growth arrest and DNA damage-inducible gene 34 (GADD34) (Han et al. 2013). GADD34 is a protein phosphatase 1 (PP1) regulatory subunit that promotes eIF2α dephosphorylation, which ends the translational block and restores protein synthesis in the ER. Phosphorylation of NRF2 promotes its dissolution from the NRF2/Keap1 complex and leads to NRF2 nuclear import and causes activation of its gene expression program. This leads to the upregulation of proteins mainly involved in cellular redox homeostasis, such as glutathione-S-transferase (GST) and heme-oxygenase (HO-1).

b. The ATF6 branch of the UPR

ATF6 is a type II transmembrane protein that contains a transcription factor in its cytoplasmic domain. Under ER stress conditions, the dissociation of BiP frees ATF6 to translocate to the Golgi apparatus, where it is cleaved by Golgi-resident proteases (Haze et al. 1999). The limited proteolysis of ATF6 releases the transcription factor domain into the cytosol and allows its migration into the nucleus, where it binds DNA and activates gene expression. ATF6-responsive genes are activated rapidly after the induction of ER stress and include the ER molecular chaperones BiP and GRP94, as well as protein disulfide isomerase (PDI). ATF6 also leads to the transcriptional upregulation of the X-box binding protein (XBP1) mRNA, which is converted into a stable transcription factor by the endonuclease activity of IRE1 (Yoshida et al. 2001).

c. The IRE1 branch of the UPR

IRE1 contains a Ser/Thr kinase domain and an endonuclease domain facing the cytosol (Yoshida et al. 2001; Urano et al. 2000; Calfon et al. 2002). Like PERK, IRE1 is a type I transmembrane receptor that is activated by oligomerization-induced trans-autophosphorylation in the ER membrane upon BiP dissociation. IRE1 autophosphorylation stimulates its endonuclease activity and, unlike PERK, does not result in the propagation of a phosphorylation cascade. Active IRE1 is able to degrade specific mRNAs based on localization and the amino acid sequences that they encode. An important result of this mRNA cleavage is the removal of a 26-nucleotide intron from the XBP1 mRNA, generating the XBP1s frameshift splice variant which encodes a stable and active transcription factor. XBP1s translocates to the nucleus and transactivates several cytoprotective genes involved in ER quality control, including but not limited to, p58IPK, DnaJ/Hsp40-like genes, ERdj4, and HEDJ/ERdj3, as well as protein disulfide isomerase (PDI)-P5 (Lee et al. 2003). XBP1s can also induce a negative feedback loop, which, by relieving the PERK-mediated translational block, returns the ER function to normal once the adaptive function of the UPR has been successful (Lee et al. 2003). In a separate process termed regulated IRE1-dependent decay of mRNA (RIDD), IRE1 can also cleave an XBP-1 consensus site in other mRNAs, including BIP/BIP1, PDIA4 which can have an effect on both a protective and pro-cell death role (Maurel et al. 2014; Han et al. 2009; Hollien et al. 2009). The IRE1 pathway is thought to be the final branch of the UPR to be activated, subsequent to the rapid induction of PERK and ATF6 signaling which have a major cytoprotective role. The IRE1 signal plays a dual role in the UPR, and if the ER stress persists XBP1s initiates pro-apoptotic protein synthesis, thereby facilitating the induction of apoptosis (Kim et al. 2008).

The main UPR pathways are shown schematically in Fig. 1.

1.3.2 When the Rescue Fails: ER Stress and Apoptosis

The above-described pathways define the so-called acute phase of ER stress, where the UPR signaling coalesces in order to promote cell survival. However, in the event that this response is inadequate and loss of ER proteostasis persists, the UPR promotes cell death, which occurs primarily, although not exclusively, through apoptosis (Jager et al. 2012).

One of the main mediators of this response is the PERK arm of the UPR. The phosphorylation of eIF2α will initially lead to a translation stop, relieving pressure on the stressed ER. However, this action also allows for the translation of ATF4, which will lead to the upregulation of among others, GADD34 and CHOP. GADD34 will complex with PP1 to dephosphorylate eIF2α, removing the brake on translation. It was long thought that this process would allow the resumption of physiological protein folding after the resolving of ER stress, but new reports have indicated that under severe ER stress, this would lead to an increased burden on the ER and oxidative stress, promoting apoptosis (Han et al. 2013). By upregulating genes encoding proteins with functions in protein synthesis (and also inducing further GADD34), the transcriptional activation of combined ATF4/CHOP leads to an increase in protein synthesis, at a time when the ER machinery is overburdened. Forced expression of both CHOP and ATF4 led to an increase in ROS, and a depletion of ATP (Han et al. 2013), eventually contributing to cell death. The tight interplay between ATP consumption, mainly involving mitochondria and ER stress, will be discussed in more detail below.

Another mediator of cell death in the UPR is IRE1. IRE1 has a dual role in promoting cell death, relying on its scaffolding function and mRNase activity. IRE1 binds to tumor necrosis factor receptor (TNFR)-associated factor 2 (TRAF2), a molecular activator of JNK signaling. JNK activation is important in ER stress-induced apoptosis, although the exact mechanism is not entirely clear (Hetz 2012; Urano et al. 2000). Moreover, JNK activation can also modulate autophagy, through the phosphorylation of Bcl2, which in turn triggers the activation of Beclin 1 (Wei et al. 2008), a signal potentially antagonizing cell death after ER stress.

IRE1 is also able to induce cell death through its mRNase activity. While the cleavage of XBP-1 is considered to be a pro-survival response, leading to the upregulation of various chaperones and pro-survival factors (Moenner et al. 2007), RIDD will target a specific subset of mRNAs. The mode of action of IRE1-mediated RIDD includes a direct pathway by cleavage of growth factor mRNA, or an indirect pathway by cleaving repressor micro-RNA (miRNA). The cleaved miRNAs include miR-17, miR-34a, miR-96, and miR-125b, which have been reported to suppress the expression of certain pro-death and pro-inflammatory mediators, such as caspase 2 and thioredoxin-interacting protein (TXNIP), a modulator of the nucleotide-binding oligomerization domain (NOD)-like receptor family 3 (NLRP3) inflammasome (Maurel et al. 2014).

2 The ER Connection: ER–Mitochondria Contact Sites

As was already discussed above in a previous section, the ER consists of a dynamic network of rough sheets and smooth tubules. The different morphologies that the ER plastically adopts couple structural changes to distinct functions, which are exerted and modulated by the formation of contact sites with other organelles and the PM. Given that the ER is the major store of Ca^{2+} in the cell, and the major site for phospholipid synthesis (as described in previous sections), it comes as no surprise that the ER will and needs to communicate with a plethora of other organelles in order to spatiotemporally coordinate Ca^{2+} signaling and transfer of lipids. Coupling elegant biochemical studies with powerful electron microscopy and tomography studies has evidenced that the tubular ER, through the action of specific tethering proteins, interfaces with the mitochondria, endosomes, lysosomes, and the PM through the formation of multiple contact sites (mostly at the distance of 3–15 nm). Among these, the most studied are the ER–mitochondria contact sites, also termed mitochondria-associated membranes (MAMs), but also ER-PM contact sites are becoming a much studied and hot topic.

2.1 A Brief Introduction on MAMs Discovery and Characterization

A tight physical relationship between the mitochondria and ER had already been suggested many decades ago by pioneering electron microscopic observations done by John Ruby and co-workers, who discovered a potential connection between the ER membrane and the mitochondrial outer membrane in 1969 (Ruby et al. 1969). MAMs, however, were not fully discovered as a separate biochemical entity until 1990, when a membrane structure Jean Vance isolated through cell fractionation was defined as a biochemical contact site between the ER and the mitochondria (Vance 1990). In comparison with mitochondria, which were already described in the nineteenth century, and the ER, which was first observed later in 1945 (Porter et al. 1945), the MAMs are a relatively new inclusion to our knowledge of subcellular organelles. At present, the best definition to characterize the MAMs is that of ER membranes which are closely apposed to mitochondria, which can be purified and characterized as distinct structures. The functional characterization of the MAMs did not come up to speed until the turn of the millennium, the work of Rizzuto and co-workers being a major driving force (Rizzuto et al. 1998). They suggested the presence of microdomains of high Ca^{2+} concentrations at the ER, enabling mitochondria to be exposed to higher Ca^{2+} concentrations than the cytosol after Ca^{2+} release from the ER; this suggests that the transfer between the ER and mitochondria requires a physical, proteinaceous link between the two organelles. It was at this time that the Vance lab followed up its initial discovery by investigating the role of the MAMs in lipid homeostasis in liver cells of the rat (Shiao et al.

1998). This finding was followed up by similar findings exploring the role of MAMs in lipid synthesis in yeast cells (Achleitner et al. 1999), demonstrating the highly conserved functions of ER–mitochondria contact sites. New advances in the last decade have estimated that the total surface area of mitochondria juxtaposed to the ER at around 5–20% (Rizzuto et al. 1998). Soon it became clear that the MAMs provide more than just a protein link between the two organelles; rather, they allow a functional transit of metabolites and signaling molecules between the ER and mitochondria, with implications in cell fate decisions. At the present time, even this claim seems to underestimate the true role of MAMs, and a growing number of studies are appearing involving this little piece of the puzzle of the cell in a vast cascade of crucial homeostatic functions both in the cell and outside of it. The effect of ER–mitochondria contact sites for Ca^{2+} transfer and mitochondrial metabolism, and the consequences of this communication between organelles in disease states such as cancer and neurodegenerative disorders have been widely discussed in excellent reviews (Hayashi et al. 2009; Giorgi et al. 2009; Verkhratsky and Toescu 2003; Schrepfer and Scorrano 2016). Here, focus will be mainly on the molecular components of the MAMs and their role in the regulation of ER stress.

2.2 MAMs—a Dynamic Connection at the ER–Mitochondria Interface Recruiting a Multitude of Proteins

Just like both organelles that they physically link, MAMs should not be interpreted as static bridges linking mitochondria and ER. In contrast, MAMs seem very flexible and entail a specialized set of proteins, which are able to support a multitude of signaling components in accordance with the needs of the cell. Despite the increasingly large number of proteins recognized to participate in this compartment, a main subgroup of ER and mitochondria-associated proteins has been identified. These proteins include Ca^{2+} ion channels located in the ER or the outer mitochondrial membrane [OMM, e.g., the inositol trisphosphate receptor, IP_3R (Furuichi et al. 1989) and voltage-dependent anion channel 1, VDAC1 (Szabadkai et al. 2006)], ER-resident chaperones [e.g., calnexin (Lynes et al. 2012)], various enzymes of the lipid biosynthetic pathways and lipid transfer proteins (Voelker 2005), enzymes involved in ER redox regulation [e.g., Ero1α (Pollard et al. 1998; Anelli et al. 2012; Gilady et al. 2010)], other chaperones [e.g., the glucose-regulated protein 75, Grp75 (Szabadkai et al. 2006), and the Sigma R1 receptor, S1R (Hayashi and Su 2003, 2007)], tumor suppressors [e.g., promyelocytic leukemia, PML (Giorgi et al. 2010) and phosphatase and tensin homolog deleted on chromosome 10, PTEN (Bononi et al. 2013)], and proteins were involved in the regulation of ER-vesicular sorting [e.g., phosphofurin acidic cluster sorting protein 2, PACS-2 (Myhill et al. 2008; Simmen et al. 2005)]. Recent research has uncovered two novel protein tethers that can function as linkers between the ER and mitochondria (Stoica et al. 2014). Vesicle-associated membrane protein-associated protein B and C (VAPB), an ER localized protein, and the mitochondrial outer membrane protein tyrosine phosphatase-interacting protein 51 (PTPIP51) have been found to interact and regulate ER–mitochondria connections. Mitochondrial motility has also been linked to the MAMs, since mitochondrial rho (Miro 1, 2; also known as mitochondrial rho GTPase RHOT1, 2) has been found in the MAMs (Fransson et al. 2003; Saotome et al. 2008; Kornmann et al. 2011). Proteins that regulate mitochondria fusion and shape, like the dynamin-related GTPase mitofusin 2 (MFN2), have also been discovered to play a key role in the MAMs (de Brito and Scorrano 2008; Naon et al. 2016). Despite MFN2 being one of the best studied MAM-resident proteins, some controversy has recently arisen concerning its exact role in the ER–mitochondria contact sites. Initial studies, which have since been reproduced, have indicated that MFN2 acts as a physical tether between the two organelles by homotypic and heterotypic (with MFN1) interactions (Naon et al. 2016; de Brito and Scorrano 2008; Hailey et al. 2010). These studies also found that MFN2 ablation resulted in increased distance between ER and mitochondria and a disturbed Ca^{2+} transfer between the two organelles (de Brito and Scorrano 2008). Furthermore, it was shown that MFN2 is able to interact with PERK to regulate the UPR, with MFN2 ablation causes increased activity of the PERK arm of the UPR (Munoz et al. 2013). However, some recent studies have, however, shown the opposite effect that MFN2 might act as an inhibitor of close ER–mitochondria apposition (Filadi et al. 2015, 2016). Future studies will have to clear out these discrepancies. This illustrates that, in spite of all the recent developments in the field of MAMs, both the molecular composition of the MAMs and the precise role of certain MAM-associated proteins, are still elusive.

2.3 ER–Mitochondria Contact Sites: Bridging Ca^{2+} Signals to the Cell's Fate

As already discussed briefly above, proper protein folding is not only a crucial process to maintain homeostasis, it is also highly complex, requiring the assistance of many folding enzymes and a specialized oxidized environment (Dinner et al. 2000). The ER has thus evolved over time to provide an ideal location for proteins to fold properly, being rich in Ca^{2+} and a variety of chaperones. ER-resident chaperones include BiP (also known as Grp78), the lectins calreticulin and calnexin, and enzymes dedicated to posttranslational modifications including glycosylation, disulfide bridge formation (PDIs) and oxidoreductases (Ero1) (Stevens and Argon 1999; Malhotra and Kaufman 2007; Pollard et al. 1998). Protein folding is an energetically demanding process where ATP is required for chaperone function as well as for maintaining the ER redox and Ca^{2+} homeostasis. Under resting conditions in most cells steady-state ER Ca^{2+} ranges between 0.1 and 1 mM (Berridge et al. 2003; Rizzuto and Pozzan 2006), although luminal ER Ca^{2+} levels are subjected to dynamic variations in response to various metabolic stimuli, cellular processes and a diverse set of endogenous and exogenous insults. Crucial for the maintenance of luminal ER Ca^{2+} steady-state levels are the sarco-endoplasmic reticulum Ca^{2+} ATPase (SERCA) pumps, which couples ATP hydrolysis with active transport of Ca^{2+} from the cytosol back into the ER lumen, and the ER leak channels, like the IP_3R and the RyR (Sammels et al. 2010). Dropping Ca^{2+} levels in the ER cause morphological changes in the ER and the formation of cortical ER (cER) (Shen et al. 2011). This cER formation goes along with the appearance of thin tubules making close contact with the plasma membrane of the cell (Shen et al. 2011). Chemical insults, physiological or pathological stimuli are triggers that can cause a drop in ER luminal Ca^{2+}, after which signals are conveyed to the plasma membrane to activate store operated Ca^{2+} entry (SOCE) to induce a Ca^{2+} influx from the extracellular matrix to the cytosol and eventually into the ER to refill its Ca^{2+} content. The proteins in the ER that are the initiators of the SOCE are the stromal interaction molecules (STIM1/2). These proteins are able to pull the ER toward the plasma membrane, where they interact with Ca^{2+} release-activated calcium channel protein 1 (Orai1), inducing its homo-dimerization. Upon dimerization, Orai1 forms a Ca^{2+} permeable channel that will cause a Ca^{2+} influx (Roberts-Thomson et al. 2010). Moreover, as a consequence of ER Ca^{2+} depletion, the folding machinery of the ER is disturbed and the following imbalance between the ER protein folding load and capacity will cause the accumulation of unfolded proteins in the ER lumen, engaging ER stress (Malhotra and Kaufman 2007; Ron and Walter 2007; Verfaillie et al. 2013).

While the role of Ca^{2+} as a versatile second messenger regulating several key cellular processes is a well-established phenomenon, the relevance of its trafficking through the MAMs as the main route for the control of mitochondrial bioenergetics and cell death has been ascertained in the last couple of decades (Berridge et al. 2003). Cytosolic microdomains of high $[Ca^{2+}]$ or "Ca^{2+} hot spots", generated proximal to the mitochondria, are necessary to efficiently convey the steep rise in local Ca^{2+} levels, required to cause the opening of the mitochondrial Ca^{2+} uniporter (MCU), the main active transporter of Ca^{2+} through the inner mitochondrial membrane to the mitochondrial matrix (Giacomello et al. 2010). Indeed one of the characteristics of the recently identified MCU is its very low affinity for Ca^{2+} uptake under physiological conditions (with a K_d of >10 μM) (Kirichok et al. 2004). Moreover, the discovery that the IP_3R, the major Ca^{2+} releasing channel at the ER (Varnai et al. 2005), not only colocalizes with the MAMs, but is kept in direct contact with VDAC1, located at the OMM (Szabadkai et al. 2006), further unraveled the role of MAMs as structural microdomains allowing high Ca^{2+} concentration transfer between the ER and the mitochondria. The IP_3R-VDAC1 interaction is maintained by the physical linkage provided by Grp75 at the OMM, which has been shown to facilitate the Ca^{2+} exchange from IP_3R to VDAC1, and validates the physiological relevance of the MAMs as physical Ca^{2+} transfer sites. Recently, thymocyte-expressed, positive selection-associated gene 1 (Tespa-1) was identified as a new member of the MAMs that was able to physically interact with Grp75 and modulate the Ca^{2+} signals from IP_3R (Matsuzaki et al. 2013).

Ca^{2+} uptake by mitochondria across the inner membrane regulates mitochondrial respiration and ATP production and indeed several components of the tricarboxylic acid cycle (TCA) cycle are known to be stimulated by an increase in the matrix Ca^{2+} concentration (McCormack and Denton 1980). Mitochondrial bioenergetics is favored by a regulated IP_3-dependent IP_3R-mediated Ca^{2+} release and mitochondria Ca^{2+} uptake (Pizzo et al. 2012). In line with this, knockdown of the mitochondrial Ca^{2+} uniporter regulator (MCUR1), which is required for MCU-dependent mitochondrial Ca^{2+} uptake, abrogated Ca^{2+} uptake, oxidative phosphorylation and stimulated AMPK-mediated pro-survival autophagy, which was triggered by a reduction in the cellular ATP content (Mallilankaraman et al. 2012). While the vital role of the MAMs in the regulation of mitochondrial Ca^{2+} uptake and metabolism appears apparent, what still remains unclear is whether besides Ca^{2+} other second messengers transferred through the MAMs (like lipids or reactive oxygen species, ROS) may also contribute to the regulation of mitochondrial metabolism by directly modulating transporters (like MCU), carriers or components of the mitochondrial electron transfer chain.

Besides their essential role in cellular metabolism and survival, it is well established that the mitochondria play a key role in the initiation and amplification of cell death (Kroemer et al. 2007). Given that excessive mitochondrial Ca^{2+} uptake can favor mitochondrial dysfunctions and trigger apoptosis, the involvement of the MAMs in cell death modulation also comes as no surprise. It has long been known that higher matrix Ca^{2+} levels sensitize mitochondria to undergo MOMP (mitochondrial outer membrane permeabilization) a process precipitating apoptosis (Pinton et al. 2008). Increased uptake of Ca^{2+} by mitochondria may result in changes in the permeability of the inner membrane of the mitochondria (IMM) caused by prolonged opening of the so-called permeability transition pore (PTP). PTP opening in turn induces mitochondrial swelling and OMM rupture, with the consequent release of caspase-activating factors and apoptosis induction (Kroemer et al. 2007). Moreover, upon MOMP, cytosolic cytochrome C release can amplify caspase activation by binding to the IP_3R and exacerbating its Ca^{2+} leak properties (Boehning et al. 2003). This is a feed-forward loop, since small amount of cytochrome C released in the early phases of apoptosis will result in higher Ca^{2+} transfer from the ER to the mitochondria, amplifying the apoptotic signal (Boehning et al. 2003).

Another study has described an inter-organellar amplification signal involving ER–mitochondria contact sites. The study of Iwasawa et al. (2011) delineated the existence of a reciprocal transfer of apoptosis signals, from mitochondria to the ER, and back to mitochondria, through a complex formed at a specific ER–mitochondrial subdomain, dubbed the ARCosome. This signal entails that in response to an apoptotic insult, a Bap31-Fis1 platform spanning the ER–mitochondrial interface would recruit caspase-8, thus enabling the cleavage of Bap31 into its pro-death fragment p20Bap31, which favors the emptying of ER Ca^{2+} stores. Upon Ca^{2+} uptake, the juxtaposed mitochondria would then undergo PTP opening and release apoptogenic factors, thereby launching apoptosis. Interestingly, disruption of MAMs by PACS-2 knockdown has been shown to lead to Bap31 cleavage into p20Bap31 in the absence of cell death (Simmen et al. 2005). This suggests that a tight association between ER and mitochondria is required for this mechanism of Bap31-mediated apoptosis.

Additionally, it has also become clear that not only do the targeting and sorting mechanisms modulate MAMs functionality, and also the length of the connections linking the ER and mitochondria is important to determine their effect on each organelle (Csordas et al. 2006). Elegant work done in the laboratory of Hajnoczky by using electron tomography and expression of synthetic linkers, addressed the issue of the relevance of a proper ER–mitochondria distance. These authors found that the ER and mitochondria are juxtaposed through trypsin-sensitive domains (thus proteinaceous in nature), which are approximately 10 nm at the smooth ER and 25 nm at the rough ER (Csordas et al. 2010). These close distances enable ER proteins to come in direct contact with proteins and lipids at the OMM (Hayashi et al. 2009). Tightening of these physical links, either artificially through synthetic linkers, or by stimulating ER Ca^{2+} release and transfer/uptake by mitochondria, led to mitochondrial Ca^{2+} overload and MOMP, precipitating apoptotic cell death. In contrast loosening without disrupting these ER–mitochondrial contact sites stimulated mitochondrial respiration and ATP production (Csordas et al. 2010).

These studies together suggest that a tight ER–mitochondria juxtaposition is integral to mechanisms controlling cellular fate and to amplify (lethal) inter-organellar Ca^{2+} signals. The identification of signaling molecules enabling the dynamic modulation of the distance between the ER and mitochondria through the MAMs according to the needs of the cell remains an active area of exploration.

2.4 ER–Mitochondria Contact Sites; a Greasy Connection

As mentioned above, the ER is also the prime site for lipid synthesis and is the central hub around which all lipids are formed. Initially phosphatidylserine (PS) is formed in a subdomain of the ER and further transported to the mitochondria where it is imported in the IMM and becomes a substrate for PS decarboxylase 1, resulting in the synthesis of PE. The newly formed PE is then transported back to the ER from the mitochondria or Golgi, where PE methyltransferases 1 and 2 convert it into PC. Upon closer inspection however, the vital location for the synthesis and shuttling of all these lipids resides in the MAMs (Voelker 2005). The importance of the MAMs as the sites for the biosynthesis and transfer of lipids between the ER and mitochondria was unraveled by the discovery that PS synthase1/2 (PSS-1/2) reside almost exclusively at the MAMs (Voelker 1989). Additionally, the transport of PS to the mitochondria has even been shown to constitute the rate-limiting step in the synthesis of PE in the mitochondria (Voelker 1989). Besides this role in the production of PS and PE, the MAMs were also found to be highly enriched for acyl–CoA:cholesterol acyltransferase (ACAT) (Rusinol et al. 1994), which catalyzes the formation of cholesterol esters, marking the ER–mitochondria juxtapositions as one of the prime locations involved in cholesterol synthesis. In line with this, the MAMs are also vital for steroidogenesis, where sterols produced in the ER are shuttled to the mitochondria. In the mitochondria cholesterol is further converted into pregnenolone, which is then transported back to the ER via the MAMs to be further converted to other steroids. The relevant lipid biosynthetic and transport function of the MAMs, and the presence of several enzymes involved in this pathway, has led some groups to use the activity of these enzymes as a marker for MAM's activity and ER–mitochondria connectivity (Area-Gomez et al. 2012).

The MAMs have long been known as a lipid raft-like structure, highly enriched in cholesterol (Hayashi and Fujimoto 2010). Furthermore, altering cholesterol levels itself has been shown to modulate the ER–mitochondrial connections. In line with this, when cholesterol is depleted from the cell, an increased association of the ER with mitochondria is observed, likely as a mechanism to improve MAMs functionality and re-establish cholesterol synthesis (Fujimoto et al. 2012).

Another indication of the importance of cholesterol synthesis at the MAMs is the presence of caveolin-1 (Cav-1) at these ER–mitochondria subdomains. Cav-1, which binds cholesterol with high affinity, is a vital component of the cholesterol-enriched caveolae of the plasma membrane and controls transport and efflux of cholesterol from the ER toward the plasma membrane and other organelles (Bosch et al. 2011). The functional relevance of this MAM-resident protein is further highlighted by the finding that upon Cav-1 depletion, cholesterol content in the ER and MAMs increases as well as its transport to the mitochondria, resulting in a decreased electron flux through the respiratory chain, increased ROS generation and sensitization toward apoptosis (Bosch et al. 2011).

Therefore, regulation of lipid shuttling at the MAMs has clear implications for mitochondrial lipid composition, mitochondrial metabolism and apoptotic signaling. Furthermore, the observations that mice lacking PS decarboxylase display aberrantly shaped mitochondria, showing a fragmented network (Steenbergen et al. 2005), and homozygous knockout of this gene is embryonically lethal (Steenbergen et al. 2005), support the critical importance of a proper functioning lipid synthesis and transport pathway at the MAMs for homeostasis.

Another interesting finding highlighting the role that MAMs play in lipid transport was recently proposed by a recent study investigating the oxysterol-binding protein (OSBP)-related protein (ORP) proteins. In mammals, ORP proteins (Osh proteins in yeast) have been identified as specialized proteins localized to the ER-PM contact sites (Chung et al. 2015; Moser von Filseck et al. 2015; Stefan et al. 2011). Earlier research had already shown that the closely related oxysterol-binding proteins (OSBP) were responsible for cholesterol—phosphatidylinositol 4-phosphate (PI4P) countertransport between the ER membrane and the Golgi (Rysman et al. 2010). ORP 5 and 8 are the prime mediators of lipid countertransport at the ER-PM contacts and harbor-specific domains that allow their action. A subset of ORPs contains a pleckstrin homology domain (PH), which binds PI4P lipids at the PM and mediates ER-PM contact site formation. All ORP proteins have one specialized domain in common, the OSBP-related domain (ORD domain), which is a conserved sterol-binding domain shown to be able to harbor both PI4P and PS in the case of ORP8 (Chung et al. 2015). ORP5 and ORP8 are unique among the ORP family for being C-terminally tail-anchored in the ER membrane. Interestingly, it has now been shown that besides regulating lipid countertransport at ER-PM contact sites, ORP5/8 may perform the same function at the ER–mitochondria contact sites (Galmes et al. 2016). Although the broad physiological consequence of this finding still needs to be fully appreciated, ORP5/8 depletion caused defects in mitochondria morphology and respiratory function (Galmes et al. 2016).

2.5 Regulation of the Mitochondrial Network Through the MAMs

The connection between the ER and mitochondria is becoming one of the major factors in controlling the dynamic changes in shape and motility of these two organelles. Similar to the ER, mitochondria are commonly organized in long tubular networks and are very dynamic organelles. Mitochondria are constantly fusing, fragmenting and branching, dictated by the various impulses which include metabolic stimuli, general calcium buffering needs or different cellular stresses.

Accumulating data highlight an important role for the MAMs in the coordination of ER and mitochondria shape and dynamics. Indeed, several proteins involved in mitochondrial movement along microtubules, like dynein and kinesin, are tightly regulated by local increases in Ca^{2+} concentration, a process regulated at the MAMs.

Indeed, some of the proteins involved in mitochondrial motion along microtubules, i.e., dynein and kinesin, are highly regulated by local Ca^{2+} concentration increases. It thus comes as no surprise that dynamic Ca^{2+} regulation at the ER–mitochondria interface can regulate mitochondrial movement, since this movement is inhibited by rises in cytosolic Ca^{2+} (Yi et al. 2004; Pizzo et al. 2012). Interestingly, diminished mitochondrial movement leads to an increase in their MAMs association, since these are subdomains of high Ca^{2+} release. This tighter connection improves their uptake of Ca^{2+} and their Ca^{2+} buffering capacity (Yi et al. 2004; Pizzo and Pozzan 2007). Important regulators of mitochondrial movement are the Miro proteins (Miro 1 and 2). Miro 1 and 2 are located at the OMM, where they are anchored through a short C-terminal domain. Both proteins can sense high levels of Ca^{2+} through their two EF-hand Ca^{2+} binding domains. Miro 1 and 2 enable mitochondrial movement by anchoring mitochondria to the cytoskeleton (Saotome et al. 2008; Fransson et al. 2006). In areas with high Ca^{2+} concentration, perceived by their EF hand domains, Miro 1 and 2 lose their connection to kinesin, resulting in immobile mitochondria. This allows them to effectively buffer (excessive) high cytosolic Ca^{2+} levels, or as a means of stimulating oxidative phosphorylation (Brough et al. 2005; Saotome et al. 2008). As mentioned earlier, the yeast ortholog of Miro, Gem1, has been shown to have an impact on the size and number of the ER–mitochondria encounter structure (ERMES) connecting both organelles, and to affect the phospholipid homeostasis

(Kornmann et al. 2011; Nguyen et al. 2012; Wideman et al. 2013). In addition to the regulation of mitochondrial motility, the MAMs are involved in the regulation of mitochondrial morphology, which is maintained through a delicate balance between the fusion and fission of mitochondria, affecting their size, length and shape, and the overall energy state of the cell (Scorrano 2013).

However, mitochondrial fragmentation due to excessive activity of the fission machinery is often stimulated in conditions of cellular stress. These stress conditions, which can lead to an increased production of reactive oxygen species, are able to compromise mitochondrial membrane potential and cause disturbances in the mitochondrial function, leading to excessive fragmentation. In order to prevent damage to the mitochondrial network caused by these stresses, compromised pieces of the network are fissioned off and isolated, ready to be processed by autophagy (see below) (Youle and van der Bliek 2012; Wang et al. 2011).

The exact molecular mechanism through which mitochondrial fission takes place has not been fully characterized. The major player in this process though is believed to be the highly conserved dynamin-related protein 1 (drp1). Drp1 (Dnm1 or yeast) consists of three functional domains, a GTPase, a smaller domain in the middle of the protein containing pleckstrin homology domain with unexplored function and a COOH-terminal assembly or GTPase effector domain (GED) (Muhlberg et al. 1997). Drp1 works by forming a spiral oligomer in a GTP-dependent manner through intra- and intermolecular interactions. Drp1, by oligomerizing in this manner, is believed to provide the required mechanochemical force that is needed to induce division of the outer and inner mitochondrial membranes (Ingerman et al. 2005). Drp1 activity is further regulated through (de)phosphorylation activity by GMM-associated kinase (PKA/AKAP1) and phosphatase (PP2A/Bβ2), in a Ca^{2+}-dependent manner. In neurons, Ser656 phosphorylation of Drp1 in mitochondria caused a significant lengthening of mitochondria, while dephosphorylation induced fission, mitochondrial fragmentation, and depolarization (Dickey and Strack 2011).

However, Drp1 is not the only protein mediating mitochondrial division, additional proteins such as mitochondrial fission protein 1 (Fis1) and mitochondrial fission factor (mff) (both located on the OMM) are necessary for proper division of mitochondria, but it remains unclear whether these proteins affect the division location. Both Mff and Fis1 are able to modulate both the number and size of Drp1 punctae at the OMM (Loson et al. 2013).

An elegant study in the Voeltz lab has shown that tubular ER is strongly associated with mitochondrial fission mechanics (Friedman et al. 2011). This study showed that the ER is necessary to induce mitochondrial fission through a tight physical interaction by "wrapping" around part of the mitochondrial network. This action induced local fragmentation and recruitment of Drp1. This interplay between tubular ER and mitochondria strongly hints at involvement of the MAMs in mitochondrial division. The exact signals that trigger and recruit the ER and the fission mechanics at these locations is still under investigation. For a schematic depiction of the MAMs, please see Fig. 2.

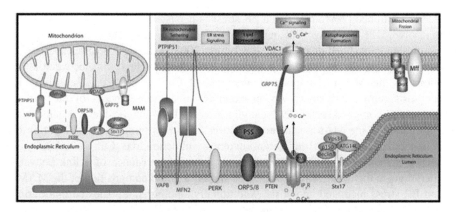

Fig. 2 Overview of the key components of the mitochondria-associated membranes (MAMs) linked to the processes they regulate. Here, the various important functions and signaling roles of the MAMs are shown. Displayed are ER–mitochondria tethering proteins, notably mitofusin 2 (MFN2) and vesicle-associated membrane protein-associated protein B and C (VAPB), an ER-localized protein, and the mitochondrial outer membrane protein tyrosine phosphatase interacting protein 51 (PTPIP51). Also shown is the link between ER stress signaling in the MAMs through the PERK-MFN2 axis. Other important functions displayed are the calcium handling of the MAMs (in the form of inositol 1, 4, 5-trisphosphate receptor (IP$_3$R) linked to voltage-dependent anion channel 1 (VDAC1) by glucose-regulated protein 75 (Grp75), PML (promyelocytic leukemia), phosphatase and tensin homolog deleted on chromosome 10 (PTEN)). Lipid homeostasis, in the form of phosphatidyl serine synthase and oxysterol-binding protein (OSBP)-related protein 5/8 (ORP5/8), is also shown. Autophagy regulation, initiated by the relocation of syntaxin 17 (Stx17) which then proceeds to recruit ATG14L, p150, Vps34, and Beclin1 (comprising the phosphoinositide 3-kinase, PI3 K, complex), is also shown. Mitochondrial fission dynamics are shown in the form of mitochondrial fission factor (Mff) and dynamin-1-like protein (Drp1)

3 UPR Signaling at the ER–Mitochondria Interface

3.1 Modulation of ER Stress Signaling at the MAMs

One of the most energy-demanding processes in the cell is the proper folding of de novo synthesized proteins. As it has been described previously, there is a tight link between the ATP produced in the mitochondria, and the demand of the stressed ER. This is highlighted by the observation that a significant proportion of mitochondria relocated to the perinuclear, rough ER after tunicamycin-induced ER stress. Relocated mitochondria exhibited a higher Ca^{2+} uptake, an increase in their transmembrane potential, increased production of ATP, a heightened reductive power and increased oxygen consumption (Bravo et al. 2011). The tightening of the ER–mitochondria contact sites under ER stress conditions points to a role in favoring a temporary increase in the intracellular ATP pool. This increased ATP is required to meet the demands of the increased amount of ER chaperones, newly expressed through the pro-survival ER stress transcriptional machinery, during the early adaptive phases of ER stress.

As mentioned previously, when ER stress is persistent or too severe for the UPR to resolve, UPR pathways will switch and turn on a pro-apoptotic signal, eventually causing apoptotic cell death (Ron and Walter 2007; Schroder and Kaufman 2005; Verfaillie et al. 2013). Several studies have disclosed a direct link between alterations in the MAMs components, deregulated Ca^{2+} transfer and sensitivity to apoptosis during ER stress. As an example, in a mouse model of the human lysosomal storage disease GM1-gangliosidosis, GM1-ganglioside was shown to accumulate in the MAMs and further causes ER stress through the stimulation of IP_3R-mediated ER Ca^{2+} depletion evoking mitochondrial Ca^{2+} overload and MOMP (Sano et al. 2009). This study suggests the existence of a link between alterations in the lipid-raft-like structures and/or the lipid composition of the MAMs and ER stress-mediated apoptosis. Additionally, another study showed that during ER stress, the expression of a truncated variant of SERCA1 (S1T) localized to the MAMs. There, SI1 led to an increasing number of contact sites, mitochondrial Ca^{2+} overload and an inhibition of mitochondrial movement, which ultimately triggered apoptosis (Chami et al. 2008). Interestingly, the PERK-ATF4 axis of the UPR was shown to be required for the induction of S1T during ER stress. This suggests that following ER stress, the activation of the PERK pathway may indirectly reinforce the ER–mitochondria contact sites through ATF4-mediated transcription of S1T (Chami et al. 2008). Other reports indicate that PERK could play a further role on the MAMs and mitochondrial physiology through ATF4 during the UPR. After ER stress and PERK activation, upregulation of ATF4 caused the transcriptional upregulation of the E3 ubiquitin ligase Parkin gene, by binding to a specific CREB/ATF site within the promotor of Parkin (Bouman et al. 2011). This higher expression of Parkin did not alleviate ER stress itself but prevented additional mitochondrial fragmentation and loss of ATP production as a result of ER stress (Bouman et al. 2011). This was highlighted in a follow-up study, where it was shown that increasing Parkin levels caused tighter functional and physical ER–mitochondrial coupling (Cali et al. 2013), which increased Ca^{2+} flux from the ER to mitochondria following stimulation with an $InsP_3$-generating agonist. This tighter inter-organellar coupling and Ca^{2+} transfer also induced a higher amount of ATP being produced through mitochondrial respiration (Cali et al. 2013), thus highlighting the importance of PERK-ATF4 induced Parkin in these pathways (Cali et al. 2013; Bouman et al. 2011).

However, besides classical UPR mechanism, PERK is able to facilitate exchanges of signals through the MAMs (Verfaillie et al. 2012). Work from our laboratory has indicated a direct role of PERK at the ER–mitochondria contact sites. Through functional studies and subcellular fractionation, we discovered that PERK is an integral member of the MAMs (Verfaillie et al. 2012). Genetic ablation of PERK in MEF cells caused a disruption of the ER–mitochondria contact sites and led to an increased protection from apoptosis that was caused by agents that simultaneously mobilize Ca^{2+} and induce ER stress through ROS (Verfaillie et al. 2012). Additionally, as mentioned before MFN2, another key component of the MAMs was found to interact with PERK directly. The PERK–MFN2 interaction was required for the progression of ER stress-mediated pathways, including UPR and apoptosis (Munoz et al. 2013). Furthermore, silencing of PERK was able to partially rescue the aberrant mitochondrial Ca^{2+} content and the fragmentation of the mitochondrial network, caused by the loss of MFN2 in MEF cells. In accordance with the important tethering role of PERK in the maintenance of the MAMs, we also uncovered that PERK inhibition/deletion reduces mitochondrial Ca^{2+} overload following release of ER Ca^{2+} by thapsigargin and improves mitochondrial oxygen consumption under ER stress conditions (van Vliet et al., unpublished data). However, more research is needed to investigate whether PERK-mediated effects on mitochondrial bioenergetics are based on its ability as a MAM-localized protein to instigate a direct transfer of essential signaling molecules/mediators between the ER and mitochondria, outside of its canonical and well-described role in UPR signaling. Moreover, further investigation is warranted whether other interactors of PERK at the ER–mitochondria contact sites, beyond MFN2, are present. Also, whether other members of the UPR are localized to or have a signaling or tethering role at the ER–mitochondria contact sites still remains to be investigated.

3.2 ER Stress, MAMs, and Autophagosome Formation

Further implicating ER–mitochondria contact sites to ER stress and UPR (and other cellular stresses), several reports have indicated that the MAMs play a central role in the initiation of autophagy and the formation of the autophagosome (Hamasaki et al. 2013; Hailey et al. 2010), the structural hallmark of macroautophagy (here called autophagy). Autophagy is the main pathway for the removal, degradation, and recycling of intracellular components through lysosomal degradation. These include superfluous or damaged organelles and proteins (Shintani and Klionsky 2004). Although autophagy is constitutively active at a basal level to preserve homeostasis, this catabolic process is typically stimulated when the cells perceive the lack of nutrients or in conditions where organellar functionality or proteostasis is compromised (Yorimitsu and Klionsky 2005; He and Klionsky 2009). Consistent with this, the stress conditions which lead to ER stress and UPR activation are also able to induce the autophagic machinery. In this context, autophagy is initiated to aid in the disposal of terminally misfolded proteins and to relieve the folding burden on the ER, and can be induced by PERK-ATF4-mediated transcriptional upregulation of selected autophagy genes (Yorimitsu et al. 2006) [for broad reviews on the molecular machinery of autophagy, see (Yorimitsu and Klionsky 2005)].

The origin of the autophagosomal membrane (also called phagophore or isolation membrane) had for many years been a mystery (Tooze and Yoshimori 2010). Now, the two major theories implicating the ER and the mitochondrial membranes as the primary source of the autophagosomal membrane may have been neatly combined in one model (Hamasaki et al. 2013). Research performed by the laboratory of Yoshimori has shown that ATG14L (autophagy-related 14 L), a pre-autophagosomal marker, is recruited to the ER–mitochondria contact sites after starvation by syntaxin 17 (Stx17) (Hamasaki et al. 2013). After disrupting the

MAMs, ATG14 recruitment and subsequent autophagosome formation were prevented. Additionally, until the process of autophagosome formation was complete, ATG5, another essential autophagy protein, was also found to localize at the ER–mitochondria contact sites (Hamasaki et al. 2013). The necessity of the autophagic machinery in maintaining a homeostatic and functioning mitochondrial network (Twig et al. 2008; Wang and Klionsky 2011) gives further credence to these recent discoveries. Interestingly, the finding that the autophagy machinery is recruited at the MAMs could link the previously discovered effect of the MAMs on the fission mechanics of the mitochondrial network to the fine-tuned regulation of mitochondrial clearance or mitophagy (Chu 2010).

4 Conclusions

Accumulating evidence highlight that the contact sites established through the MAMs between the ER and mitochondria constitute a dynamic and complex subcellular signaling node regulating a variety of cellular processes. These range not only from classical ER-associated processes such as lipid synthesis/transport, Ca^{2+} signaling, and ER stress, but also to mitochondria shape, motility, bioenergetics, autophagy/mitophagy, and cell death. The myriad of functions that MAMs coordinate may imply a certain molecular plasticity in regulatory signaling mechanisms enabling crucial cellular decisions to be taken at the ER–mitochondria interface. Indeed, an increasing number of proteins have been shown to localize at the ER–mitochondria contact sites and modulate interorganellar crosstalk in both physiological and pathological conditions. This plasticity, moreover, seems particularly important under the conditions of ER stress, where the UPR has the ability to promote/re-establish ER proteostasis and survival or to incite cell death. In this context, it is emerging that certain ER stress sensors, such as PERK, have the ability to directly couple loss of ER folding capacity and Ca^{2+} signaling functions to the mitochondria, by associating with the MAMs. Finally, how ER–mitochondria contact sites integrate and modulate UPR signaling during ER stress remains a subject of future investigations.

References

Achleitner G, Gaigg B, Krasser A, Kainersdorfer E, Kohlwein SD, Perktold A, Zellnig G, Daum G (1999) Association between the endoplasmic reticulum and mitochondria of yeast facilitates interorganelle transport of phospholipids through membrane contact. Eur J Biochem 264:545–553

Anelli T, Bergamelli L, Margittai E, Rimessi A, Fagioli C, Malgaroli A, Pinton P, Ripamonti M, Rizzuto R, Sitia R (2012) Ero1α regulates Ca(2+) fluxes at the endoplasmic reticulum-mitochondria interface (MAM). Antioxid Redox Signal 16:1077–1087

Area-Gomez E, Castillo MDCL, Tambini MD, Guardia-Laguarta C, De Groof AJ, Madra M, Ikenouchi J, Umeda M, Bird TD, Sturley SL, Schon EA (2012) Upregulated function of mitochondria-associated ER membranes in Alzheimer disease. EMBO J 31:4106–4123

Berridge MJ (2002) The endoplasmic reticulum: a multifunctional signaling organelle. Cell Calcium 32:235–249

Berridge MJ, Bootman MD, Roderick HL (2003) Calcium signalling: dynamics, homeostasis and remodelling. Nat Rev Mol Cell Biol 4:517–529

Boehning D, Patterson RL, Sedaghat L, Glebova NO, Kurosaki T, Snyder SH (2003) Cytochrome c binds to inositol (1, 4, 5) trisphosphate receptors, amplifying calcium-dependent apoptosis. Nat Cell Biol 5:1051–1061

Bononi A, Bonora M, Marchi S, Missiroli S, Poletti F, Giorgi C, Pandolfi PP, Pinton P (2013) Identification of PTEN at the ER and MAMs and its regulation of Ca(2+) signaling and apoptosis in a protein phosphatase-dependent manner. Cell Death Differ 20:1631–1643

Bosch M, Mari M, Herms A, Fernandez A, Fajardo A, Kassan A, Giralt A, Colell A, Balgoma D, Barbero E, Gonzalez-Moreno E, Matias N, Tebar F, Balsinde J, Camps M, Enrich C, Gross SP, Garcia-Ruiz C, Perez-Navarro E, Fernandez-Checa JC, Pol A (2011) Caveolin-1 deficiency causes cholesterol-dependent mitochondrial dysfunction and apoptotic susceptibility. Curr Biol 21:681–686

Bouman L, Schlierf A, Lutz AK, Shan J, Deinlein A, Kast J, Galehdar Z, Palmisano V, Patenge N, Berg D, Gasser T, Augustin R, Trumbach D, Irrcher I, Park DS, Wurst W, Kilberg MS, Tatzelt J, Winklhofer KF (2011) Parkin is transcriptionally regulated by ATF4: evidence for an interconnection between mitochondrial stress and ER stress. Cell Death Differ 18:769–782

Bravo R, Vicencio JM, Parra V, Troncoso R, Munoz JP, Bui M, Quiroga C, Rodriguez AE, Verdejo HE, Ferreira J, Iglewski M, Chiong M, Simmen T, Zorzano A, Hill JA, Rothermel BA, Szabadkai G, Lavandero S (2011) Increased ER-mitochondrial coupling promotes mitochondrial respiration and bioenergetics during early phases of ER stress. J Cell Sci 124:2143–2152

Brough D, Schell MJ, Irvine RF (2005) Agonist-induced regulation of mitochondrial and endoplasmic reticulum motility. Biochem J 392:291–297

Calfon M, Zeng H, Urano F, Till JH, Hubbard SR, Harding HP, Clark SG, Ron D (2002) IRE1 couples endoplasmic reticulum load to secretory capacity by processing the XBP-1 mRNA. Nature 415:92–96

Cali T, Ottolini D, Negro A, Brini M (2013) Enhanced parkin levels favor ER-mitochondria crosstalk and guarantee Ca(2+) transfer to sustain cell bioenergetics. Biochim Biophys Acta 1832:495–508

Chami M, Oules B, Szabadkai G, Tacine R, Rizzuto R, Paterlini-Brechot P (2008) Role of SERCA1 truncated isoform in the proapoptotic calcium transfer from ER to mitochondria during ER stress. Mol Cell 32:641–651

Chu CT (2010) A pivotal role for PINK1 and autophagy in mitochondrial quality control: implications for Parkinson disease. Hum Mol Genet 19:R28–R37

Chung J, Torta F, Masai K, Lucast L, Czapla H, Tanner LB, Narayanaswamy P, Wenk MR, Nakatsu F, de Camilli P (2015) Intracellular Transport. PI4P/phosphatidylserine countertransport at ORP5- and ORP8-mediated ER-plasma membrane contacts. Science 349:428–432

Csordas G, Renken C, Varnai P, Walter L, Weaver D, Buttle KF, Balla T, Mannella CA, Hajnoczky G (2006) Structural and functional features and significance of the physical linkage between ER and mitochondria. J Cell Biol 174:915–921

Csordas G, Varnai P, Golenar T, Roy S, Purkins G, Schneider TG, Balla T, Hajnoczky G (2010) Imaging interorganelle contacts and local calcium dynamics at the ER-mitochondrial interface. Mol Cell 39:121–132

Cullinan SB, Zhang D, Hannink M, Arvisais E, Kaufman RJ, Diehl JA (2003) Nrf2 is a direct PERK substrate and effector of PERK-dependent cell survival. Mol Cell Biol 23:7198–7209

de Brito OM, Scorrano L (2008) Mitofusin 2 tethers endoplasmic reticulum to mitochondria. Nature 456:605–610

Dickey AS, Strack S (2011) PKA/AKAP1 and PP2A/Bbeta2 regulate neuronal morphogenesis via Drp1 phosphorylation and mitochondrial bioenergetics. J Neurosci 31:15716–15726

Dinner AR, Sali A, Smith LJ, Dobson CM, Karplus M (2000) Understanding protein folding via free-energy surfaces from theory and experiment. Trends Biochem Sci 25:331–339

Filadi R, Greotti E, Turacchio G, Luini A, Pozzan T, Pizzo P (2015) Mitofusin 2 ablation increases endoplasmic reticulum-mitochondria coupling. Proc Natl Acad Sci U S A 112:E2174–E2181

Filadi R, Greotti E, Turacchio G, Luini A, Pozzan T, Pizzo P (2016) Presenilin 2 modulates endoplasmic reticulum-mitochondria coupling by tuning the antagonistic effect of mitofusin 2. Cell Rep 15:2226–2238

Fransson A, Ruusala A, Aspenstrom P (2003) Atypical Rho GTPases have roles in mitochondrial homeostasis and apoptosis. J Biol Chem 278:6495–6502

Fransson S, Ruusala A, Aspenstrom P (2006) The atypical Rho GTPases miro-1 and miro-2 have essential roles in mitochondrial trafficking. Biochem Biophys Res Commun 344:500–510

Friedman JR, Lackner LL, West M, Dibenedetto JR, Nunnari J, Voeltz GK (2011) ER tubules mark sites of mitochondrial division. Science 334:358–362

Fujimoto M, Hayashi T, Su TP (2012) The role of cholesterol in the association of endoplasmic reticulum membranes with mitochondria. Biochem Biophys Res Commun 417:635–639

Furuichi T, Yoshikawa S, Miyawaki A, Wada K, Maeda N, Mikoshiba K (1989) Primary structure and functional expression of the inositol 1, 4, 5-trisphosphate-binding protein P400. Nature 342:32–38

Galmes R, Houcine A, Van Vliet AR, Agostinis P, Jackson CL, Giordano F (2016) ORP5/ORP8 localize to endoplasmic reticulum-mitochondria contacts and are involved in mitochondrial function. EMBO Rep 17(6):800–810

Giacomello M, Drago I, Bortolozzi M, Scorzeto M, Gianelle A, Pizzo P, Pozzan T (2010) Ca^{2+} hot spots on the mitochondrial surface are generated by Ca^{2+} mobilization from stores, but not by activation of store-operated Ca^{2+} channels. Mol Cell 38:280–290

Gilady SY, Bui M, Lynes EM, Benson MD, Watts R, Vance JE, Simmen T (2010) Ero1α requires oxidizing and normoxic conditions to localize to the mitochondria-associated membrane (MAM). Cell Stress Chaperones 15:619–629

Giorgi C, de Stefani D, Bononi A, Rizzuto R, Pinton P (2009) Structural and functional link between the mitochondrial network and the endoplasmic reticulum. Int J Biochem Cell Biol 41:1817–1827

Giorgi C, Ito K, Lin HK, Santangelo C, Wieckowski MR, Lebiedzinska M, Bononi A, Bonora M, Duszynski J, Bernardi R, Rizzuto R, Tacchetti C, Pinton P, Pandolfi PP (2010) PML regulates apoptosis at endoplasmic reticulum by modulating calcium release. Science 330:1247–1251

Gorlach A, Klappa P, Kietzmann T (2006) The endoplasmic reticulum: folding, calcium homeostasis, signaling, and redox control. Antioxid Redox Signal 8:1391–1418

Hailey DW, Rambold AS, Satpute-Krishnan P, Mitra K, Sougrat R, Kim PK, Lippincott-Schwartz J (2010) Mitochondria supply membranes for autophagosome biogenesis during starvation. Cell 141:656–667

Hamasaki M, Furuta N, Matsuda A, Nezu A, Yamamoto A, Fujita N, Oomori H, Noda T, Haraguchi T, Hiraoka Y, Amano A, Yoshimori T (2013) Autophagosomes form at ER-mitochondria contact sites. Nature 495:389–393

Han D, Lerner AG, Vande Walle L, Upton JP, Xu W, Hagen A, Backes BJ, Oakes SA, Papa FR (2009) IRE1alpha kinase activation modes control alternate endoribonuclease outputs to determine divergent cell fates. Cell 138:562–75

Han J, Back SH, Hur J, Lin YH, Gildersleeve R, Shan J, Yuan CL, Krokowski D, Wang S, Hatzoglou M, Kilberg MS, Sartor MA, Kaufman RJ (2013) ER-stress-induced transcriptional regulation increases protein synthesis leading to cell death. Nat Cell Biol 15:481–490

Harding HP, Zhang Y, Ron D (1999) Protein translation and folding are coupled by an endoplasmic-reticulum-resident kinase. Nature 397:271–274

Hayashi T, Fujimoto M (2010) Detergent-resistant microdomains determine the localization of sigma-1 receptors to the endoplasmic reticulum-mitochondria junction. Mol Pharmacol 77:517–528

Hayashi T, Rizzuto R, Hajnoczky G, Su TP (2009) MAM: more than just a housekeeper. Trends Cell Biol 19:81–88

Hayashi T, Su TP (2003) Sigma-1 receptors (sigma(1) binding sites) form raft-like microdomains and target lipid droplets on the endoplasmic reticulum: roles in endoplasmic reticulum lipid compartmentalization and export. J Pharmacol Exp Ther 306:718–725

Hayashi T, Su TP (2007) Sigma-1 receptor chaperones at the ER-mitochondrion interface regulate Ca(2+) signaling and cell survival. Cell 131:596–610

Haze K, Yoshida H, Yanagi H, Yura T, Mori K (1999) Mammalian transcription factor ATF6 is synthesized as a transmembrane protein and activated by proteolysis in response to endoplasmic reticulum stress. Mol Biol Cell 10:3787–3799

He CC, Klionsky DJ (2009) Regulation mechanisms and signaling pathways of autophagy. Annu Rev Genet 43:67–93

Hendershot LM (2004) The ER function BiP is a master regulator of ER function. Mt Sinai J Med 71:289–297

Hetz C (2012) The unfolded protein response: controlling cell fate decisions under ER stress and beyond. Nat Rev Mol Cell Biol 13:89–102

Hollien J, Lin JH, Li H, Stevens N, Walter P, Weissman JS (2009) Regulated Ire1-dependent decay of messenger RNAs in mammalian cells. J Cell Biol 186:323–331

Ingerman E, Perkins EM, Marino M, Mears JA, McCaffery JM, Hinshaw JE, Nunnari J (2005) Dnm1 forms spirals that are structurally tailored to fit mitochondria. J Cell Biol 170:1021–1027

Iwasawa R, Mahul-Mellier AL, Datler C, Pazarentzos E, Grimm S (2011) Fis1 and Bap31 bridge the mitochondria-ER interface to establish a platform for apoptosis induction. EMBO J 30:556–568

Jager R, Bertrand MJ, Gorman AM, Vandenabeele P, Samali A (2012) The unfolded protein response at the crossroads of cellular life and death during endoplasmic reticulum stress. Biol Cell 104:259–270

Kim I, Xu W, Reed JC (2008) Cell death and endoplasmic reticulum stress: disease relevance and therapeutic opportunities. Nat Rev Drug Discov 7:1013–1030

Kirichok Y, Krapivinsky G, Clapham DE (2004) The mitochondrial calcium uniporter is a highly selective ion channel. Nature 427:360–364

Kornmann B, Osman C, Walter P (2011) The conserved GTPase Gem1 regulates endoplasmic reticulum-mitochondria connections. Proc Natl Acad Sci U S A 108:14151–14156

Kroemer G, Galluzzi L, Brenner C (2007) Mitochondrial membrane permeabilization in cell death. Physiol Rev 87:99–163

Lee AH, Iwakoshi NN, Glimcher LH (2003) XBP-1 regulates a subset of endoplasmic reticulum resident chaperone genes in the unfolded protein response. Mol Cell Biol 23:7448–7459

Loson OC, Song Z, Chen H, Chan DC (2013) Fis1, Mff, MiD49, and MiD51 mediate Drp1 recruitment in mitochondrial fission. Mol Biol Cell 24:659–667

Lynes EM, Bui M, Yap MC, Benson MD, Schneider B, Ellgaard L, Berthiaume LG, Simmen T (2012) Palmitoylated TMX and calnexin target to the mitochondria-associated membrane. EMBO J 31:457–470

Malhotra JD, Kaufman RJ (2007) The endoplasmic reticulum and the unfolded protein response. Semin Cell Dev Biol 18:716–731

Mallilankaraman K, Cardenas C, Doonan PJ, Chandramoorthy HC, Irrinki KM, Golenar T, Csordas G, Madireddi P, Yang J, Muller M, Miller R, Kolesar JE, Molgo J, Kaufman B, Hajnoczky G, Foskett JK, Madesh M (2012) MCUR1 is an essential component of mitochondrial Ca^{2+} uptake that regulates cellular metabolism. Nat Cell Biol 14:1336–1343

Matsuzaki H, Fujimoto T, Tanaka M, Shirasawa S (2013) Tespa1 is a novel component of mitochondria-associated endoplasmic reticulum membranes and affects mitochondrial calcium flux. Biochem Biophys Res Commun 433:322–326

Maurel M, Chevet E, Tavernier J, Gerlo S (2014) Getting RIDD of RNA: IRE1 in cell fate regulation. Trends Biochem Sci 39:245–254

McCormack JG, Denton RM (1980) Role of calcium ions in the regulation of intramitochondrial metabolism. Properties of the Ca^{2+}-sensitive dehydrogenases within intact uncoupled mitochondria from the white and brown adipose tissue of the rat. Biochem J 190:95–105

McCracken AA, Brodsky JL (2003) Evolving questions and paradigm shifts in endoplasmic-reticulum-associated degradation (ERAD). BioEssays 25:868–877

Moenner M, Pluquet O, Bouchecareilh M, Chevet E (2007) Integrated endoplasmic reticulum stress responses in cancer. Cancer Res 67:10631–10634

Muhlberg AB, Warnock DE, Schmid SL (1997) Domain structure and intramolecular regulation of dynamin GTPase. EMBO J 16:6676–6683

Munoz JP, Ivanova S, Sanchez-Wandelmer J, Martinez-Cristobal P, Noguera E, Sancho A, Diaz-Ramos A, Hernandez-Alvarez MI, Sebastian D, Mauvezin C, Palacin M, Zorzano A (2013) Mfn2 modulates the UPR and mitochondrial function via repression of PERK. EMBO J 32:2348–2361

Myhill N, Lynes EM, Nanji JA, Blagoveshchenskaya AD, Fei H, Simmen KC, Cooper TJ, Thomas G, Simmen T (2008) The subcellular distribution of calnexin is mediated by PACS-2. Mol Biol Cell 19:2777–2788

Naon D, Zaninello M, Giacomello M, Varanita T, Grespi F, Lakshminaranayan S, Serafini A, Semenzato M, Herkenne S, Hernandez-Alvarez MI, Zorzano A, De Stefani D, Dorn GW 2nd, Scorrano L (2016) Critical reappraisal confirms that Mitofusin 2 is an endoplasmic reticulum-mitochondria tether. Proc Natl Acad Sci U S A 113:11249–11254

Nguyen TT, Lewandowska A, Choi JY, Markgraf DF, Junker M, Bilgin M, Ejsing CS, Voelker DR, Rapoport TA, Shaw JM (2012) Gem1 and ERMES do not directly affect phosphatidylserine transport from ER to mitochondria or mitochondrial inheritance. Traffic 13:880–890

Noack J, Molinari M (2014) Protein trafficking: RESETting proteostasis. Nat Chem Biol 10:881–882

Ogata M, Hino S, Saito A, Morikawa K, Kondo S, Kanemoto S, Murakami T, Taniguchi M, Tanii I, Yoshinaga K, Shiosaka S, Hammarback JA, Urano F, Imaizumi K (2006) Autophagy is activated for cell survival after endoplasmic reticulum stress. Mol Cell Biol 26:9220–9231

Park SH, Blackstone C (2010) Further assembly required: construction and dynamics of the endoplasmic reticulum network. EMBO Rep 11:515–521

Penno A, Hackenbroich G, Thiele C (2013) Phospholipids and lipid droplets. Biochim Biophys Acta 1831:589–594

Pinton P, Giorgi C, Siviero R, Zecchini E, Rizzuto R (2008) Calcium and apoptosis: ER-mitochondria Ca^{2+} transfer in the control of apoptosis. Oncogene 27:6407–6418

Pizzo P, Drago I, Filadi R, Pozzan T (2012) Mitochondrial Ca(2+) homeostasis: mechanism, role, and tissue specificities. Pflugers Arch 464:3–17

Pizzo P, Pozzan T (2007) Mitochondria-endoplasmic reticulum choreography: structure and signaling dynamics. Trends Cell Biol 17:511–517

Pollard MG, Travers KJ, Weissman JS (1998) Ero1p: a novel and ubiquitous protein with an essential role in oxidative protein folding in the endoplasmic reticulum. Mol Cell 1:171–182

Porter KR, Claude A, Fullam EF (1945) A study of tissue culture cells by electron microscopy: methods and preliminary observations. J Exp Med 81:233–246

Preissler S, Rato C, Chen R, Antrobus R, Ding S, Fearnley IM, Ron D (2015) AMPylation matches BiP activity to client protein load in the endoplasmic reticulum. Elife 4:e12621

Rizzuto R, Pinton P, Carrington W, Fay FS, Fogarty KE, Lifshitz LM, Tuft RA, Pozzan T (1998) Close contacts with the endoplasmic reticulum as determinants of mitochondrial Ca2+ responses. Science 280:1763–1766

Rizzuto R, Pozzan T (2006) Microdomains of intracellular Ca2+: molecular determinants and functional consequences. Physiol Rev 86:369–408

Roberts-Thomson SJ, Peters AA, Grice DM, Monteith GR (2010) ORAI-mediated calcium entry: mechanism and roles, diseases and pharmacology. Pharmacol Ther 127:121–130

Ron D, Walter P (2007) Signal integration in the endoplasmic reticulum unfolded protein response. Nat Rev Mol Cell Biol 8:519–529

Ruby JR, Dyer RF, Skalko RG (1969) Continuities between mitochondria and endoplasmic reticulum in the mammalian ovary. Z Zellforsch Mikrosk Anat 97:30–37

Rusinol AE, Cui Z, Chen MH, Vance JE (1994) A unique mitochondria-associated membrane fraction from rat liver has a high capacity for lipid synthesis and contains pre-Golgi secretory proteins including nascent lipoproteins. J Biol Chem 269:27494–27502

Rysman E, Brusselmans K, Scheys K, Timmermans L, Derua R, Munck S, van Veldhoven PP, Waltregny D, Daniels VW, Machiels J, Vanderhoydonc F, Smans K, Waelkens E, Verhoeven G, Swinnen JV (2010) De novo lipogenesis protects cancer cells from free radicals and chemotherapeutics by promoting membrane lipid saturation. Cancer Res 70:8117–8126

Sammels E, Parys JB, Missiaen L, de Smedt H, Bultynck G (2010) Intracellular Ca^{2+} storage in health and disease: a dynamic equilibrium. Cell Calcium 47:297–314

Sano R, Annunziata I, Patterson A, Moshiach S, Gomero E, Opferman J, Forte M, D'Azzo A (2009) GM1-ganglioside accumulation at the mitochondria-associated ER membranes links ER stress to Ca(2+)-dependent mitochondrial apoptosis. Mol Cell 36:500–511

Saotome M, Safiulina D, Szabadkai G, Das S, Fransson A, Aspenstrom P, Rizzuto R, Hajnoczky G (2008) Bidirectional Ca^{2+}-dependent control of mitochondrial dynamics by the Miro GTPase. Proc Natl Acad Sci U S A 105:20728–20733

Schrepfer E, Scorrano L (2016) Mitofusins, from Mitochondria to Metabolism. Mol Cell 61:683–694

Schroder M, Kaufman RJ (2005) The mammalian unfolded protein response. Annu Rev Biochem 74:739–789

Scorrano L (2013) Keeping mitochondria in shape: a matter of life and death. Eur J Clin Invest 43:886–893

Shen WW, Frieden M, Demaurex N (2011) Remodelling of the endoplasmic reticulum during store-operated calcium entry. Biol Cell 103:365–380

Shiao YJ, Balcerzak B, Vance JE (1998) A mitochondrial membrane protein is required for translocation of phosphatidylserine from mitochondria-associated membranes to mitochondria. Biochem J 331(Pt 1):217–223

Shibata Y, Shemesh T, Prinz WA, Palazzo AF, Kozlov MM, Rapoport TA (2010) Mechanisms determining the morphology of the peripheral ER. Cell 143:774–788

Shintani T, Klionsky DJ (2004) Autophagy in health and disease: a double-edged sword. Science 306:990–995

Simmen T, Aslan JE, Blagoveshchenskaya AD, Thomas L, Wan L, Xiang Y, Feliciangeli SF, Hung CH, Crump CM, Thomas G (2005) PACS-2 controls endoplasmic reticulum-mitochondria communication and Bid-mediated apoptosis. EMBO J 24:717–729

Steenbergen R, Nanowski TS, Beigneux A, Kulinski A, Young SG, Vance JE (2005) Disruption of the phosphatidylserine decarboxylase gene in mice causes embryonic lethality and mitochondrial defects. J Biol Chem 280:40032–40040

Stefan CJ, Manford AG, Baird D, Yamada-Hanff J, Mao Y, Emr SD (2011) Osh proteins regulate phosphoinositide metabolism at ER-plasma membrane contact sites. Cell 144:389–401

Stevens FJ, Argon Y (1999) Protein folding in the ER. Semin Cell Dev Biol 10:443–454

Stoica R, de Vos KJ, Paillusson S, Mueller S, Sancho RM, Lau KF, Vizcay-Barrena G, Lin WL, Xu YF, Lewis J, Dickson DW, Petrucelli L, Mitchell JC, Shaw CE, Miller CC (2014) ER-mitochondria associations are regulated by the VAPB-PTPIP51 interaction and are disrupted by ALS/FTD-associated TDP-43. Nat Commun 5:3996

Szabadkai G, Bianchi K, Varnai P, de Stefani D, Wieckowski MR, Cavagna D, Nagy AI, Balla T, Rizzuto R (2006) Chaperone-mediated coupling of endoplasmic reticulum and mitochondrial Ca^{2+} channels. J Cell Biol 175:901–911

Tooze SA, Yoshimori T (2010) The origin of the autophagosomal membrane. Nat Cell Biol 12:831–835

Twig G, Elorza A, Molina AJA, Mohamed H, Wikstrom JD, Walzer G, Stiles L, Haigh SE, Katz S, Las G, Alroy J, Wu M, Py BF, Yuan J, Deeney JT, Corkey BE, Shirihai OS (2008) Fission and selective fusion govern mitochondrial segregation and elimination by autophagy. EMBO J 27:433–446

Urano F, Wang X, Bertolotti A, Zhang Y, Chung P, Harding HP, Ron D (2000) Coupling of stress in the ER to activation of JNK protein kinases by transmembrane protein kinase IRE1. Science 287:664–666

Vance JE (1990) Phospholipid synthesis in a membrane fraction associated with mitochondria. J Biol Chem 265:7248–7256

Varnai P, Balla A, Hunyady L, Balla T (2005) Targeted expression of the inositol 1, 4, 5-triphosphate receptor (IP3R) ligand-binding domain releases Ca^{2+} via endogenous IP3R channels. Proc Natl Acad Sci U S A 102:7859–7864

Verfaillie T, Garg AD, Agostinis P (2013) Targeting ER stress induced apoptosis and inflammation in cancer. Cancer Lett 332:249–264

Verfaillie T, Rubio N, Garg AD, Bultynck G, Rizzuto R, Decuypere JP, Piette J, Linehan C, Gupta S, Samali A, Agostinis P (2012) PERK is required at the ER-mitochondrial contact sites to convey apoptosis after ROS-based ER stress. Cell Death Differ 19:1880–1891

Verkhratsky A, Toescu EC (2003) Endoplasmic reticulum Ca(2+) homeostasis and neuronal death. J Cell Mol Med 7:351–361

Voelker DR (1989) Reconstitution of phosphatidylserine import into rat liver mitochondria. J Biol Chem 264:8019–8025

Voelker DR (2005) Bridging gaps in phospholipid transport. Trends Biochem Sci 30:396–404

Voeltz GK, Prinz WA, Shibata Y, Rist JM, Rapoport TA (2006) A class of membrane proteins shaping the tubular endoplasmic reticulum. Cell 124:573–586

Voeltz GK, Rolls MM, Rapoport TA (2002) Structural organization of the endoplasmic reticulum. EMBO Rep 3:944–950

von Filseck JM, Copic A, Delfosse V, Vanni S, Jackson CL, Bourguet W, Drin G (2015) Intracellular Transport. Phosphatidylserine transport by ORP/Osh proteins is driven by phosphatidylinositol 4-phosphate. Science 349:432–436

Wang K, Klionsky DJ (2011) Mitochondria removal by autophagy. Autophagy 7:297–300

Wang M, Kaufman RJ (2016) Protein misfolding in the endoplasmic reticulum as a conduit to human disease. Nature 529:326–335

Wang X, Winter D, Ashrafi G, Schlehe J, Wong YL, Selkoe D, Rice S, Steen J, Lavoie MJ, Schwarz TL (2011) PINK1 and Parkin target Miro for phosphorylation and degradation to arrest mitochondrial motility. Cell 147:893–906

Wei Y, Pattingre S, Sinha S, Bassik M, Levine B (2008) JNK1-mediated phosphorylation of Bcl-2 regulates starvation-induced autophagy. Mol Cell 30:678–688

Wideman JG, Lackey SW, Srayko MA, Norton KA, Nargang FE (2013) Analysis of mutations in Neurospora crassa ERMES components reveals specific functions related to beta-barrel protein assembly and maintenance of mitochondrial morphology. PLoS ONE 8:e71837

Yi M, Weaver D, Hajnoczky G (2004) Control of mitochondrial motility and distribution by the calcium signal: a homeostatic circuit. J Cell Biol 167:661–672

Yorimitsu T, Klionsky DJ (2005) Autophagy: molecular machinery for self-eating. Cell Death Differ 12(Suppl 2):1542–1552

Yorimitsu T, Nair U, Yang Z, Klionsky DJ (2006) Endoplasmic reticulum stress triggers autophagy. J Biol Chem 281:30299–30304

Yoshida H, Matsui T, Yamamoto A, Okada T, Mori K (2001) XBP1 mRNA is induced by ATF6 and spliced by IRE1 in response to ER stress to produce a highly active transcription factor. Cell 107:881–891

Youle RJ, van der Bliek AM (2012) Mitochondrial fission, fusion, and stress. Science 337:1062–1065

Coordinating Organismal Metabolism During Protein Misfolding in the ER Through the Unfolded Protein Response

Vishwanatha K. Chandrahas, Jaeseok Han and Randal J. Kaufman

Abstract The endoplasmic reticulum (ER) is a cellular organelle responsible for folding of secretory and membrane proteins. Perturbance in ER homeostasis caused by various intrinsic/extrinsic stimuli challenges the protein-folding capacity of the ER, leading to an ER dysfunction, called ER stress. Cells have developed a defensive response to adapt and/or survive in the face of ER stress that may be detrimental to cell function and survival. When exposed to ER stress, the cell activates a complex and elaborate signaling network that includes translational modulation and transcriptional induction of genes. In addition to these autonomous responses, recent studies suggest that the stressed tissue secretes peptides or unknown factors that transfer the signal to other cells in the same or different organs, leading the organism as a whole to cope with challenges in a non-autonomous manner. In this review, we discuss the mechanisms by which cells adapt to ER stress challenges autonomously and transfer the stress signal to non-stressed cells in different organs.

Contents

1 ER Stress and the Unfolded Protein Response (UPR) 104
2 The Autonomous UPR in Metabolism 105
    2.1 The UPR in Pancreatic β Cells 106
    2.2 The UPR in the Liver 109
    2.3 The UPR in Adipose Tissue 112

V.K. Chandrahas · R.J. Kaufman (✉)
Degenerative Diseases Program, Sanford_Burnham_Prebys Medical
Discovery Institute, 92037 La Jolla, CA, USA
e-mail: rkaufman@sbpdiscovery.org

J. Han (✉)
Soonchunhyang Institute of Med-bio Science (SIMS), Soonchunhyang University,
31151 Cheonan-si, Chungcheongnam-do, Republic of Korea
e-mail: hanjs015@sch.ac.kr

2.4 The UPR in the Nervous System.. 114
3 The Non-autonomous UPR 115
   3.1 Proximal Effects of the Non-autonomous UPR.. 117
   3.2 Distal Effects of the Non-autonomous UPR... 118
   3.3 Potential Mechanism of the Non-autonomous UPR... 121
4 Conclusions.. 123
References.. 124

1 ER Stress and the Unfolded Protein Response (UPR)

The ER in eukaryotes is a membranous organelle that provides diverse cellular processes such as protein folding, protein glycosylation, lipid biosynthesis besides serving as a calcium repository for its controlled release during calcium signaling. Newly synthesized proteins, prior to their secretion from the ER lumen, require proper folding into their native structure and into multisubunit complexes. The rate of protein synthesis at the ER is continually regulated in accordance with the protein-folding capacity. However, occasional physiological insults perturb the protein-folding capacity of the ER causing an accumulation of unfolded and/or misfolded proteins, a situation termed ER stress. To counteract the ER stress, an adaptive response called the unfolded protein response (UPR) is executed to restore protein-folding homeostasis (Wang and Kaufman 2016). These responses include expression of chaperones to increase protein-folding capacity of the ER, elimination of unfolded/misfolded proteins by ER-associated degradation (ERAD) and autophagy, and attenuation of global protein synthesis to reduce the protein-folding load on the ER, altogether leading to improved protein homeostasis and cell survival, whereas chronic ER stress activates cell death pathways. There are three known ER transmembrane proteins that sense protein misfolding in the ER. These include two type I transmembrane proteins, a protein kinase/endoribonuclease called inositol requiring protein-1α (Ire1α) and a protein kinase R-like ER kinase (PERK), and a type II transmembrane protein, activating transcription factor 6α (ATF6α). Ire1α and PERK possess functionally interchangeable luminal domains that sense ER stress, whereas ATF6α has a C-terminal ER stress sensing domain inside the ER lumen and an N-terminal cytosolic domain harboring a basic leucine zipper (bZIP) transcription factor domain. In the basal state, the luminal domains of all sensors are bound to the chaperone GRP78/BiP and reside in the ER. Accumulation of unfolded proteins causes release of BiP from the luminal domains of the sensors because the unfolded proteins bind and sequester the peptide-binding activity of BiP. As a consequence, the luminal domains of IRE1α and PERK undergo dimerization/oligomerization, autophosphorylation, and subsequent activation. Activated Ire1α initiates its intrinsic endoribonuclease activity to remove a 26 base intron from *Xbp1* mRNA (or a 256 base intron in *Hac1* mRNA in yeast) that encodes a transcription factor XBP1s (s = spliced). XBP1s enters the nucleus and upregulates transcription of genes involved in ER homeostasis. The substrate for PERK is the alpha subunit of eukaryotic translation initiation factor 2 (eIF2α). Activated PERK phosphorylates a critical serine residue (S51) on eIF2α and deters its ability to initiate mRNA translation to reduce global protein synthesis. The transient decreased load of proteins in the ER then provides an opportunity to restore protein homeostasis in the ER. The release of BiP from ATF6α leads its translocation from ER to the Golgi in a coat protein complex II (COPII) vesicle-dependent manner (Schindler and Schekman 2009). At the Golgi, two proteolytic cleavage reactions on ATF6α by Golgi-resident Site I and Site II proteases (S1P and S2P) liberate the N-terminal bZIP domain (ATF6 (N)), which enters the nucleus to affect transcription of genes involved in the UPR to relieve ER stress. For an extensive review on the mechanism of UPR activation, see (Schröder and Kaufman 2005; Wang and Kaufman 2014, 2016). In this review, we focus on the role of the UPR in metabolic organs and in the brain. The role of the UPR in lung (Osorio et al. 2013), heart (Zhou and Tabas 2013), inflammation (Grootjans et al. 2016), and cancer (Wang and Kaufman 2014) was recently reviewed.

2 The Autonomous UPR in Metabolism

The functions of the three membrane-bound UPR sensors manifest independently of each other upon ER stress. The cross talk between their signaling in conditions of disease, such as inflammatory and metabolic signaling pathways, has also been demonstrated extensively. Although chronic low levels of ER stress and a constitutively active state of the UPR in cells may not necessarily reflect a disease rather a normal physiological response of the cell, persistent ER stress may lead to cell death. Identification of glucose-regulated protein (GRP) as targets of the UPR describes the physiological functions of UPR in glucose/energy metabolism. Subsequently, several studies were performed to establish the role of UPR in glucose metabolism that in addition uncovered the role of the UPR in lipid metabolism, cell proliferation, and differentiation as well as disease. In this section, we elaborate our knowledge on autonomous functions of the UPR in several cell types and tissues such as pancreatic β cells, hepatocytes, adipocytes, and neurological tissue. We describe a role of the UPR in different cell types and tissues and the physiological role of UPR in cellular metabolism and diseases emphasizing glucose and lipid metabolism, insulin action, as well as metabolic disorders (Fig. 1). Since the UPR is becoming the target for the treatment of diabetes, obesity, liver associated as well as tumorigenesis, our review of the understanding of the cross talk of UPR signaling pathways and between cells will be significant.

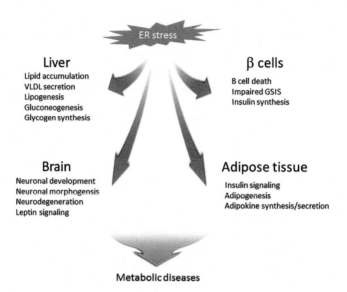

Fig. 1 Role of the autonomous UPR in glucose and lipid metabolism and in neuron development. A basal level of UPR operates in liver, pancreatic β cells, and adipose tissue to regulate glucose and lipid metabolism. Increased proinsulin synthesis in β cells causes autonomous activation of UPR to accommodate their folding and secretion. Consequently, dysfunction in UPR sensors and chaperones is known to cause metabolic disorders like type 2 diabetes and obesity. Besides energy metabolism, UPR activation is required during brain development. The process of neuronal differentiation and morphogenesis elicits the UPR due to increased protein synthesis. The reduced UPR in adult brains further emphasize the requirement of the UPR in brain development

2.1 The UPR in Pancreatic β Cells

The pancreas is a vital endocrine organ capable of producing several hormones including insulin, glucagon, and somatostatin that circulate in blood. Insulin, which is secreted by β cells in the islets of Langerhans in the pancreas, facilitates glucose absorption from the blood into cells. Pancreatic β cells contain a large pool of mRNAs representing 20% of total mRNAs in β cells that encode the inactive precursor proinsulin to be stored in secretory granules. In response to increased blood glucose levels following diet, proinsulin is processed into insulin, which is released to decrease blood glucose levels. The need for the biosynthesis, folding, and processing of insulin imposes heavy burden on β cells necessitating the constant deployment of an active UPR to maintain glucose metabolism (Back and Kaufman 2012).

Glucose metabolism largely depends on the mass of pancreatic β cells and their insulin secretion capacity. The failure of β cells to efficiently secrete insulin is the major cause of type 2 diabetes. An increased demand for insulin due to insulin resistance leads to an expansion of β cell mass to compensate and stabilize blood glucose concentrations. Studies showed that insulin demand itself signals enhanced insulin secretion and the expansion of β cell mass is accomplished through the activation of UPR. Therefore, β cells that engage mild ER stress with an active UPR are more likely to proliferate (Sharma et al. 2015). These studies underscore the importance of physiological UPR pathways in health and proliferation of β cell mass.

Besides the canonical role of IRE1α and XBP1s in the restoration of ER homeostasis during UPR, they were shown to be important in insulin receptor signaling through activation of c-Jun N-terminal kinase (JNK). Through the regulation of JNK, XBP1s was demonstrated as the key regulator of insulin receptor signaling, obesity, and glucose homeostasis. Experiments performed in mice and cell culture models demonstrated that pharmacological induction of ER stress or complete/partial reduction in the function of XBP1s ($Xbp1$ $-/-$ or $Xbp1$ $-/+$) without inducing ER stress impairs insulin receptor signaling. Additionally, $Xbp1$ $-/-$ mice exhibited impaired glucose homeostasis (Özcan et al. 2004). The observation of impaired glucose homeostasis and insulin signaling under conditions of diminished XBP1s function in the absence of ER stress leads to the speculation that components of the UPR play pivotal roles in glucose metabolism. Besides insulin signaling, Ire1α regulates proinsulin biosynthesis and folding upon exposure of pancreatic β cells to high glucose levels independent of XBP1s; however, chronic exposure of β cells to high glucose levels elicits an ER stress response followed by suppression of insulin gene expression. These results were corroborated by the fact that conditional knockdown of Ire1α signaling resulted in diminished insulin biosynthesis (Lipson et al. 2006). Furthermore, an ER stress-independent function of Ire1α in blood glucose homeostasis was demonstrated from experiments that showed that a DNA-binding-defective mutant of XBP1s was able to interact with Forkhead boxO1 (FoxO1) transcription factor. FoxO1 in pancreatic β cells was known to regulate their differentiation, proliferation, and apoptosis, and inhibition of FoxO1 protects pancreatic β cells from apoptosis (Martinez et al. 2008). Due to the impaired DNA-binding ability of XBP1s mutant, it was unable to participate in the UPR. However, its ability to interact with FoxO1 was not affected. The interaction between XBP1s and FoxO1 enhanced degradation of the FoxO1 in a proteasome-dependent pathway and hence diminished the activity of FoxO1 improving serum glucose concentration without affecting the ER-folding capacity. Accordingly, depletion of XBP1s followed by high-fat diet caused accumulation of FoxO1 and an improved glucose intolerance (Zhou et al. 2011). In addition, β cell-specific ablation of $Xbp1$ was shown to cause β cell failure suggesting the role of the IRE1α-XBP1s pathway for optimal insulin secretion from pancreatic β cells (Lee et al. 2011).

ER stress-dependent activation of PERK followed by eIF2α phosphorylation ensues a temporarily reduced global translation rate with the concomitant reduction in protein load into ER. However, PERK activation followed by eIF2α phosphorylation upregulates the translation of selective mRNAs, e.g., ATF4, ATF5, and CHOP. PERK in addition to ATF4, ATF5, and CHOP is important in the regulation of glucose metabolism, obesity, type 2 diabetes, as well as inflammatory signaling. Studies performed using animal models suggested a critical role for the PERK pathway in pancreatic β cell function (Lee and Ozcan 2014). Using *Perk* knockout mouse models, Gupta et al. demonstrated that PERK was a critical component in maintaining intact ERAD, ER morphology, ER to Golgi anterograde translocation, as well as an overall β cell morphology (Gupta et al. 2010). Consistently, suppression of PERK by expressing dominant negative PERK mutants or *Perk* deletion caused excessive accumulation of proinsulin and ubiquitinated proteins in pancreatic β cells and an abnormal β cell morphology. Similar results were observed with respect to insulin secretion and blood glucose accumulation in *Akita* models that exhibit hyperglycemia and hypoinsulinemia; a condition similar to Wolcott–Rallison syndrome in humans results from loss of PERK function. The deletion of *Perk* in *Akita* mice causes expansion of the β cell mass albeit with reduced accumulation of insulin and proinsulin and a concomitant increase in glucagon expression in adjacent alpha cells; eventually, these mice rapidly develop diabetes (Gupta et al. 2010). Another study by Zhang et al. showed that PERK is required for the maintenance of β cell mass and the deletion of PERK-impaired β cell differentiation and glucose-stimulated insulin secretion (Zhang et al. 2006). A more detailed analysis of the PERK pathway in insulin secretion revealed that eIF2α phosphorylation is a critical requirement in the growth of pancreatic β cells. Studies performed in homozygous mice harboring a mutant eIF2α (S51A) that cannot perform PERK-mediated phosphorylation exhibited severe hypoglycemia at 4–6 h after birth (Scheuner et al. 2001). Furthermore, the pancreas from these mice showed a β cell deficiency with a concomitant reduction in serum insulin content as low as 35–50% of that from their wild-type counterparts. These results suggest that an intact PERK–eIF2α pathway is central for glucose homeostasis (Scheuner et al. 2001). Although a lack of the PERK pathway impaired glucose homeostasis and caused hyperglycemia at two weeks of life, defective eIF2α phosphorylation caused hypoglycemia at 6 h postnatally. It is likely that eIF2α phosphorylation through a different eIF2α protein kinase regulates gluconeogenesis in the liver early after birth where PERK-mediated eIF2α phosphorylation is important at two weeks postnatally to maintain β cell function to prevent hyperglycemia and diabetes. These results suggest that PERK and eIF2α may control glucose homeostasis independently of each other and warrant further studies to understand their roles in glucose homeostasis.

Unlike the mode of action for IRE1α and PERK upon UPR activation, ATF6α leaves the ER where it is then processed in the Golgi apparatus to release a cytosolic fragment which enters the nucleus to induce transcription of genes involved in ER stress management. Besides its undisputed function in the UPR, the role of ATF6α in glucose homeostasis and pancreatic β cell integrity has been demonstrated. A lower pancreatic insulin content as well as impaired insulin secretion was observed in *Atf6α* −/− mice upon high-fat diet feeding. These mice also exhibited a swollen ER in β cells with increased *Xbp1* splicing in liver implicating increased ER stress. However, pancreas from mice that were fed with regular diet exhibited normal insulin secretion, suggesting that ATF6α is important in β cell physiology under circumstances of metabolic overload (Usui et al. 2012). In addition to its effect on proinsulin secretion, ATF6α also downregulates gluconeogenesis and hence decreases blood glucose levels. In obese mammals, pancreatic glucagon in association with CREB-regulated transcription coactivator 2 (CRTC2) induces hepatic gluconeogenesis leading to increased blood glucose levels. However, ATF6α in hepatic cells inhibits the activity of CRTC2 and reduces gluconeogenesis, as well as blood glucose levels. Consequently, RNAi-mediated depletion of ATF6α results in increased glucose production, whereas overexpression of ATF6α reverses the effects of CRTC2 (Wang et al. 2009). In addition to its effect on glucose metabolism, ATF6α was identified as an important UPR sensor to signal β cell proliferation in response to increased glucose. While depletion of ATF6α activity using small molecule inhibitors abrogated glucose-induced β cell proliferation in mouse islet cultures, overexpression of ATF6α was sufficient to induce their proliferation (Sharma et al. 2015). These studies underscore the importance of ATF6α pathway in type 2 diabetes.

2.2 The UPR in the Liver

The liver is a vital organ known to carry out synthesis and storage of cellular fatty acids and cholesterol as well as their metabolism contributing to lipid homeostasis in the body. Total lipid concentration in the liver is a function of fatty acid uptake from the circulating pool or from dietary intake, de novo lipogenesis as facilitated by sterol regulatory element-binding proteins (SREBPs), fatty acid secretion, and fatty acid oxidation. Perturbation in one or more of these processes due to genetic and/or environmental factors is thought to be the cause of metabolic disorders such as non-alcoholic fatty liver disease (NAFLD), dyslipidemia, obesity. Characterization of ER stress and UPR induction by the three sensors IRE1α, PERK, and ATF6α in these metabolic disorders has revealed that each UPR pathway independently regulates lipid homeostasis through different mechanisms (Basseri and Austin 2012; Han and Kaufman 2016; Rutkowski et al. 2008).

The IRE1α/XBP1 pathway was shown to affect lipid synthesis independent of their role in UPR as well as during ER stress as seen by an elevated level of phospholipids due to overexpression of XBP1s (Sriburi et al. 2004). Postnatal liver-specific deletion of *Xbp1* resulted in reduced expression of lipogenic genes including acetyl-CoA carboxylase (ACC), stearoyl-CoA desaturase (SCD), diacylglycerol acetyl-CoA transferase (DGAT). Furthermore, liver-specific deletion of *Xbp1* resulted in hypocholesterolemia and hypotriglyceridemia in livers or in plasma in addition to a reduced very low-density lipoproteins (VLDL) content, suggesting that XBP1 regulates both de novo lipid biogenesis and their secretion (Lee et al. 2008). Deletion of *Xbp1* in these mice activates IRE1α by a feedback mechanism that in turn triggers IRE1α-dependent decay (RIDD) of cytosolic mRNA molecules. Accordingly, suppression of RIDD or IRE1α ablation in *Xbp1*-deficient mice reversed the hypolipidemic conditions suggesting the contribution of XBP1 in regulating lipid metabolism through RIDD (So et al. 2012). Studies have demonstrated a role for IRE1α in preventing fatty liver under conditions of ER stress by limiting lipid accumulation and facilitating lipid secretion. Consequently, *Ire1α* null hepatocytes upon chemical induction of ER stress exhibited higher lipid content with reduced plasma lipid levels as a result of compromised lipid secretion (Zhang et al. 2011). These hepatocytes also contained an elevated level of lipogenic transcription factors including peroxisome proliferator-activated receptor γ (PPARγ), liver X receptor α (LXRα), and ACC indicating that IRE1α is required to regulate hepatic lipid accumulation (Zhang et al. 2011). More detailed analysis of hepatocyte-specific Ire1α deletion did not find a reduction in expression of lipogenic genes, but rather defined a defect in microsomal triglyceride transfer protein (MTP)-mediated transfer of triglycerides into the smooth ER, where assembly of VLDL particles is known to occur. The defect was due to reduction in protein disulfide isomerase (PDI), a cofactor required for MTP activity. PDI is a transcriptional target of XBP1s (Wang et al. 2012, 2015b). Thus, reduction in IRE1α and XBP1s blocks VLDL secretion and causes cytosolic accumulation of triglycerides.

PERK activation due to ER stress phosphorylates eIF2α at Ser51, which in turn decreases global protein synthesis. Apart from ER stress, eIF2α phosphorylation in response to nutrient depletion, fasting, or a high-fat diet implies a role for eIF2α phosphorylation in lipid metabolism (Basseri and Austin 2012). In order to further understand the role of eIF2α in the liver, studies were performed by hepatocyte-specific overexpression of GADD34, which associates with protein phosphatase 1 and dephosphorylates eIF2α to reestablish mRNA translation and hence mitigates eIF2α phosphorylation. This approach diminished hepatosteatosis in high-fat diet-fed mice and lowered glycogen levels in the liver (Oyadomari et al. 2008). Additionally, diminished eIF2α phosphorylation also lowered adipogenic receptor PPARγ and its downstream regulators in transgenic mouse liver. In contrast, normal eIF2α phosphorylation levels were known to upregulate these factors (Scheuner et al. 2001). Further studies using mammary-specific PERK deletion showed that expression level of lipogenic enzymes including fatty acid synthase (FAS), ACC, SCD, ATP citrate lyase (ACL), as well as SREBP was declined. Consequently, the amount of free fatty acids in milk obtained from these mice was lower due to reduced triglyceride synthesis (Bobrovnikova-Marjon et al. 2008). These studies besides revealing a role for eIF2α phosphorylation in lipid metabolism also suggest it may increase expression of adipogenic factors, possible ATF4 and CHOP, that are selectively upregulated after eIF2α phosphorylation, owing to upstream open reading frames. Therefore, an extensive analysis of upstream ORFs of lipogenic genes that are induced after eIF2α phosphorylation may shed light on the mechanism by which translation of these genes is regulated. This hypothesis is further supported by the fact that ATF4 was shown to be required for cholesterol metabolism in the liver. Liver-specific deletion of *Atf4* decreased expression of genes CYP7A1, encoding cholesterol 7α-hydroxylase, and CYP27A1, encoding sterol 27-hydoxylase. These hepatocyte-specific ATF4 knockout mice exhibited higher cholesterol content in liver and lower serum cholesterol levels (Fusakio et al. 2016). Thus, the use of eIF2α phosphorylation inhibitors may contribute to the greater understanding of role of UPR in metabolic regulation and disorders.

SREBPs are ER-resident transcription factors that induce expression of genes required for cholesterol (SREBP2) and lipid (SREBP1a/1c) synthesis. In sterol-rich conditions, SREBP2 is embedded in the ER membrane in a complex with SREBP cleavage-activating protein (SCAP) and insulin-induced gene 1 (INSIG1). Upon sensing low levels of cholesterol, SCAP disengages SREBP2 from INSIG1 and escorts it to the Golgi. In the Golgi apparatus, SREBP2 undergoes sequential intramembrane-regulated proteolysis by S1P and S2P proteases that liberate the N-terminal transactivation domain (SREBP2(N)). Subsequently, SREBP2(N) enters the nucleus and regulates transcription of genes required for cholesterol biosynthesis and uptake. Moreover, SREBP activation has been observed upon ER stress (Basseri and Austin 2012). Striking similarity in the mode of processing of ATF6α and SREBP by S1P and S2P proteases (Bobrovnikova-Marjon et al. 2008; Rutkowski and Kaufman 2004) as well as their reliance on PERK for processing (Bobrovnikova-Marjon et al. 2008; Teske et al. 2011) implicates the role of ATF6α pathway in lipid metabolism. Biochemical studies to understand the role of ATF6α and SREBP in lipid metabolism revealed a direct interaction between ATF6α and SREBP (Zeng et al. 2004). Overexpression of SREBP(N) alone or with ATF6α(N) in adenoviral vectors was performed in HepG2 cells, and two target genes of SREBP, HMG-CoA reductase and HMG-CoA synthase mRNA levels were reduced upon ATF6α(N) coexpression in a dose-dependent manner. Additionally, SREBP expression induced accumulation of lipids in Hep2G cells that was greatly diminished by coexpression of ATF6α(N) (Zeng et al. 2004) suggesting that the ATF6α reduces lipid homeostasis by antagonizing SREBP. Although the inhibitory effects of ATF6α(N) overexpression on cholesterol synthesis are apparent, Bommiasamy et al. observed that ATF6α(N) overexpression increased the synthesis of phosphatidyl ethanolamine and phosphatidylcholine that are major components of biological membranes (Bommiasamy et al. 2009). Consistent with this, ATF6α(N) overexpression caused ER membrane expansion with a distinct ER morphology that was not evident in cells expressing DNA-binding defective mutant of ATF6α(N). These results support the role of ATF6α in augmenting ER biogenesis (Bommiasamy et al. 2009). Further studies were performed by Wu et al. and Yamamoto et al. employing *Atf6α* knockout mouse model to understand the role of ATF6α in lipid metabolism in liver (Wu et al. 2007; Yamamoto et al. 2010). Although both *Atf6α* +/+ and *Atf6α* −/− mice behave similarly in the absence of ER stress, injection of tunicamycin (Tm) for 48 h elicited steatosis and reduced the fitness of *Atf6α* −/− mice, and death of these mice was ensued three days after Tm injection. Whereas serum examination of these mice showed higher levels of alanine aminotransferase (ALT) in *Atf6α* −/− mice, a clear indication of hepatic injury, histological examination of livers from *Atf6α* +/+ and *Atf6α* −/− mice revealed morphological differences. A closer examination of these livers showed that livers from *Atf6α* −/− mice had higher content of triglycerides, cholesterol, and accumulation of lipid droplets with an associated reduction in the concentration of β oxidation enzymes carnitine palmitoyl transferases (CPTs)-I and -II and acyl-CoA oxidase (Acox1) (Wu et al. 2007; Yamamoto et al. 2010). Therefore, abnormal liver symptoms were ascribed to defective catabolism of lipids in which ATF6α is a critical component. Studies to date show the dual role of ATF6α in both lipid metabolism (catabolism) and lipogenesis based on cellular needs. ATF6α negatively regulates triglyceride and cholesterol synthesis preventing their storage and promotes lipogenesis to aid membrane biogenesis by promoting the synthesis structural membrane lipids (Bommiasamy et al. 2009; Yamamoto et al. 2010). Analogous to ATF6α, CREBH (cyclic AMP-responsive element-binding protein H) is activated by sequential cleavage by S1P and S2P proteases and liberates an N-terminal transcription factor into the nucleus. Activation of CREBH by lipoic acids in hepatic tissues was shown to inhibit the SREBP1-c and hence reduce transcription of fatty acid synthesis genes. Consequently, Zucker diabetic fatty (ZDF) rats that were fed lipoic acids demonstrated reduced triglycerides in their livers, indicating an improvement in severe hypertriglyceridemia owing to CREBH activation (Tong et al. 2015). Subsequent studies demonstrated that accumulation of triglycerides activates CREBH which then induces expression of fibroblast growth factor 21 (FGF21) which prevents adipose tissue lipolysis and hence ameliorates hepatic steatosis. Consistently, CREBH-deficient mice exhibited impaired FGF21 production and symptoms of hepatic steatosis which was improved by adenoviral expression of FGF21 (Park et al. 2016). Although roles played independently by ATF6 and CREBH were demonstrated in liver metabolism, their ability to heterodimerize through their bZIP domains and bind to a conserved DNA motif warrants further investigation of ATF6–CREBH dimers in liver metabolism. Additionally, due to the regulation of lipid metabolism and lipogenesis by ATF6α and CREBH, identification of small molecule inhibitors as well as activators of the ATF6α pathway will have profound therapeutic influences.

2.3 The UPR in Adipose Tissue

Adipose tissue is a type of connective tissue primarily composed of adipocytes specialized in storing excess fat in the form of triglycerides in lipid droplets. Apart from storing energy-rich fatty acids, adipose tissue is highly flexible in their storage and release by hormonal signals and energetic needs of the organism and thus maintains energy homeostasis. Insulin plays an important role in adipocyte metabolism and promotes synthesis and storage of triglycerides. When there is ample glucose supply in the circulation, high levels of circulating insulin bind to their receptors on adipocytes and signal cytosol to cell membrane translocation of glucose transporters (GLUT-4), which in turn enhances glucose flux into cells. Glucose undergoes glycolysis and produces dihydroxyacetone phosphate as intermediates, which serves as a substrate for triglycerides to be stored in adipocytes.

Whereas the uptake, esterification of fatty acids into triglycerides, their storage, and the consequential expansion of adipose tissue represent an adaptive response to over nutrition, its limitless expansion leads to obesity-related symptoms such as cardiovascular disease, insulin resistance, and type 2 diabetes (Gregor and Hotamisligil 2007). Furthermore, the IRE1α–XBP1, PERK–eIF2α, and ATF6α pathways play roles in adipogenesis and lipid accumulation. Among these UPR pathways, the inhibitory role of the PERK–eIF2α pathway was demonstrated, while IRE1α and ATF6α were shown to promote adipogenesis (Rutkowski et al. 2015). The presence of insulin resistance in obese conditions, which is a prominent symptom during ER stress, suggests that ER stress may be one cause of obesity. An examination of genetic or diet-induced (high-fat diet) obese mouse model indicated the activation of PERK and IRE1α by their phosphorylation, JNK activation, and elevated GRP78/BiP expression in adipose tissue of these mice compared to the lean mice suggested that obesity induces ER stress. In order to test the role of XBP1 in insulin signaling, XBP1 gain or loss of function models was established in mouse embryonic fibroblasts. Subsequent to insulin stimulation, a reduced level of tyrosine phosphorylation on the insulin receptor substrate 1 (IRS1) was evident in *Xbp1* −/− fibroblasts with the concomitant increase in serine phosphorylation indicative of a compromised insulin signaling (Özcan et al. 2004). Besides modulating insulin signaling, XBP1 was also shown to control adipogenesis by regulating the WNT10b and β catenin (WNT/β) signaling. In preadipocytes, WNT/β-catenin suppresses adipogenic factors C/EBPα and PPARγ and hence maintains cells in an undifferentiated state. However, during adipogenesis, overexpression of XBP1s and C/EBPα is observed. Subsequently, XBP1s inhibits the transcription of WNT10b and simultaneously antagonizes the anti-adipogenic β-catenin signaling pathway, leading to an unrestrained adipogenesis (Cho et al. 2013).

Han et al. demonstrated the upregulation of canonical UPR markers in 3T3-L1 cells and suggested a physiological role of the UPR during adipogenesis (Han et al. 2013). During day 4 of adipogenesis, in addition to the adipocyte-specific gene expression, there was an increased level of phosphorylated and total eIF2α, phosphorylated and total IRE1α, spliced and native *Xbp1*, as well as CHOP. However, pharmacological induction of ER stress by Tm or by hypoxic conditions attenuated adipogenesis in spite of elevated eIF2α phosphorylation (Zhang et al. 2013). Although the UPR was induced and CHOP levels were elevated, inducible CHOP expression was shown to inhibit adipocyte differentiation by binding to C/EBP transcription factors (Batchvarova et al. 1995). These studies implicate the ER stress-independent physiological role of the PERK–eIF2α pathway in adipogenesis.

The role of ATF6α in adipose tissue has not been studied intensely. Limited studies performed in wild-type and shRNA-mediated ATF6α knockdown C3H10T1/2 cell lines implicated a role for ATF6α in adipogenesis. Upon induction of adipogenesis, expression of adipogenic transcription factors C/EBPβ, SREBP1c, GLUT4, and fatty acid-binding protein were significantly reduced in ATF6α knockdown cells with an associated reduced oil O red staining (Lowe et al. 2012). Although dimerization of ATF6α with XBP1 has been demonstrated previously, the role of this complex in regulating C/EBP transcription factors or WNT/β-catenin signaling needs further investigation.

2.4 The UPR in the Nervous System

The process of neurogenesis, proliferation, and differentiation of neurons in vertebrates greatly increases protein synthesis, thereby increasing the demand for efficient protein folding, necessitating the UPR activation to maintain protein homeostasis. Accordingly, in vitro and in vivo studies demonstrated that activation of UPR sensors occurs during neuronal differentiation of mouse embryonic stem cells as well as differentiation of bone marrow stem cells into neurons (Cho et al. 2009; Godin et al. 2016). Zhang et al. demonstrated an upregulation of ER-resident chaperones such as GRP78/BiP, calreticulin, GRP94, and protein disulfide isomerase in embryonic brains (Zhang et al. 2007). Additionally, signs of UPR such as phosphorylated eIF2α levels and XBP1s was enriched in embryonic brains, but not detected in adult brains. The enrichment of these markers in embryonic brain in comparison with adult brain implicates a physiological role of UPR during differentiation of progenitor cells into newborn neurons (Zhang et al. 2007). Furthermore, an absolute requirement for PERK–eIF2α in the stabilization of olfactory receptor (OR) choice was elucidated. In mammals, generation of a particular olfactory receptor requires transcriptional activation of one in thousands of alleles of olfactory receptors followed by a feedback signal that stabilizes this olfactory receptor choice. Dalton et al. showed that olfactory receptor expression activates PERK which then phosphorylates eIF2α (Dalton et al. 2013). Consequential paucity in global translation selectively increases expression of activating transcription factor 5 (ATF5) in olfactory sensory neurons. ATF5 in turn increases the expression of ORs and adenylyl cyclase 3 which relieves ER stress to restore translation and stabilize the OR choice (Dalton et al. 2013). Consistently, ATF5 knockout mice exhibit unstable OR expression.

During neurogenesis, neurons undergo morphological change by acquiring dendrites by giving rise to branches. This process required increased protein synthesis, and hence, it can be envisaged that ER stress and UPR are physiological phenomena during neuron morphogenesis. To test this hypothesis, Murakami et al. observed an upregulation of ER chaperone GRP78/BiP during murine neuron morphogenesis by analysis of dendrite sprouting. In addition, increased phosphorylation of eIF2α and IRE1α as well as localization of XBP1s at the proximal dendrites was observed (Murakami et al. 2007). These studies suggest the UPR is spatially and temporally regulated during neuronal morphogenesis. Similar studies were performed in *C. elegans* sensory neurons, and an elevated level of chaperone HSP-4, the mammalian homolog of BiP, was seen during morphogenesis. Consistent with these results, loss of IRE1α produced defects in dendritic morphogenesis which could be reversed by overexpression of HSP-4 (Wei et al. 2015). Although these studies emphasize the importance of IRE1α dendrite growth, upregulation of BiP that is predominantly induced by ATF6α further warrants investigation of ATF6α in dendritic sprouting. In a study performed in *C. elegans*, trafficking of glutamate receptors from the ER was shown to be dependent on a functional UPR. In the absence of IRE1α or XBP1s, receptor subunits accumulated in the ER. However, previous studies suggested a role for PERK in non-neuronal cells in facilitating ER to Golgi translocation of vesicles including ATF6α (Gupta et al. 2010) which depends on COPII for ER to Golgi translocation (Schindler and Schekman 2009). Therefore, it can be anticipated that trafficking of glutamate receptors may depend on more than one UPR component and further studies are required to understand the role of UPR in receptor trafficking.

Another study conducted by Hayashi et al. isolated hippocampal neurons exhibit demonstrated that during neuronal development, increased UPR signaling measured by activation of IRE1α and splicing of *Xbp1* in response to brain-derived neurotropic factor (BDNF) (Hayashi et al. 2007). In contrast to XBP1s that was localized to dendrite during neurogenesis, BDNF induces translocation of *Xbp1s* to cell nucleus suggesting that UPR conveys signaling from neurites to the nucleus. Accordingly, BDNF-induced neurite outgrowth was attenuated in *Xbp1* −/− neurons. Additionally, BDNF also induces eIF2α phosphorylation via PERK and selectively upregulates expression of ATF4 (Hayashi et al. 2007). ATF4 inhibits the function of cyclic AMP-responsive element-binding protein (CREB) and hence CREB-mediated attenuation of long-term potentiation and long-term memory. Experiments performed in heterozygous $eIF2\alpha^{+/S51A}$ mice demonstrated reduced eIF2α phosphorylation, indicating that lower eIF2α phosphorylation enhanced memory and learning. Consistent with this, pharmacological treatment with Sal003 that induces eIF2α phosphorylation decreased memory. These results suggest ER stress-independent physiological role of eIF2α in memory and learning (Costa-Mattioli et al. 2007).

In addition to neurogenesis, cognitive function, UPR induction is also evident in human neurodegenerative diseases, such as Parkinson's disease, Alzheimer's disease, Huntington's disease, and amyotrophic lateral sclerosis (ALS). In several instances of neurological diseases, upregulation of one or more of UPR sensors was shown to improve disease conditions; e.g., forced expression of ATF6α in mouse model improved the stroke outcome and ATF6α was found to be neuroprotective (Yu et al. 2016). Another clinical study showed the linkage between homozygous ATF6α mutations and cone dysfunction in humans. *Atf6α* −/− mice were also shown to develop cone and rod dysfunction with age (Kohl et al. 2015). Although direct contribution of UPR sensors or genetic mutations in UPR genes was not known in these diseases, their function in abating the onset of diseases reveals unidentified selective physiological functions of each UPR pathway.

3 The Non-autonomous UPR

In the previous section, we discussed how ER stress is involved in the development or dysfunction of individual metabolic target organs. However, as an organism, the cells or tissues in our body communicate with each other closely to maintain homeostasis or signal danger/needs to nearby and/or remotely separated cells. For example, adipocytes alter their secretory profile of substances to communicate with inflammatory immune cells in adipose tissue or even affect the function of distal target organs such as the brain and muscle. Therefore, the autonomous effect of ER stress on certain tissues might have a systemic influence on other organs either proximal or distal to the cell that experiences ER stress. While the cell autonomous regulation of the UPR has been extensively studied, the concept that the UPR can affect distal tissues is relatively new. Recently, however, several lines of evidence suggest the non-autonomous effect of ER stress and the UPR. The communication is achieved by cell-to-cell interaction or secretion of soluble factor(s) from the stressed cells. In the following section, we discuss the systemic regulation of the ER stress and the UPR and potential mechanisms by which cell/tissue can communicate with each other (Fig. 2).

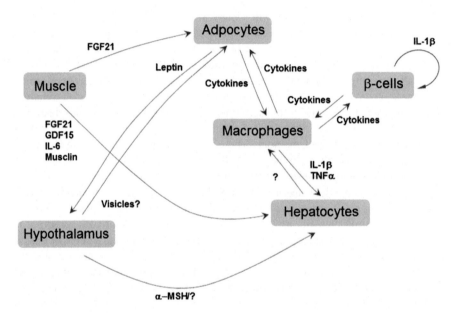

Fig. 2 Non-autonomous regulation of the UPR upon ER stress. Stressed macrophages communicate with various tissues including pancreatic β cells, hepatocytes, and adipocytes by secreting pro-inflammatory cytokines, and vice versa, they are recruited or activated by ER-stressed tissues. ER stress in adipocytes modulates leptin secretion to affect hypothalamus function and increased secretion of cytokines to recruit macrophages to exacerbate the inflammatory response. The stressed hypothalamus secretes substances that modulate the function of adipocytes and hepatocytes. Upon ER stress, myocytes secrete several substances to transfer the signal to hepatocytes and adipocytes. ER-stressed β cells enhance secretion of IL-1β which in turn leads β cells to dysfunction or death. In addition, stressed β cells secrete cytokines to recruit pro-inflammatory macrophages that exacerbate inflammatory signals in islets. Although it is becoming clear that there is communication between tissues upon ER stress, the mediators of this communication are not well defined

3.1 Proximal Effects of the Non-autonomous UPR

One feature of metabolic disease is elevated levels of circulating glucose and lipids, which might cause ER stress-mediated dysfunction in several target organs/cells including pancreatic β cells, macrophages, hepatocytes, neuronal tissues, and adipose tissues. In addition to this autonomous effect of ER stress, cells also seem to communicate with each other by altering expression pattern of surface proteins or secretory protein profiles from ER-stressed cells. For example, secretion of pro-inflammatory cytokines from macrophages is greatly enhanced by ER stress-induced ATF4 (Iwasaki et al. 2013) or XBP1s activation (Martinon et al. 2010). In addition, the accumulation of inflammatory macrophages observed in obesity-induced diabetic animals and T2D individuals strongly suggests that enhanced pro-inflammatory cytokines from ER-stressed macrophages might be crucially involved in β cell dysfunction (Ehses et al. 2007; Richardson et al. 2009). This is supported by the observation that β cells displayed reduced expression of β cell-specific genes and impaired glucose-stimulated insulin secretion (GSIS) when they were cocultured with inflammatory macrophages or were introduced into conditioned medium (CM) from inflammatory macrophages experiencing ER stress (Eguchi et al. 2012). Consistently, neutralizing IL-1β and TNF-α suppressed the interaction between insulinoma cells and macrophages (Eguchi et al. 2012). The mediator for this cross talk might be pro-inflammatory cytokines since they are known to induce apoptosis and dysfunction in pancreatic β cells. In fact, combination of IL-1β, TNFα, and IFNγ induced β cell apoptosis through nitric oxide (NO)– or Ca^{2+} depletion-mediated ER stress (Cardozo et al. 2005; Kharroubi et al. 2004; Oyadomari et al. 2001). In particular, IL-1β is one of the most important pro-apoptotic and pro-inflammatory cytokines, responsible for disorders in β cell activity and subsequent pathogenesis of T2D. Consistently, an anti-IL-1β antibody or a recombinant human IL-1β receptor antagonist improved glycemia and β cell secretory function in T2D individuals (Larsen et al. 2007). Interestingly, IL-1β production and secretion were greatly enhanced by β cells when they were exposed to high glucose, and IL-1β-producing β cells were observed in pancreatic sections of T2D individuals but not in non-diabetic subjects (Maedler et al. 2002). It is of note that ER stress can promote IL-1β secretion from the β cell through induction of TXNIP and subsequent activation of the inflammasome (Lerner et al. 2012; Oslowski et al. 2012), suggesting IL-1β has an autocrine effect to exacerbate islet inflammation and ER stress. The liver is another organ that is greatly influenced by metabolic abnormalities. Mounting evidence suggests that ER stress induces cellular dysfunction in hepatocytes and Kupffer cells autonomously. However, recent studies suggest cross talk between hepatocytes and other components of the liver including Kupffer cell or other inflammatory cells. For example, conditioned medium (CM) from hepatocytes treated with chemical ER stress inducers enhanced secretion of pro-inflammatory cytokines including IL-1β and TNF-α from macrophages (Xiu et al. 2015). In addition, CM from ER-stressed hepatocytes caused induction of the UPR in macrophages as indicated by elevated expression of BiP and CHOP. These results suggest that not only inflammatory signals but also ER stress can be transferred from hepatocytes to macrophages (Xiu et al. 2015). Similar transmission of ER stress is also reported in other studies where CM from ER-stressed cancer cells activated the UPR gene induction in macrophages and dendritic cells (Cullen et al. 2013; Mahadevan et al. 2012, 2011). Vice versa, hepatocytes incubated in CM from immortalized macrophages as well as Kupffer cells experiencing palmitate-induced ER stress also displayed upregulation of UPR marker genes (Pardo et al. 2015). It is of note that pro-inflammatory cytokines are known to induce insulin resistance in hepatic cells through the activation of ER stress as well as an inflammatory response (Hotamisligil 2010). Since palmitate treatment of macrophages enhanced secretion of pro-inflammatory cytokines including IL-6, IL-1β, TNFα, and MCP1 through ER stress and the UPR activation (Namgaladze et al. 2014), it is likely that enhanced secretion of pro-inflammatory cytokines from ER-stressed macrophages transmits the stress signal from the donor cell to target cells. However, the exact mechanism remains to be elucidated.

3.2 Distal Effects of the Non-autonomous UPR

To cope with dynamic changes in the metabolic state as an organism, certain parts of the body influence other parts to modify their function. Many of these effects seem to be controlled by only a subset of cells, suggesting systemic regulation in a non-autonomous manner.

First, the central nervous system (CNS), in particular arcuate nucleus (ARC) of the hypothalamus, plays a critical role in energy homeostasis regulation through the modulation of complex and distributed neuronal networks. Two specific subpopulations of ARC are responsible for regulation of whole-body energy balance and metabolism: (i) AgRP neurons expressing orexigenic neuropeptides agouti-related protein (AgRP) and (ii) neighboring POMC neurons expressing anorexigenic neuropeptides alpha-melanocyte-stimulating hormone (α-MSH, a product of proopiomelanocortin (POMC) processing). Accumulating evidence indicates the existence of a causal link between hypothalamic ER stress and the development of leptin resistance by impairing leptin signaling in POMC neurons (Ramirez and Claret 2015). In fact, diet-induced or pharmacologically induced ER stress in POMC neurons impaired processing of POMC to α-MSH through reduction in proconverting enzyme 2 (PC2) (Cakir et al. 2013). Given that α-MSH plays a major anorexigenic role in the regulation of feeding and energy expenditure, ER stress-mediated reduction in α-MSH might lead animals into less energy consumption, exacerbating obesity and insulin resistance. Interestingly, reduced α-MSH secretion from ER-stressed POMC neurons impaired glucose homeostasis due to enhanced gluconeogenesis mediated by CREB in the liver (Schneeberger et al. 2015). It was also recently shown that IRE1α in POMC neurons is essential for thermogenesis and glucose homeostasis (Yao et al. 2016). These observations suggest ER stress in POMC neurons affects hepatocyte function non-autonomously, possibly through the endocrinal effect of α-MSH. Similarly, another study showed that constitutive induction of XBP1s in POMC neurons enhanced metabolic rate and thermogenesis in brown adipose tissue (BAT), resulting in improved body-weight homeostasis as well as insulin sensitivity against diet-induced obesity (DIO) (Williams et al. 2014). In addition, the effect of XBP1s in POMC neurons improved glucose homeostasis in DIO context through decreased glucose production in hepatocytes. These results suggest that effect of XBP1s in POMC neuron might be transferred to other organs in the body including BAT and hepatocytes. Consistently, activation of XBP1s and its downstream target genes in the liver were observed in mice that expressed elevated levels of XBP1s in POMC neurons (Williams et al. 2014). A similar phenomenon was also observed in *C. elegans*, where activation of IRE1 and subsequent induction of XBP1s in neuronal tissue were able to transmit a metabolic signal to other organs at a distance in a non-autonomous manner (Taylor and Dillin 2013). In this study, Taylor et al. found that constitutive expression XBP1s in neurons of *C. elegans* activated the UPR in their intestinal cells. The transmission of the stress signal was associated with release of small clear vesicle (SCV) containing signaling molecules from neuronal cells. Not only in mammals and *C. elegans*, transmission of the UPR from one cell to distal cells was also observed in the *Drosophila*, where ER-stressed intestinal epithelial cells (IECs) induced PERK activity and eIF2α phosphorylation in intestinal stem cells, resulting in proliferation of those cells (Wang et al. 2015a). Given that lifespan is generally extended when intestinal stem cell (ISC) proliferation is limited (Biteau et al. 2010), transmission of ER stress from stressed IECs to ISCs might explain how ER stress affects the aging process.

Another example of distal dissemination of ER stress is cross talk between skeletal myocyte (SM) and other tissues. As a major site of glucose utilization, SMs play an important role in metabolic homeostasis. Chronic metabolic disease (Han and Kaufman 2016) or exercise (Wu et al. 2011) activates UPR pathways, suggesting a potential role of ER stress in SM function during stress conditions. The altered function or physiological condition of SM is known to affect other organs through secretion of certain substances called myokines (Pedersen and Febbraio 2012). Among them, the best well-known example is fibroblast growth factor 21 (FGF21) that was induced during metabolic status or exercise. FGF21 as a myokine seems to exert its effect on other tissues mainly through endocrine actions because β-klotho, the crucial coreceptor for FGF21 action, is expressed in FGF21 target organs including adipose tissue, the liver, the hypothalamus, but not in SM (Fisher and Maratos-Flier 2016). Interestingly, FGF21 expression and secretion from SM are induced by eIF2α phosphorylation and subsequent activation of ATF4 due to mitochondrial or ER perturbations (Keipert et al. 2014; Kim et al. 2013; Tyynismaa et al. 2010). FGF21 released from SM has endocrine effects leading to expansion of WAT and reduced hepatic lipid accumulation, leading to improved insulin sensitivity in DIO mice (Kim et al. 2013). Consistently, pharmacological activation of eIF2α phosphorylation in SM promoted secretion of FGF21 and surpassed fat accumulation by enhancing energy expenditure in BAT despite their increased food intake compared to wild-type mice (Miyake et al. 2016). Besides FGF21, there are several other ER stress-induced myokines such as growth differentiation factor 15 (GDF15) (Park et al. 2012) and IL-6 (Welc and Clanton 2013), and Musclin (Gu et al. 2015), most of which are also known to be induced by ER stress in several tissues (Ost et al. 2016). Therefore, more investigation is needed to elucidate the exact role of ER stress into how these myokines secreted from SM have a non-autonomous effect on other cells/organs.

Another distal effect of ER stress was observed between adipocytes and other organs during metabolic disease. It is generally accepted that ER stress and chronic inflammation in adipose tissue contribute to obesity-induced insulin resistance in rodents and humans (Boden et al. 2008; Ozcan et al. 2006). One of underlying mechanisms might be pro-inflammatory macrophage infiltration into adipose tissue by ER-stressed adipocytes. Supporting this notion, secretion of macrophage-recruiting cytokines was greatly enhanced from adipocytes experiencing ER stress, resulting in increased pro-inflammatory macrophages in epididymal fat (Li et al. 2013; Nguyen et al. 2012). Consistently, macrophage infiltration in diet- or genetically induced obese mice was significantly reduced by decreasing ER stress or by *Chop* deletion, both of which improved insulin sensitivity (Dong et al. 2014; Maris et al. 2012). Similarly, palmitate-induced ER stress promoted pro-inflammatory cytokine secretion from pancreatic β cells to recruit macrophages to the islet that enhanced islet inflammation, suggesting that distal effects of ER stress from damaged/stressed cells may exert actions on immune cells that may be common in the body (Eguchi et al. 2012). Besides pro-inflammatory cytokines, adipocytes secrete various adipocyte-derived substances including adiponectin, leptin, and resistin, which function as classic circulating hormones to communicate with other distal organs (Kwon and Pessin 2013). For example, adiponectin has potent anti-diabetic properties by enhancing fatty acid oxidation and glucose uptake in muscle and suppressing gluconeogenesis in hepatocytes (Yamauchi et al. 2002). Leptin, the satiety hormone, suppresses food intake and induces weight loss, improving metabolic abnormalities (Ouchi et al. 2011). The relationship between adipokines and ER stress in adipocytes was demonstrated previously. In human and murine adipocytes, adiponectin synthesis and multimerization were greatly reduced upon ER stress (Kim et al. 2006; Mondal et al. 2012). Leptin secretion was reduced upon ER stress (Xu et al. 2010). Given previous observations and the importance of these hormones in regulating systemic energy homeostasis, it is likely that the decreased secretion of these adipocyte hormones upon ER stress exacerbates metabolic dysfunction in adipose tissue as well as in other cell types in the body (Xu et al. 2010). However, it was also reported that TUDCA, a chemical chaperone that reduces protein misfolding in the ER, improved hepatic and muscle insulin sensitivity in obese human subjects but not in adipose tissues, suggesting reduced ER stress in adipose tissues may not be related to insulin resistance (Kars et al. 2010). Therefore, the exact role of ER stress in adipose tissue remains to be elucidated.

3.3 Potential Mechanism of the Non-autonomous UPR

As discussed above, mounting evidence suggests that ER-stressed cells communicate with other cells or organs at proximal or distal distances. These observations question how this communication is accomplished. Although the exact mechanism is not clear yet, several studies suggest that this interaction could be mediated by direct contact between cells or by secreted substances in an endocrine manner.

A number of studies suggest that ER stress dislocates ER chaperone proteins into other cellular organelles. For example, upon ER stress, BiP is detected in nuclei, mitochondria, the cytosol, the plasma membrane, and the extracellular space (Kern et al. 2009; Matsumoto and Hanawalt 2000; Ni et al. 2009; Sun et al. 2006; Zhang et al. 2010). Secreted BiP from ER-stressed cells can bind to cell surface receptors of adjacent endothelial cells and activate ERK and AKT pathways to protect cells from ER stress (Kern et al. 2009). Another ER chaperone calreticulin (CRT) is also known to be on the cell surface or secreted from ER-stressed cells. The relocation of CRT to the surface upon ER stress enhances uptake of stressed cells by phagocytic dendritic cells (DCs) to induce production and secretion of pro-inflammatory cytokines (Peters and Raghavan 2011). In addition, cell surface exposure of CRT on ER-stressed tumor cells is significantly correlated with increased accumulation of anti-tumor DC cells in human individuals (Fucikova et al. 2016). Significantly, exposure of CRT on the cell surface of tumor cells might act as an 'eat me' signal for DCs (Feng et al. 2015; Martins et al. 2010). Although it is not clear how ER stress is transferred from cancer cells to immune cells, it is evident that ER stress evokes an initial danger signal from damaged cells to a surveillance system to prepare the organism for subsequent challenges.

Cells experiencing ER stress are also known to secrete substances such as peptide hormones or pro-inflammatory cytokines. For example, pancreatic β cells promote secretion of IL-1β through ER stress-induced inflammasome activation (Lerner et al. 2012; Oslowski et al. 2012) as well as other pro-inflammatory cytokines (Igoillo-Esteve et al. 2010). Similar phenomenon was also observed in other tissues including differentiated adipocytes (Kim et al. 2014), myocytes (Ost et al. 2016), and hepatocytes (Lawless and Greene 2012). The secreted cytokines from these tissues are likely to affect either their own tissue (autocrine), adjacent tissues (paracrine), or tissues at a distance (endocrine). Chemokines secreted from ER-stressed cells might recruit inflammatory cells into the affected region, where recruited macrophages secrete pro-inflammatory cytokines to exacerbate inflammatory response on adjacent cells. This is observed in the pancreas, liver, and WAT. Besides pro-inflammatory cytokines, secretion of substances from affected tissues changes upon ER stress to transfer the signal to distal regions. For example, ER stress modulates secretion patterns of myokines from SM or adipokines from adipocytes that influence the function of other organs.

Another potential mechanism to disseminate signals from the ER-stressed cell might be mediated by extracellular vesicles (EVs) including exosomes, which are suggested as crucial mediators of intercellular communications (Javeed et al. 2015;

Robbins and Morelli 2014; Yoon et al. 2014). The exosome is known to contain plasma membrane and cytoplasmic proteins as well as mRNAs and microRNAs. Toxic insults or stresses were reported to facilitate the generation of these vesicles (Raposo and Stoorvogel 2013). Recent studies suggest that ER stress can induce secretion of extracellular vesicles in various cell types. Hyperoxia in lung epithelial cells caused ER stress, which in turn exerted EV production and secretion through the ATF6α pathway (Moon et al. 2015). ER stress-induced EVs from lung epithelial cells stimulated the production of IL-6, TNFα, and other cytokines in macrophages (Moon et al. 2015). Similarly, exosomes secreted from pancreatic cancer provoked induction of the UPR genes in pancreatic β cells in islets (Javeed et al. 2015), suggesting that EVs can transfer the stress signal from one cell to another cell. In hepatocytes, enhanced secretion of EVs upon ER stress is also observed, in which PA enhanced significant EV release through the IRE1α/XBP1 pathway (Kakazu et al. 2016). Consistently, neuronal cells in *C. elegans* undergoing ER stress promoted XBP1-mediated secretion of small clear vesicles which remotely induced the UPR induction in intestinal cells (Taylor and Dillin 2013). These results suggest that extracellular vesicles including exosomes can be a potential mediator of the stress signal.

The most intriguing phenomenon for a non-autonomous effect of ER stress is that the signal from an ER-stressed cell might induce the UPR in target cells, suggesting transmission of ER stress. For example, the signal from stressed cancer cells was transmitted to macrophages through TLR4 signaling, resulting in induction of the UPR and enhanced secretion of pro-inflammatory cytokines through the IRE1α/XBP1 pathway (Mahadevan et al. 2011). DCs also appear to receive ER stress signals from stressed cancer cells, leading to enhanced production of pro-inflammatory cytokines as well as suppressive phenotypes that impair T cell proliferation and facilitate tumor growth (Mahadevan et al. 2012). Transmission of ER stress from tumor to immune cells was also observed in other studies (Cullen et al. 2013; Peters and Raghavan 2011). Transmission of ER stress is not limited to the interplay between cancer cell and immune cells. In ovarian cancer, impaired MHC class I-restricted antigen presentation to CD8 + T cells was linked to reactive oxygen species (ROS)-induced ER stress and increased lipid metabolism in DCs induced by an unknown transmissible stress factor in the tumor microenvironment. Although the concept of lipid accumulation-dependent dysfunction of antigen presentation by DCs was previously described (Herber et al. 2010), these findings suggest ER stress-induced *Xbp1* splicing causes lipid accumulation in DCs that reduce antigen presentation (Cubillos-Ruiz et al. 2015). Indeed, deletion of *Xbp1* in DCs abolished lipid accumulation in DCs and increased T cell-mediated anti-tumor immunity, resulting in decreased tumor burden and improved survival (Cubillos-Ruiz et al. 2015; Grootjans et al. 2016).

The neuronal cells can transmit the ER stress signal to other tissues including intestine in *C. elegans* (Taylor and Dillin 2013). In addition, ER-stressed intestinal epithelial cells caused PERK activation in intestinal stem cells to promote proliferation of the target cells in Drosophila (Wang et al. 2015a). Recently, it was reported that constitutive expression of XBP1s in POMC neurons induced the

Xbp1s pathway in the liver, causing improved hepatic insulin sensitivity by suppressing hepatic glucose production (Williams et al. 2014). These results strongly suggest that transmission of ER stress might be evolutionally shared in *C. elegans* to mammals to alarm other cells in the organism for upcoming challenges to ER homeostasis.

4 Conclusions

In this review, we describe how ER stress upon metabolic challenge provokes the UPR within the stressed cell and some of the mechanisms by which tissues communicate under stress conditions. For a while, it was regarded that the UPR is activated to increase stress resistance in the stressed cell with challenges to maintain homeostasis and survival. However, recent findings suggest the stress signal might be conveyed from stressed cells/tissues to unstressed organs in the body. It is of note that damaged cells need to transfer danger signals to unstressed cells in the same organism to prepare for the threat to the organism. Consistently, it is known that different cells communicate and now similar concepts can be applied to transferring signals from ER-stressed cells to unstressed cells. As we are at the beginning of understanding the mechanism of communication of stress signals from the ER, there remain many questions unanswered. First, although there are several candidates responsible for this transmission, it is still not clarified what molecules mediate transfer signals from cells experiencing ER stress. Second, studies are needed to identify which arms of the UPR are involved in transmission of ER stress. Several studies suggest XBP1s is responsible for this process, but it is not clear whether other UPR pathways are involved or not. Finally, not only transmission from stressed to non-stressed cells but also how do reciprocal interactions between ER-stressed cells upon metabolic challenges in different tissues of an organism affect different cell types, as they might be exposed to stress at the same time in the body. For example, islet β cells with ER stress during obesity signal to recruit inflammatory cells experiencing ER stress, which might exacerbate the inflammatory response in islets to progress β cell death. Understanding the mechanisms by which the UPR copes with ER stress autonomously as well as non-autonomously will provide new therapeutic approaches to cure metabolic diseases.

Acknowledgements We apologize to those who we were unable to reference due to space limitations. RJK is supported by NIH grants DK042394, DK103185, DK110973, and CA198103. This work was supported by Basic Science Research Program through the National Research Foundation of Korea (NRF) funded by the Ministry of Education (NRF-2015R1D1A1A01058846, NRF-2017033069).

References

Back SH, Kaufman RJ (2012) Endoplasmic reticulum stress and type 2 diabetes. Annu Rev Biochem 81:767–793

Basseri S, Austin RC (2012) Endoplasmic reticulum stress and lipid metabolism: mechanisms and therapeutic potential. Biochem Res Int 2012:13

Batchvarova N, Wang XZ, Ron D (1995) Inhibition of adipogenesis by the stress-induced protein CHOP (Gadd153). EMBO J 14:4654–4661

Biteau B, Karpac J, Supoyo S, Degennaro M, Lehmann R, Jasper H (2010) Lifespan extension by preserving proliferative homeostasis in Drosophila. PLoS Genet 6:e1001159

Bobrovnikova-Marjon E, Hatzivassiliou G, Grigoriadou C, Romero M, Cavener DR, Thompson CB, Diehl JA (2008) PERK-dependent regulation of lipogenesis during mouse mammary gland development and adipocyte differentiation. Proc Natl Acad Sci USA 105:16314–16319

Boden G, Duan X, Homko C, Molina EJ, Song W, Perez O, Cheung P, Merali S (2008) Increase in endoplasmic reticulum stress-related proteins and genes in adipose tissue of obese, insulin-resistant individuals. Diabetes 57:2438–2444

Bommiasamy H, Back SH, Fagone P, Lee K, Meshinchi S, Vink E, Sriburi R, Frank M, Jackowski S, Kaufman RJ, Brewer JW (2009) ATF6α induces XBP1-independent expansion of the endoplasmic reticulum. J Cell Sci 122:1626–1636

Cakir I, Cyr NE, Perello M, Litvinov BP, Romero A, Stuart RC, Nillni EA (2013) Obesity induces hypothalamic endoplasmic reticulum stress and impairs proopiomelanocortin (POMC) post-translational processing. J Biol Chem 288:17675–17688

Cardozo AK, Ortis F, Storling J, Feng YM, Rasschaert J, Tonnesen M, Van Eylen F, Mandrup-Poulsen T, Herchuelz A, Eizirik DL (2005) Cytokines downregulate the sarcoendoplasmic reticulum pump Ca_2 + ATPase 2b and deplete endoplasmic reticulum Ca^{2+}, leading to induction of endoplasmic reticulum stress in pancreatic beta-cells. Diabetes 54:452–461

Cho YM, Jang Y-S, Jang Y-M, Chung S-M, Kim H-S, Lee J-H, Jeong S-W, Kim I-K, Kim JJ, Kim K-S, Kwon O-J (2009) Induction of unfolded protein response during neuronal induction of rat bone marrow stromal cells and mouse embryonic stem cells. Exp Mol Med 41:440–452

Cho YM, Kim DH, Kwak S-N, Jeong S-W, Kwon O-J (2013) X-box binding protein 1 enhances adipogenic differentiation of 3T3-L1 cells through the downregulation of Wnt10b expression. FEBS Lett 587:1644–1649

Costa-Mattioli M, Gobert D, Stern E, Gamache K, Colina R, Cuello C, Sossin W, Kaufman R, Pelletier J, Rosenblum K, Krnjević K, Lacaille J-C, Nader K, Sonenberg N (2007) eIF2α phosphorylation bidirectionally regulates the switch from short- to long-term synaptic plasticity and memory. Cell 129:195–206

Cubillos-Ruiz JR, Silberman PC, Rutkowski MR, Chopra S, Perales-Puchalt A, Song M, Zhang S, Bettigole SE, Gupta D, Holcomb K, Ellenson LH, Caputo T, Lee AH, Conejo-Garcia JR, Glimcher LH (2015) ER stress sensor XBP1 controls anti-tumor immunity by disrupting dendritic cell Homeostasis. Cell 161:1527–1538

Cullen SJ, Fatemie S, Ladiges W (2013) Breast tumor cells primed by endoplasmic reticulum stress remodel macrophage phenotype. Am J Cancer Res 3:196–210

Dalton Ryan P, Lyons David B, Lomvardas S (2013) Co-opting the unfolded protein response to elicit olfactory receptor feedback. Cell 155:321–332

Dong H, Huang H, Yun X, Kim DS, Yue Y, Wu H, Sutter A, Chavin KD, Otterbein LE, Adams DB, Kim YB, Wang H (2014) Bilirubin increases insulin sensitivity in leptin-receptor deficient and diet-induced obese mice through suppression of ER stress and chronic inflammation. Endocrinology 155:818–828

Eguchi K, Manabe I, Oishi-Tanaka Y, Ohsugi M, Kono N, Ogata F, Yagi N, Ohto U, Kimoto M, Miyake K, Tobe K, Arai H, Kadowaki T, Nagai R (2012) Saturated fatty acid and TLR signaling link beta cell dysfunction and islet inflammation. Cell Metab 15:518–533

Ehses JA, Perren A, Eppler E, Ribaux P, Pospisilik JA, Maor-Cahn R, Gueripel X, Ellingsgaard H, Schneider MK, Biollaz G, Fontana A, Reinecke M, Homo-Delarche F, Donath MY (2007) Increased number of islet-associated macrophages in type 2 diabetes. Diabetes 56:2356–2370

Feng M, Chen JY, Weissman-Tsukamoto R, Volkmer JP, Ho PY, McKenna KM, Cheshier S, Zhang M, Guo N, Gip P, Mitra SS, Weissman IL (2015) Macrophages eat cancer cells using their own calreticulin as a guide: roles of TLR and Btk. Proc Natl Acad Sci USA 112:2145–2150

Fisher FM, Maratos-Flier E (2016) Understanding the physiology of FGF21. Annu Rev Physiol 78:223–241

Fucikova J, Becht E, Iribarren K, Goc J, Remark R, Damotte D, Alifano M, Devi P, Biton J, Germain C, Lupo A, Fridman WH, Dieu-Nosjean MC, Kroemer G, Sautes-Fridman C, Cremer I (2016) Calreticulin expression in human non-small cell lung cancers correlates with increased accumulation of antitumor immune cells and favorable prognosis. Cancer Res 76:1746–1756

Fusakio ME, Willy JA, Wang Y, Mirek ET, Al Baghdadi RJT, Adams CM, Anthony TG, Wek RC (2016) Transcription factor ATF4 directs basal and stress-induced gene expression in the unfolded protein response and cholesterol metabolism in the liver. Mol Biol Cell 27:1536–1551

Godin JD, Creppe C, Laguesse S, Nguyen L (2016) Emerging roles for the unfolded protein response in the developing nervous system. Trends Neurosci 39:394–404

Gregor MF, Hotamisligil GS (2007) Thematic review series: adipocyte biology. Adipocyte stress: the endoplasmic reticulum and metabolic disease. J Lipid Res 48:1905–1914

Grootjans J, Kaser A, Kaufman RJ, Blumberg RS (2016) The unfolded protein response in immunity and inflammation. Nat Rev Immunol 16:469–484

Gu N, Guo Q, Mao K, Hu H, Jin S, Zhou Y, He H, Oh Y, Liu C, Wu Q (2015) Palmitate increases musclin gene expression through activation of PERK signaling pathway in C2C12 myotubes. Biochem Biophys Res Commun 467:521–526

Gupta S, McGrath B, Cavener DR (2010) PERK (EIF2AK3) regulates proinsulin trafficking and quality control in the secretory pathway. Diabetes 59:1937–1947

Han J, Kaufman RJ (2016) The role of ER stress in lipid metabolism and lipotoxicity. J Lipid Res 57:1329–1338

Han J, Murthy R, Wood B, Song B, Wang S, Sun B, Malhi H, Kaufman RJ (2013) ER stress signalling through eIF2α and CHOP, but not IRE1alpha, attenuates adipogenesis in mice. Diabetologia 56:911–924

Hayashi A, Kasahara T, Iwamoto K, Ishiwata M, Kametani M, Kakiuchi C, Furuichi T, Kato T (2007) The role of brain-derived neurotrophic factor (BDNF)-induced XBP1 splicing during brain development. J Biol Chem 282:34525–34534

Herber DL, Cao W, Nefedova Y, Novitskiy SV, Nagaraj S, Tyurin VA, Corzo A, Cho HI, Celis E, Lennox B, Knight SC, Padhya T, McCaffrey TV, McCaffrey JC, Antonia S, Fishman M, Ferris RL, Kagan VE, Gabrilovich DI (2010) Lipid accumulation and dendritic cell dysfunction in cancer. Nat Med 16:880–886

Hotamisligil GS (2010) Endoplasmic reticulum stress and the inflammatory basis of metabolic disease 140:900–917

Igoillo-Esteve M, Marselli L, Cunha DA, Ladriere L, Ortis F, Grieco FA, Dotta F, Weir GC, Marchetti P, Eizirik DL, Cnop M (2010) Palmitate induces a pro-inflammatory response in human pancreatic islets that mimics CCL2 expression by beta cells in type 2 diabetes. Diabetologia 53:1395–1405

Iwasaki Y, Suganami T, Hachiya R, Shirakawa I, Kim-Saijo M, Tanaka M, Hamaguchi M, Takai-Igarashi T, Nakai M, Miyamoto Y, Ogawa Y (2013) Activating transcription factor 4 links metabolic stress to Interleukin-6 expression in Macrophages. *Diabetes*

Javeed N, Sagar G, Dutta SK, Smyrk TC, Lau JS, Bhattacharya S, Truty M, Petersen GM, Kaufman RJ, Chari ST, Mukhopadhyay D (2015) Pancreatic cancer-derived exosomes cause paraneoplastic beta-cell dysfunction. Clin Cancer Res 21:1722–1733

Kakazu E, Mauer AS, Yin M, Malhi H (2016) Hepatocytes release ceramide-enriched pro-inflammatory extracellular vesicles in an IRE1α-dependent manner. J Lipid Res 57:233–245

Kars M, Yang L, Gregor MF, Mohammed BS, Pietka TA, Finck BN, Patterson BW, Horton JD, Mittendorfer B, Hotamisligil GS, Klein S (2010) Tauroursodeoxycholic acid may improve liver and muscle but not adipose tissue insulin sensitivity in obese men and women. Diabetes 59:1899–1905

Keipert S, Ost M, Johann K, Imber F, Jastroch M, van Schothorst EM, Keijer J, Klaus S (2014) Skeletal muscle mitochondrial uncoupling drives endocrine cross-talk through the induction of FGF21 as a myokine. Am J Physiol Endocrinol Metab 306:E469–E482

Kern J, Untergasser G, Zenzmaier C, Sarg B, Gastl G, Gunsilius E, Steurer M (2009) GRP-78 secreted by tumor cells blocks the antiangiogenic activity of bortezomib. Blood 114:3960–3967

Kharroubi I, Ladriere L, Cardozo AK, Dogusan Z, Cnop M, Eizirik DL (2004) Free fatty acids and cytokines induce pancreatic {β}-cell apoptosis by different mechanisms: role of nuclear factor-{κ}B and endoplasmic reticulum stress. Endocrinology 145:5087–5096

Kim HB, Kong M, Kim TM, Suh YH, Kim WH, Lim JH, Song JH, Jung MH (2006) NFATc4 and ATF3 negatively regulate adiponectin gene expression in 3T3-L1 adipocytes. Diabetes 55:1342–1352

Kim KH, Jeong YT, Oh H, Kim SH, Cho JM, Kim YN, Kim SS, Kim DH, Hur KY, Kim HK, Ko T, Han J, Kim HL, Kim J, Back SH, Komatsu M, Chen H, Chan DC, Konishi M, Itoh N, Choi CS, Lee MS (2013) Autophagy deficiency leads to protection from obesity and insulin resistance by inducing Fgf21 as a mitokine. Nat Med 19:83–92

Kim S, Joe Y, Jeong SO, Zheng M, Back SH, Park SW, Ryter SW, Chung HT (2014) Endoplasmic reticulum stress is sufficient for the induction of IL-1β production via activation of the NF-κB and inflammasome pathways. Innate Immun 20:799–815

Kohl S, Zobor D, Chiang W-C, Weisschuh N, Staller J, Menendez IG, Chang S, Beck SC, Garrido MG, Sothilingam V, Seeliger MW, Stanzial F, Benedicenti F, Inzana F, Heon E, Vincent A, Beis J, Strom TM, Rudolph G, Roosing S, den Hollander AI, Cremers FPM, Lopez I, Ren H, Moore AT, Webster AR, Michaelides M, Koenekoop RK, Zrenner E, Kaufman RJ, Tsang SH, Wissinger B, Lin JH (2015) Mutations in the unfolded protein response regulator ATF6 cause the cone dysfunction disorder achromatopsia. Nat Genet 47:757–765

Kwon H, Pessin JE (2013) Adipokines mediate inflammation and insulin resistance. Front Endocrinol (Lausanne) 4:71

Larsen CM, Faulenbach M, Vaag A, Volund A, Ehses JA, Seifert B, Mandrup-Poulsen T, Donath MY (2007) Interleukin-1-Receptor antagonist in type 2 diabetes mellitus. N Engl J Med 356:1517–1526

Lawless MW, Greene CM (2012) Toll-like receptor signalling in liver disease: ER stress the missing link? Cytokine 59:195–202

Lee A-H, Heidtman K, Hotamisligil GS, Glimcher LH (2011) Dual and opposing roles of the unfolded protein response regulated by IRE1α and XBP1 in proinsulin processing and insulin secretion. Proc Natl Acad Sci 108:8885–8890

Lee A-H, Scapa EF, Cohen DE, Glimcher LH (2008) Regulation of hepatic lipogenesis by the transcription factor XBP1. Science 320:1492–1496

Lee J, Ozcan U (2014) Unfolded protein response signaling and metabolic diseases. J Biol Chem 289:1203–1211

Lerner AG, Upton JP, Praveen PV, Ghosh R, Nakagawa Y, Igbaria A, Shen S, Nguyen V, Backes BJ, Heiman M, Heintz N, Greengard P, Hui S, Tang Q, Trusina A, Oakes SA, Papa FR (2012) IRE1α induces thioredoxin-interacting protein to activate the NLRP3 inflammasome and promote programmed cell death under irremediable ER stress. Cell Metab 16:250–264

Li Y, Zhang H, Jiang C, Xu M, Pang Y, Feng J, Xiang X, Kong W, Xu G, Li Y, Wang X (2013) Hyperhomocysteinemia promotes insulin resistance by inducing endoplasmic reticulum stress in adipose tissue. J Biol Chem 288:9583–9592

Lipson KL, Fonseca SG, Ishigaki S, Nguyen LX, Foss E, Bortell R, Rossini AA, Urano F (2006) Regulation of insulin biosynthesis in pancreatic beta cells by an endoplasmic reticulum-resident protein kinase IRE1. Cell Metab 4:245–254

Lowe CE, Dennis RJ, Obi U, O'Rahilly S, Rochford JJ (2012) Investigating the involvement of the ATF6α pathway of the unfolded protein response in adipogenesis. Int J Obes 36:1248–1251

Maedler K, Sergeev P, Ris F, Oberholzer J, Joller-Jemelka HI, Spinas GA, Kaiser N, Halban PA, Donath MY (2002) Glucose-induced beta cell production of IL-1β contributes to glucotoxicity in human pancreatic islets. J Clin Invest 110:851–860

Mahadevan NR, Anufreichik V, Rodvold JJ, Chiu KT, Sepulveda H, Zanetti M (2012) Cell-extrinsic effects of tumor ER stress imprint myeloid dendritic cells and impair CD8(+) T cell priming. PLoS ONE 7:e51845

Mahadevan NR, Rodvold J, Sepulveda H, Rossi S, Drew AF, Zanetti M (2011) Transmission of endoplasmic reticulum stress and pro-inflammation from tumor cells to myeloid cells. Proc Natl Acad Sci USA 108:6561–6566

Maris M, Overbergh L, Gysemans C, Waget A, Cardozo AK, Verdrengh E, Cunha JP, Gotoh T, Cnop M, Eizirik DL, Burcelin R, Mathieu C (2012) Deletion of C/EBP homologous protein (Chop) in C57Bl/6 mice dissociates obesity from insulin resistance. Diabetologia 55:1167–1178

Martinez SC, Tanabe K, Cras-Méneur C, Abumrad NA, Bernal-Mizrachi E, Permutt MA (2008) Inhibition of Foxo1 protects pancreatic islet β-cells against fatty acid and endoplasmic reticulum stress-induced apoptosis. Diabetes 57:846–859

Martinon F, Chen X, Lee AH, Glimcher LH (2010) TLR activation of the transcription factor XBP1 regulates innate immune responses in macrophages. Nat Immunol 11:411–418

Martins I, Kepp O, Galluzzi L, Senovilla L, Schlemmer F, Adjemian S, Menger L, Michaud M, Zitvogel L, Kroemer G (2010) Surface-exposed calreticulin in the interaction between dying cells and phagocytes. Ann N Y Acad Sci 1209:77–82

Matsumoto A, Hanawalt PC (2000) Histone H3 and heat shock protein GRP78 are selectively cross-linked to DNA by photoactivated gilvocarcin V in human fibroblasts. Cancer Res 60:3921–3926

Miyake M, Nomura A, Ogura A, Takehana K, Kitahara Y, Takahara K, Tsugawa K, Miyamoto C, Miura N, Sato R, Kurahashi K, Harding HP, Oyadomari M, Ron D, Oyadomari S (2016) Skeletal muscle-specific eukaryotic translation initiation factor 2α phosphorylation controls amino acid metabolism and fibroblast growth factor 21-mediated non-cell-autonomous energy metabolism. FASEB J 30:798–812

Mondal AK, Das SK, Varma V, Nolen GT, McGehee RE, Elbein SC, Wei JY, Ranganathan G (2012) Effect of endoplasmic reticulum stress on inflammation and adiponectin regulation in human adipocytes. Metab Syndr Relat Disord 10:297–306

Moon HG, Cao Y, Yang J, Lee JH, Choi HS, Jin Y (2015) Lung epithelial cell-derived extracellular vesicles activate macrophage-mediated inflammatory responses via ROCK1 pathway. Cell Death Dis 6:e2016

Murakami T, Hino SI, Saito A, Imaizumi K (2007) Endoplasmic reticulum stress response in dendrites of cultured primary neurons. Neuroscience 146:1–8

Namgaladze D, Lips S, Leiker TJ, Murphy RC, Ekroos K, Ferreiros N, Geisslinger G, Brune B (2014) Inhibition of macrophage fatty acid β-oxidation exacerbates palmitate-induced inflammatory and endoplasmic reticulum stress responses. Diabetologia 57:1067–1077

Nguyen MT, Chen A, Lu WJ, Fan W, Li PP, Oh DY, Patsouris D (2012) Regulation of chemokine and chemokine receptor expression by PPARγ in adipocytes and macrophages. PLoS ONE 7:e34976

Ni M, Zhou H, Wey S, Baumeister P, Lee AS (2009) Regulation of PERK signaling and leukemic cell survival by a novel cytosolic isoform of the UPR regulator GRP78/BiP. PLoS ONE 4:e6868

Oslowski CM, Hara T, O'Sullivan-Murphy B, Kanekura K, Lu S, Hara M, Ishigaki S, Zhu LJ, Hayashi E, Hui ST, Greiner D, Kaufman RJ, Bortell R, Urano F (2012)

Thioredoxin-interacting protein mediates ER stress-induced β cell death through initiation of the inflammasome. Cell Metab 16:265–273

Osorio F, Lambrecht B, Janssens S (2013) The UPR and lung disease. Semin Immunopathol 35:293–306

Ost M, Coleman V, Kasch J, Klaus S (2016) Regulation of myokine expression: role of exercise and cellular stress. Free Radic Biol Med 98:78–89

Ouchi N, Parker JL, Lugus JJ, Walsh K (2011) Adipokines in inflammation and metabolic disease. Nat Rev Immunol 11:85–97

Oyadomari S, Harding HP, Zhang Y, Oyadomari M, Ron D (2008) Dephosphorylation of translation initiation factor 2α enhances glucose tolerance and attenuates hepatosteatosis in mice. Cell Metab 7:520–532

Oyadomari S, Takeda K, Takiguchi M, Gotoh T, Matsumoto M, Wada I, Akira S, Araki E, Mori M (2001) Nitric oxide-induced apoptosis in pancreatic beta cells is mediated by the endoplasmic reticulum stress pathway. Proc Natl Acad Sci USA 98:10845–10850

Özcan U, Cao Q, Yilmaz E, Lee A-H, Iwakoshi NN, Özdelen E, Tuncman G, Görgün C, Glimcher LH, Hotamisligil GS (2004) Endoplasmic reticulum stress links obesity, insulin action, and type 2 diabetes. Science 306:457–461

Ozcan U, Yilmaz E, Ozcan L, Furuhashi M, Vaillancourt E, Smith RO, Gorgun CZ, Hotamisligil GS (2006) Chemical chaperones reduce ER stress and restore glucose homeostasis in a mouse model of type 2 diabetes. Science 313:1137–1140

Pardo V, Gonzalez-Rodriguez A, Guijas C, Balsinde J, Valverde AM (2015) Opposite cross-talk by oleate and palmitate on insulin signaling in hepatocytes through macrophage activation. J Biol Chem 290:11663–11677

Park JG, Xu X, Cho S, Hur KY, Lee MS, Kersten S, Lee AH (2016) CREBH-FGF21 axis improves hepatic steatosis by suppressing adipose tissue lipolysis. Sci Rep 6:27938

Park SH, Choi HJ, Yang H, Do KH, Kim J, Kim HH, Lee H, Oh CG, Lee DW, Moon Y (2012) Two in-and-out modulation strategies for endoplasmic reticulum stress-linked gene expression of pro-apoptotic macrophage-inhibitory cytokine 1. J Biol Chem 287:19841–19855

Pedersen BK, Febbraio MA (2012) Muscles, exercise and obesity: skeletal muscle as a secretory organ. Nat Rev Endocrinol 8:457–465

Peters LR, Raghavan M (2011) Endoplasmic reticulum calcium depletion impacts chaperone secretion, innate immunity, and phagocytic uptake of cells. J Immunol 187:919–931

Ramirez S, Claret M (2015) Hypothalamic ER stress: a bridge between leptin resistance and obesity. FEBS Lett 589:1678–1687

Raposo G, Stoorvogel W (2013) Extracellular vesicles: exosomes, microvesicles, and friends. J Cell Biol 200:373–383

Richardson SJ, Willcox A, Bone AJ, Foulis AK, Morgan NG (2009) Islet-associated macrophages in type 2 diabetes. Diabetologia 52:1686–1688

Robbins PD, Morelli AE (2014) Regulation of immune responses by extracellular vesicles. Nat Rev Immunol 14:195–208

Rutkowski DT, Kaufman RJ (2004) A trip to the ER: coping with stress. Trends Cell Biol 14:20–28

Rutkowski DT, Wu J, Back SH, Callaghan MU, Ferris SP, Iqbal J, Clark R, Miao H, Hassler JR, Fornek J, Katze MG, Hussain MM, Song B, Swathirajan J, Wang J, Yau GD, Kaufman RJ (2008) UPR pathways combine to prevent hepatic steatosis caused by ER stress-mediated suppression of transcriptional master regulators. Dev Cell 15:829–840

Rutkowski JM, Stern JH, Scherer PE (2015) The cell biology of fat expansion. J Cell Biol 208:501–512

Scheuner D, Song B, McEwen E, Liu C, Laybutt R, Gillespie P, Saunders T, Bonner-Weir S, Kaufman RJ (2001) Translational control is required for the unfolded protein response and in vivo glucose homeostasis. Mol Cell 7:1165–1176

Schindler AJ, Schekman R (2009) In vitro reconstitution of ER-stress induced ATF6 transport in COPII vesicles. Proc Natl Acad Sci 106:17775–17780

Schneeberger M, Gomez-Valades AG, Altirriba J, Sebastian D, Ramirez S, Garcia A, Esteban Y, Drougard A, Ferres-Coy A, Bortolozzi A, Garcia-Roves PM, Jones JG, Manadas B, Zorzano A, Gomis R, Claret M (2015) Reduced α-MSH underlies hypothalamic ER-stress-induced hepatic gluconeogenesis. Cell Rep 12:361–370

Schröder M, Kaufman RJ (2005) The mammalian unfolded protein response. Annu Rev Biochem 74:739–789

Sharma RB, O'Donnell AC, Stamateris RE, Ha B, McCloskey KM, Reynolds PR, Arvan P, Alonso LC (2015) Insulin demand regulates β cell number via the unfolded protein response. J Clin Invest 125:3831–3846

So J-S, Hur Kyu Y, Tarrio M, Ruda V, Frank-Kamenetsky M, Fitzgerald K, Koteliansky V, Lichtman Andrew H, Iwawaki T, Glimcher Laurie H, Lee A-H (2012) Silencing of lipid metabolism genes through IRE1α-mediated mRNA decay lowers plasma lipids in mice. Cell Metab 16:487–499

Sriburi R, Jackowski S, Mori K, Brewer JW (2004) XBP1. a link between the unfolded protein response, lipid biosynthesis, and biogenesis of the endoplasmic reticulum 167:35–41

Sun FC, Wei S, Li CW, Chang YS, Chao CC, Lai YK (2006) Localization of GRP78 to mitochondria under the unfolded protein response. Biochem J 396:31–39

Taylor RC, Dillin A (2013) XBP-1 is a cell-nonautonomous regulator of stress resistance and longevity. Cell 153:1435–1447

Teske BF, Wek SA, Bunpo P, Cundiff JK, McClintick JN, Anthony TG, Wek RC (2011) The eIF2 kinase PERK and the integrated stress response facilitate activation of ATF6 during endoplasmic reticulum stress. Mol Biol Cell 22:4390–4405

Tong X, Christian P, Zhao M, Wang H, Moreau R, Su Q (2015) Activation of hepatic CREBH and Insig signaling in the anti-hypertriglyceridemic mechanism of R-α-lipoic acid. J Nutr Biochem 26:921–928

Tyynismaa H, Carroll CJ, Raimundo N, Ahola-Erkkila S, Wenz T, Ruhanen H, Guse K, Hemminki A, Peltola-Mjosund KE, Tulkki V, Oresic M, Moraes CT, Pietilainen K, Hovatta I, Suomalainen A (2010) Mitochondrial myopathy induces a starvation-like response. Hum Mol Genet 19:3948–3958

Usui M, Yamaguchi S, Tanji Y, Tominaga R, Ishigaki Y, Fukumoto M, Katagiri H, Mori K, Oka Y, Ishihara H (2012) Atf6α-null mice are glucose intolerant due to pancreatic β-cell failure on a high-fat diet but partially resistant to diet-induced insulin resistance. Metabolism 61:1118–1128

Wang L, Ryoo HD, Qi Y, Jasper H (2015a) PERK limits drosophila lifespan by promoting intestinal stem cell proliferation in response to ER stress. PLoS Genet 11:e1005220

Wang M, Kaufman RJ (2014) The impact of the endoplasmic reticulum protein-folding environment on cancer development. Nat Rev Cancer 14:581–597

Wang M, Kaufman RJ (2016) Protein misfolding in the endoplasmic reticulum as a conduit to human disease. Nature 529:326–335

Wang S, Chen Z, Lam V, Han J, Hassler J, Finck BN, Davidson NO, Kaufman RJ (2012) IRE1α-XBP1s induces PDI expression to increase MTP activity for hepatic VLDL assembly and lipid homeostasis. Cell Metab 16:473–486

Wang S, Park S, Kodali VK, Han J, Yip T, Chen Z, Davidson NO, Kaufman RJ (2015b) Identification of protein disulfide isomerase 1 as a key isomerase for disulfide bond formation in apolipoprotein B100. Mol Biol Cell 26:594–604

Wang Y, Vera L, Fischer WH, Montminy M (2009) The CREB coactivator CRTC2 links hepatic ER stress and fasting gluconeogenesis. Nature 460:534–537

Wei X, Howell AS, Dong X, Taylor CA, Cooper RC, Zhang J, Zou W, Sherwood DR, Shen K (2015) The unfolded protein response is required for dendrite morphogenesis. eLife 4, e06963

Welc SS, Clanton TL (2013) The regulation of interleukin-6 implicates skeletal muscle as an integrative stress sensor and endocrine organ. Exp Physiol 98:359–371

Williams KW, Liu T, Kong X, Fukuda M, Deng Y, Berglund ED, Deng Z, Gao Y, Liu T, Sohn JW, Jia L, Fujikawa T, Kohno D, Scott MM, Lee S, Lee CE, Sun K, Chang Y,

Scherer PE, Elmquist JK (2014) Xbp1s in Pomc neurons connects ER stress with energy balance and glucose homeostasis. Cell Metab 20:471–482

Wu J, Ruas JL, Estall JL, Rasbach KA, Choi JH, Ye L, Bostrom P, Tyra HM, Crawford RW, Campbell KP, Rutkowski DT, Kaufman RJ, Spiegelman BM (2011) The unfolded protein response mediates adaptation to exercise in skeletal muscle through a PGC-1α/ATF6α complex. Cell Metab 13:160–169

Wu J, Rutkowski DT, Dubois M, Swathirajan J, Saunders T, Wang J, Song B, Yau GD, Kaufman RJ (2007) ATF6α optimizes long-term endoplasmic reticulum function to protect cells from chronic stress. Dev Cell 13:351–364

Xiu F, Catapano M, Diao L, Stanojcic M, Jeschke MG (2015) Prolonged endoplasmic reticulum-stressed hepatocytes drive an alternative macrophage polarization. Shock 44:44–51

Xu L, Spinas GA, Niessen M (2010) ER stress in adipocytes inhibits insulin signaling, represses lipolysis, and alters the secretion of adipokines without inhibiting glucose transport. Horm Metab Res 42:643–651

Yamamoto K, Takahara K, Oyadomari S, Okada T, Sato T, Harada A, Mori K (2010) Induction of liver steatosis and lipid droplet formation in ATF6α-knockout mice burdened with pharmacological endoplasmic reticulum stress. Mol Biol Cell 21:2975–2986

Yamauchi T, Kamon J, Minokoshi Y, Ito Y, Waki H, Uchida S, Yamashita S, Noda M, Kita S, Ueki K, Eto K, Akanuma Y, Froguel P, Foufelle F, Ferre P, Carling D, Kimura S, Nagai R, Kahn BB, Kadowaki T (2002) Adiponectin stimulates glucose utilization and fatty-acid oxidation by activating AMP-activated protein kinase. Nat Med 8:1288–1295

Yao T, Deng Z, Gao Y, Sun J, Kong X, Huang Y, He Z, Xu Y, Chang Y, Yu KJ, Findley BG, Berglund ED, Wang RT, Guo H, Chen H, Li X, Kaufman RJ, Yan J, Liu T, Williams KW (2016) Ire1α in Pomc neurons is required for thermogenesis and glycemia. *Diabetes*

Yoon YJ, Kim OY, Gho YS (2014) Extracellular vesicles as emerging intercellular communicasomes. BMB Rep 47:531–539

Yu Z, Sheng H, Liu S, Zhao S, Glembotski CC, Warner DS, Paschen W, Yang W (2016) Activation of the ATF6 branch of the unfolded protein response in neurons improves stroke outcome. J Cereb Blood Flow Metab

Zeng L, Lu M, Mori K, Luo S, Lee AS, Zhu Y, Shyy JYJ (2004) ATF6 modulates SREBP2-mediated lipogenesis. EMBO J 23:950–958

Zhang K, Wang S, Malhotra J, Hassler JR, Back SH, Wang G, Chang L, Xu W, Miao H, Leonardi R, Chen YE, Jackowski S, Kaufman RJ (2011) The unfolded protein response transducer IRE1α prevents ER stress-induced hepatic steatosis. EMBO J 30:1357–1375

Zhang Q, Yu J, Liu B, Lv Z, Xia T, Xiao F, Chen S, Guo F (2013) Central activating transcription factor 4 (ATF4) regulates hepatic insulin resistance in mice via S6K1 signaling and the vagus nerve. Diabetes 62:2230–2239

Zhang W, Feng D, Li Y, Iida K, McGrath B, Cavener DR (2006) PERK EIF2AK3 control of pancreatic β cell differentiation and proliferation is required for postnatal glucose homeostasis. Cell Metab 4:491–497

Zhang X, Szabo E, Michalak M, Opas M (2007) Endoplasmic reticulum stress during the embryonic development of the central nervous system in the mouse. Int J Dev Neurosci 25:455–463

Zhang Y, Liu R, Ni M, Gill P, Lee AS (2010) Cell surface relocalization of the endoplasmic reticulum chaperone and unfolded protein response regulator GRP78/BiP. J Biol Chem 285:15065–15075

Zhou AX, Tabas I (2013) The UPR in atherosclerosis. Semin Immunopathol 35:321–332

Zhou Y, Lee J, Reno CM, Sun C, Park SW, Chung J, Lee J, Fisher SJ, White MF, Biddinger SB, Ozcan U (2011) Regulation of glucose homeostasis through a XBP-1-FoxO1 interaction. Nat Med 17:356–365

ER Stress and Neurodegenerative Disease: A Cause or Effect Relationship?

Felipe Cabral-Miranda and Claudio Hetz

Abstract The accumulation of protein aggregates has a fundamental role in the patophysiology of distinct neurodegenerative diseases. This phenomenon may have a common origin, where disruption of intracellular mechanisms related to protein homeostasis (here termed proteostasis) control during aging may result in abnormal protein aggregation. The unfolded protein response (UPR) embodies a major element of the proteostasis network triggered by endoplasmic reticulum (ER) stress. Chronic ER stress may operate as possible mechanism of neurodegenerative and synaptic dysfunction, and in addition contribute to the abnormal aggregation of key disease-related proteins. In this article we overview the most recent findings suggesting a causal role of ER stress in neurodegenerative diseases.

F. Cabral-Miranda · C. Hetz
Faculty of Medicine, Biomedical Neuroscience Institute, University of Chile, Santiago, Chile

F. Cabral-Miranda · C. Hetz
Faculty of Medicine, Center for Geroscience, Brain Health and Metabolism, University of Chile, Santiago, Chile

F. Cabral-Miranda · C. Hetz (✉)
Program of Cellular and Molecular Biology, Institute of Biomedical Sciences, University of Chile, Independencia 1027, P.O.BOX 70086, Santiago, Chile
e-mail: chetz@med.uchile.cl; chetz@hsph.harvard.edu
URL: http://www.hetzlab.cl

F. Cabral-Miranda
Instituto de Ciências Biomédicas, Universidade Federal do Rio de Janeiro, Rio de Janeiro, Brazil

C. Hetz
Buck Institute for Research on Aging, Novato, CA 94945, USA

C. Hetz
Department of Immunology and Infectious Diseases, Harvard School of Public Health, Boston, MA 02115, USA

Current Topics in Microbiology and Immunology (2018) 414:131–158
DOI 10.1007/82_2017_52
© Springer International Publishing AG 2017
Published Online: 02 September 2017

Contents

1 Introduction.. 132
2 PMDs and ER Stress—Towards Causality?... 133
3 Alzheimer's Disease 136
4 Parkinson's Disease... 140
5 Amyotrophic Lateral Sclerosis 143
6 Huntington's Disease... 145
7 Prion-Related Disorders.. 147
8 Closing Remarks 148
References.. 149

1 Introduction

The unfolded protein response (UPR) is a well-conserved intracellular signalling pathway responsive to different types of cellular stress. Activation of the UPR triggers a collection of molecular pathways related to an increased capacity of protein folding, transport and degradation by the activation of three types of ER-resident sensors: IRE1, PERK and ATF6 (Fig. 1). However, stressful stimuli that cannot be counterbalanced by the UPR direct cell fate towards cell death (Hetz 2012; Walter and Ron 2011). The UPR mediates a constant homeostatic surveillance of the proteome in highly secretory cells where the demand for high protein synthesis triggers physiological (non-apoptotic) levels of ER stress. However, chronic ER stress is emerging as an important driver of diverse diseases including cancer, diabetes, autoimmune diseases and neurodegenerative diseases (Oakes and Papa 2015; Wang and Kaufman 2016).

In vertebrates, the UPR has evolved towards the establishment of a complex network of interconnected signaling pathways initiated by the stimulation of three mayor types of transducers located at the ER known as IRE1 alpha and beta, activating transcription factor-6 (ATF6) alpha and beta and protein kinase RNA (PKR)-like ER kinase (PERK) (Walter and Ron 2011; Wang and Kaufman 2016). Under ER stress conditions PERK oligomerizes and trans-autophosphorylates, inhibiting general protein translation through the phosphorylation of eukaryotic translation initiator factor-2 (eIF2α) at serine 51 (Hetz 2012; Walter and Ron 2011). In addition to reducing the overload of proteins entering into the ER of a stressed cell, eIF2α phosphorylation allows the selective translation of the mRNA encoding the transcription factor ATF4, contributing to the reinforcement of the antioxidant response, the folding capacity of the ER, in addition to enhance macroautophagy levels (Oakes and Papa 2015). IRE1 initiates the most conserved UPR signaling branch, where more of the advances in terms of understanding its regulation have been provided. IRE1α is a kinase and endoribonuclease (RNase) domain on its cytosolic region. In response to luminal stimulation, IRE1α dimerizes and trans-autophosphorylates, inducing a conformational change that activates the RNase domain, catalysing the excision of a 26-nucleotide intron within the XBP1 mRNA (Hetz 2012). This unconventional splicing event shifts the coding reading frame of the mRNA to generate a stable and active transcription factor known as XBP1s. The RNase domain of IRE1α also regulates RNA stability through a direct cleavage involving specific sequences and secondary structure, a process known as Regulated IRE1-dependent decay or RIDD (Ron and Walter 2011). ATF6α is also an ER transmembrane protein that contains a bZIP transcription factor on its cytosolic domain. Under ER stress, ATF6 translocates to the Golgi apparatus, where it is cleaved by the proteases S1P and S2P releasing its cytosolic domain (ATF6f) (Oakes and Papa 2015) ATF6f induces the upregulation of a selective set of UPR genes, highlighting the reinforcement of the ER-associated degradation (ERAD) pathway. Overall, the UPR operates as a complex signaling network that engages multiple outputs to restore ER proteostasis and sustain cellular function.

For decades, researchers established the notion that different neurodegenerative diseases share similar molecular signatures, one of which is the formation and accumulation of protein aggregates, an event that partially correlates with disrupted function of synapses and in some cases, with neuronal death (Scheper and Hoozemans 2015; Henstridge et al. 2016). These diseases are now classified as protein misfolding disorders (PMDs) and include Alzheimer's disease (AD), Parkinson´s disease (PD), amyotrophic lateral sclerosis (ALS), Huntington's disease (HD), and prion-related disorders (PrDs), among others (Soto 2003). Accumulating evidence suggests that UPR mediators are directly involved in the physiopathology of PMDs impacting synaptic function and neuronal physiology (Scheper and Hoozemans 2015). Accordingly, the presence of ER stress markers have been described in tissue derived from patients affected by AD (Hoozemans et al. 2005; Cornejo and Hetz 2013), PD (Mercado et al. 2016), HD (Jiang et al. 2016), ALS (Rozas et al. 2016), and PrDs (Mays and Soto 2016). In fact, the experimental expression of PMD-related proteins is sufficient to trigger the UPR response, as shown in multiple disease models using cell culture or animal models (Hoozemans et al. 2012; Hetz Flores and Mollereau 2014; Smith and Mallucci 2016). In this chapter we aim to discuss the most recent findings relating PMDs and ER stress, in addition to speculating about how the UPR may regulate the physiopathology and outcomes of such devastating diseases, thus shedding light into their complex etiology.

2 PMDs and ER Stress—Towards Causality?

Protein homeostasis or proteostasis is supported by a complex set of intracellular responses highly conserved across evolution which includes a collection of mechanisms related to protein translation, folding, trafficking, secretion and degradation distributed in different compartments inside the cell (Balch et al. 2008). Those pathways are collectively referred as the proteostasis network and are comprised of different sub-networks with superposed functions and molecular effectors (Powers and Balch 2013; Labbadia and Morimoto 2014). Proteostasis

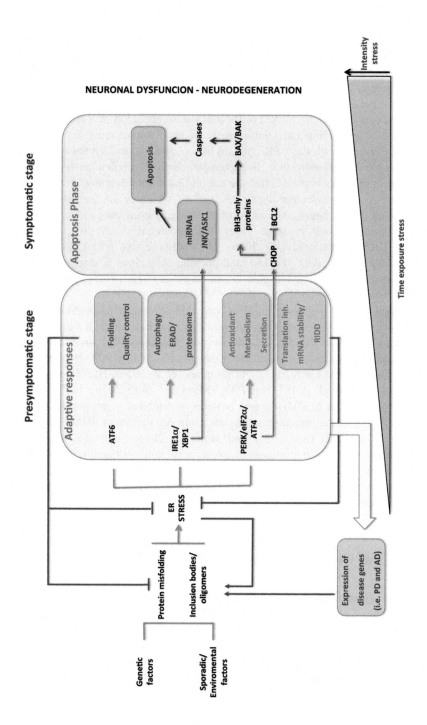

◀Fig. 1 UPR signalling outputs and neurodegeneration we propose a scenario in which mutation of specific disease-related genes and environmental factors triggers the misfolding of a particular protein, which forms different types of aggregates ranging from small oligomeric species to inclusion bodies. This pathological event triggers ER stress, which activates UPR sensors such as ATF6f, IRE1 and PERK that activate distinct UPR-related responses are observed over time. Early UPR responses attenuate protein synthesis at the ER by transiently shutting down translation, mRNA decay, and autophagy. These are adaptive responses, promoted by UPR transcription factors, that work to restore ER function and maintain cell survival. Prolonged ER stress overwhelms the adaptive responses of the UPR and apoptosis is induced that eliminates irreversibly damaged cells. Various apoptotic pathways have been described including upregulation of CHOP through ATF4. This pathway also modulates the expression of BCL-2 family members where BH3-only proteins regulate the activation of BAK and/or BAX to trigger apoptosis. CHOP also induces the expression GADD34, augmenting the levels of ROS and enhanced protein synthesis. Altered calcium homeostasis due to IP3R or RYR activation may also contribute to cell death. IRE1 also induces the activation of NJK and ASK1, contributing to cell death. In addition, IRE1 can degrade micro RNAs (miRNAs, RIDD activity) that negatively control the expression of caspases. In addition, ER stress may feedback toward enhancing abnormal protein aggregation of disease-related proteins. UPR signaling events also can modulate the expression of a variety of genes involved in the etiology of the disease

represents a coordinated quality control system that ensures that unfolded or misfolded proteins do not accumulate forming cytotoxic aggregates (proteotoxicity) that may alter cell normal physiology (Di Domenico et al. 2014; Kaushik and Cuervo 2015). When detected, misfolded proteins are managed by distinct protein chaperones in order to be properly refolded or targeted to the ubiquitin proteasome system (UPS) or the autophagy-lysosome pathway for degradation. Since more than 30% of the proteome is synthesized through the ER and Golgi, the UPR represents a major contributor of proteostasis as its arms and molecular effectors are directly involved with almost every aspect of the secretory pathway (Kaushik and Cuervo 2015).

Inclusion bodies formed by misfolded protein aggregates have long been identified as potential inducers of abnormal neuronal function as shown by observations that date back to the 1980s (Glenner and Wong 2012; Masters et al. 1985; Caughey and Lansbury 2003; Kaushik and Cuervo 2015; Di Domenico et al. 2014). It is proposed that altered proteostasis may result in abnormal protein aggregation since most proteins that accumulate and aggregate do not contain specific mutations as in familial cases of patients affected by PMDs. Overall, different neurodegenerative diseases are marked by distinct misfolded protein aggregates that arise in divergent brain regions inducing synaptic dysfunction and degeneration in vulnerable neuronal populations (Walker et al. 2015). However, the thesis that inclusion bodies represent the primary cause of neuronal toxicity is still under debate (Ross and Poirier 2005; Tompkins and Hill 1997; Brown 1998; De Strooper and Karran 2016), and soluble oligomers are the most relevant neurotoxic species altering neuronal function. In addition, loss-of-function of the affected protein due to the conformational changes driving aggregation may also contribute to neurodegeneration (Balch et al. 2008). Importantly, a vast proportion of human population will develop amyloid deposits and mild cognitive impairments late in life to AD and amyloid β aggregates can be encountered in the brain of non-demented elderly (Kern and Behl 2009; Saxena and Caroni 2011). Those findings sustain that other molecular factors must interact with misfolded proteins hence promoting their toxicity (Endres and Reinhardt 2013). Remarkably, protein aggregates are also present in the brain of subjects submitted to acute deleterious stimuli such as brain ischemia (Hu et al. 2001), sustaining that proteostasis disruption is a key event related to neuronal damage also following acute stimuli (Fawcett et al. 2015).

Compelling evidence indicate that ER stress markers are up-regulated in PMDs and often co-localize with protein aggregates in the brain of patients (Hetz and Mollereau 2014; Scheper and Hoozemans 2015; Smith and Mallucci 2016). Those findings could be interpreted as (i) neuroprotective signals to sustain proteostasis, (ii) a pro-degenerative stimuli due to chronic ER dysfunction or (iii) an epiphenomena that is irrelevant to the disease process (Salminen et al. 2009; Freeman and Mallucci 2016). In the next sections we will discuss the involvement of the UPR in different PMDs highlighting its implication to synaptic dysfunction and neuronal degeneration.

3 Alzheimer's Disease

Dementia affects about 24 million people worldwide and the majority of the cases are AD patients (Ballard et al. 2011; Alzheimer's Association 2015). AD represents one of the biggest disease-related burdens for modern society with an estimated health-care cost of UD$172 billion per year only in USA (Ballard et al. 2011; Cavedo et al. 2014) and by 2050, one new case of AD is expected to arise every 33 s in USA, resulting in almost 1 million new cases per year (Alzheimer's Association 2015). Its symptoms are characterized by a progressive and irreversible degeneration of the brain that leads to a gradual memory loss and decline of cognition and eventual death (Ballard et al. 2011). Alois Alzheimer characterized it in 1906, and noticed distinctive plaques and neurofibrillary tangles in the brain of the first described clinical case (Hippius and Neundörfer 2003).

Remarkably, those histopathological markers are used until this day to the confirmation of the diagnosis in *post mortem* tissue (Scheff et al. 2016; Ballard et al. 2011). Subsequent decades of intense study of AD patient brains led to the exposure of the nature of both senile plaques (also known as amyloid plaques) and neurofibrillary tangles composed respectively by misfolded amyloid β protein aggregates and hyperphosphorylated Tau (p-Tau), a microtubule assembly protein (Brandt et al. 2005; Glenner and Wong 2012). Senile plaques are amyloid deposits with a dense central core, surrounded by dystrophic neurites and a severe inflammatory reaction while neurofibrillary tangles are abnormal filamentous inclusions found in neuronal somata composed principally by abnormally folded Tau protein (Brandt et al. 2005; Glenner and Wong 2012; Oakley et al. 2006; Vossel et al. 2010).

The origin of amyloid β peptides arises from the cleavage of ubiquitous type I transmembrane protein known as amyloid precursor protein (APP) which may be submitted to a normal processing known as the non-amyloidogenic pathway or an abnormal processing known as the amyloidogenic pathway. Its cleavage is mediated by signal peptidases during its translocation from the ER to the Golgi apparatus, where the protein gains its maturation. Following the non-amyloidogenic pathway, enzyme alpha-secretase cleaves APP and leads to the formation of APPs-alpha fragment, which presents no toxicity (Luo et al. 2001; Dovey et al. 2001). In the amyloidogenic pathway, however, APP may be alternatively degraded by beta-secretase BACE-1 (beta-site APP cleaving enzyme 1) and gamma-secretase, which leads to the formation of neurotoxic amyloid β peptides (Luo et al. 2001; Dovey et al. 2001; Haass 2004). Mutations in coding regions for enzyme preselinin-1 favour the amyloidogenic pathway and are the primarily cause of autosomal dominant familial Alzheimer's disease (FAD) (Berezovska et al. 2005). The disturbance in this metabolic cascade leads to the accumulation of neurotoxic oligomers that forms intra and extracellular deposits, which account for disruption of synaptic transmission and neuronal survival hence leading to the devastating symptoms associated to the disease (Reitz 2012; McGeer and McGeer 2013). Small oligomers or even intraneuronal amyloid β peptides rather than amyloid plaques may represent the most relevant neurotoxic species (Endres and Reinhardt 2013).

The characteristic histopathology of AD brains suggests that proteostasis dysfunction shall exert a fundamental role in the physiopathology of the disease (Walker et al. 2015; Caughey and Lansbury 2003; Glenner and Wong 2012). Consistently, upregulation of several chaperones is observed in the brain of AD patients including heat shock protein-27 (HSP-27) and the ER stress-associated chaperone BiP/GRP78 (Hamos et al. 1991). Interestingly, authors stressed that BiP presented increased expression in neurons that did not show deterioration, suggesting that this chaperone may exert a neuroprotective role prior to degeneration in AD. Moreover, mutations in presilinin-1, a gene related to an increased incidence of FAD have been shown to downregulate BiP expression (Katayama et al. 1999), thus suggesting that presenilin mutations were capable of altering the UPR firing and ER homeostasis, although opposite results showed that neither loss of presenilin-1 nor expression of its variants is sufficient to trigger UPR (Sato et al. 2000). Analysing *post-mortem* tissue obtained from patients in different Braak stages, Hoozemans and colleagues reported an up-regulation of BiP in the temporal cortex and hippocampus of AD patients that positively correlated with increased proportion of neurofibrillary tangles and amyloid pathology (Hoozemans et al. 2005). Remarkably, using immunohistochemistry to identify the neuronal populations positive for BiP in AD patients compared to non-demented elderly, authors found that neurons positive for BiP did not co-localize with neurofibrillary tangles, suggesting that BiP expression preceded tangle formation. Those first reports made clear that ER stress was activated during the course of AD, thus driving researchers to evaluate the role of other UPR mediators in the physiopathology of the disease. Accordingly, neurons in CA1 region in the hippocampus of AD patients exhibited an increased expression of phospho-PERK when compared to non-demented controls (Hoozemans et al. 2009), a result that positively correlated with Braak stage. Those findings are also consistent with animal models for AD (Abisambra et al. 2013; Page et al. 2006). Besides the indication of increased phosphorylation of PERK and expression of BiP, this work also presents consistent evidence supporting that tau interferes with ERAD.

The PERK and eIF2α pathway is another arm of the UPR with important implications to AD pathology (Lourenco et al. 2015). Its phosphorylation was shown to modulate BACE-1 expression in AD mice (Vassar 2009). Likely, ATF4 was also shown to regulate gamma-secretase activity during amino acid imbalance, which triggers ER stress, also favouring amyloidogenic pathway (Mitsuda et al. 2007). Cellular models present evidence consistent with PERK activation following exogenous amyloid β treatment. Neuronal cultures treated with amyloid β develop increased phosphorylation of PERK, which confers cellular protection, as its silencing increases neuronal death following amyloid β treatment (Lee et al. 2010a). Therefore, it is possible that ER stress response is activated to protect neuron homeostasis and prevent neurodegeneration linked to AD pathogenesis. Finally, IRE1 may trigger JNK-3 in the brain following amyloid β exposure in animal models and its deletion restores the translational block induced by oligomeric amyloid β (Yoon et al. 2012).

XBP1s is another pivotal UPR component implicated in AD pathophysiology. Transgenic flies expressing amyloid β showed increased XBP1 mRNA splicing, which prevented oligomers toxicity in both models (Casas-Tinto et al. 2011). Consistently, brain samples from AD patients presented augmented XBP1 when compared to age matched-controls, an observation extended to its downstream mediator protein disulfide isomerase (PDI) but not to BiP, indicating disrupted UPR function in AD pathogenesis (Lee et al. 2010b). Consistently, XBP1 mRNA splicing is observed after amyloid β treatment in neuronal cell lines (Castillo-Carranza et al. 2012) and also in the temporal cortex of AD patients (Katayama et al. 2004). Remarkably, genome wide screening to identify XBP1s-target genes uncovered a cluster of genes related to AD (Acosta-Alvear et al. 2007). Finally, a direct genetic association between XBP1 promoter polymorphism and AD increased incidence was revealed in the Chinese population, suggesting that XBP1 may stand as a UPR etiological component for AD (Liu et al. 2013) and the expression of XBP1s downstream target PDI was also shown to co-localize with neurofibrillary tangles in AD patients brain (Honjo et al. 2010). A recent study indicated that delivering the active form of XBP1s into the hippocampus of AD mice restored synaptic function although findings from our group indicate that ablating IRE1 in the CNS ameliorates AD pathology (Duran-Aniotz et al. 2017). This is in agreement with recent findings indicating that XBP1s has a relevant role in controlling learning and memory-related processes (Martínez et al. 2016). Importantly, targeting PERK expression in the context of AD restored synaptic function due to the recovery of protein synthesis of synaptic proteins, improving memory capacity and synaptic transmission (Ma et al. 2013; Duran-Aniotz et al. 2014). Similarly, targeting of other eIF2α kinases like PKR and

Table 1 Functional studies linking ER stress with neurodegeneration

Disease	Model	UPR manipulation	Phenotype	Reference
AD	P301L mice	Thapsigargin	Increased phosphorylation of tau, increased caspase-3 cleavage	Ho et al. (2012)
	APP/PS1 Tg mice	JNK3-/-	Reduced amyloid β, neuronal loss and cognitive dysfunction	Yoon et al. (2012)
		PERK CNS KO	Improved learning and memory and LTP	Ma et al. (2013)
		AAV-XBP1s	Rescued spine density, synaptic plasticity and memory function	Cissé et al. (2016)
		IRE1 CNS KO	Improved learning and LTP, reduced amyloid b	Duran-Aniotz et al. (2017)
PD	α-Synuclein over expression	Salubrinal	Attenuates disease manifestation	Colla et al. (2012a, b)
	Neurotoxins	AAV-XBP1s	Increased dopaminergic survival	Unterberger et al. (2006)
		ATF6 KO	Increased neurodegeneration	Egawa et al. (2011); Hashida et al. (2012)
		CHOP KO	Neuroprotection	Silva et al. (2005)
		AAV-BiP	Dopaminergic survival, decreased αSyn aggregation	Gorbatyuk et al. (2012)
		AAV-XBP1s	Neuroprotection, reduced striatal denervation	Valdés et al. (2014)
ALS	SOD1 overexpressing mice	PERK ±	Disease exacerbation, enhancement SOD1 aggregation	Wang et al. (2011)
		Salubrinal	Increased lifespan	Saxena et al. (2009)
		ATF4 KO	Partial embryonic lethality, protection against disease progression	Matus et al. (2013)
		XBP1 CNS KO	Extended life span, decreased SOD1	Hetz et al. (2009)
		Blockage of ASK-1 binding to SOD1	Motorneuron death	Nishitoh et al. (2008)
		AAV-SIL-1	Delayed muscle denervation and prolonged survival	Filézac de L'Etang et al. (2015)
		Guanabenz	Accelerated disease progression	Vieira et al. (2015)
		Guanabenz	Delayed disease onset, increased survival, less accumulation of aggregates, improved motor performance and attenuated motor neuron loss	Wang et al. (2014); Jiang et al (2014)

A summary of selected most recent findings regarding the causality between UPR genetic or pharmacological manipulation and aggregation/toxicity of classical misfolded proteins in distinct NDDs is presented

GCN-2 impact synaptic function in AD models (Lourenco et al. 2015; Ma et al. 2013). ATF4 expression was also recently shown to control axonal degeneration in AD models through a cell-nonautonomous mechanism (Baleriola et al. 2014) suggesting that XBP1s and ATF4 may represent opposite forces in AD progression.

Another study indicated that neurons containing neurofibrillary tangles have increased phosphorylation of eIF2α and PERK (Culmsee and Landshamer 2006). Increased levels of p-PERK and p-eIF2α were also described in the hippocampus of aged P301L mutant Tau transgenic mice (Ho et al. 2012). Accordingly, Abisambra and colleagues reported that ERAD machinery is disrupted by Tau accumulation, leading to ER stress (Abisambra et al. 2013). Nevertheless, a recent report described that treatment with both the chemical chaperone TUDCA or a specific small-molecule inhibitor of the PERK pathway prevents phosphorylation of Tau mediated by metabolic stress (van der Harg et al. 2014). Alternatively, Ho and colleagues offered a different explanation, presenting data sustaining that induction of Tau phosphorylation is sufficient to activate the UPR. Pharmacological induction of ER stress may also induce phosphorylation of Tau (Ho et al. 2012). Such findings sustain that the UPR and Tau hyperphosphorylation may act as self-amplifying molecular events, thus creating a vicious cycle of neuronal degeneration signalling. Moreover, XBP1 unconventional splicing has been demonstrated in genetic manipulated *Drosophila* that overexpress Tau, an animal model that recapitulates most molecular signatures of AD pathology (Loewen and Feany 2010).

Taken together, cumulative evidence suggests the existence of a complex regulatory molecular network that mediate AD pathology by the ER stress response although the number of studies that directly manipulate the UPR in AD models using mammalian systems remains erratic (Cornejo and Hetz 2013) (please refer to Table 1). Such approaches will certainly shed light into its complicated relation in order to better establish a solid impact of UPR in AD pathogenesis.

4 Parkinson's Disease

PD is characterized by severe motor dysfunction composed by rest tremor, slowness of movement, rigidity and postural instability (Jankovic and Aguilar 2008). Following AD, PD is the second most common age-related neurodegenerative disorder affecting 0.6% of the population who are 65–69 years of age and 2.6% of the population between 85 and 89 years of age (De Lau and Breteler 2006). PD symptoms are dictated by a massive loss of dopaminergic neurons in the *substantia nigra pars compacta* leading to a severe decrease in the content of dopamine in this brain circuit. Remarkably, just like AD, its physiopathology is marked by the presence of neurofibrillary deposits of misfolded proteins known as Lewy bodies composed by the misfolded protein α-Synuclein which colocalizes with ubiquitin (Varma and Sen 2015). Despite the fact that most cases of PD are sporadic, mutations in several genes including SNCA (codifies for α-Synuclein), Leucine-rich repeat kinase 2 (LRRK2), parkin (PRKN), Parkinson Disease protein

7 (PARK7/DJ1), PTEN-induced putative kinase 1 (PINK1) and Parkinson Disease protein 9 (PARK9/ATP13A2) trigger familial parkinsonian syndromes which account for less than 10% of the total PD cases (Duvoisin 1995) with mutations in leucine-rich repeat serine/threonine kinase 2 gene (LRRK2/PARK8) being the most frequent cause of familial PD occurring in 1–2% of genetic cases (Varma and Sen 2015).

Different groups have implemented cellular models of PD, relying on drugs that elicit specific neuronal death for dopaminergic neurons such as 6-hydroxydopamine (6-OHDA), 1-methyl-4-phenyl-pyridinium (MPP+), and rotenone in order to understand the molecular pathways responsible for neuronal degeneration in PD (Betarbet et al. 2000; Ryu et al. 2002). In 2002, Ryu and colleagues described that PC12 cells exposed to 6-hydroxydopamine presented an increased expression of transcripts associated with the UPR, suggesting that ER stress mediators may regulate neuronal death associated to PD (Ryu et al. 2002). Those results were confirmed by the observation of augmented phosphorylation of IRE1 and PERK besides an increased expression of its downstream targets in that model. Consistently, Hoozemans and collaborators were the first to describe the presence of UPR mediators in post mortem tissue from PD patients (Hoozemans et al. 2007). Authors showed increased phosphorylation of PERK and eIF2α by immunohistochemistry in the substantia nigra of PD patients compared to age-matched controls and remarkably, that those UPR mediators were in co-localization with α-Synuclein inclusions. Other studies were also able to identify other ER stress markers in brain tissue from PD patients such as PDI (Conn et al. 2004) and components of ERAD like Herp (Slodzinski et al. 2009) in co-localization with Lewy bodies, suggesting that the UPR may be a prominent regulator of PD pathology.

Genetic models of PD that include transgenic mice overexpressing various human variants of α-Synuclein in the brain also point to UPR triggering as a molecular target of neuronal degeneration (Mercado et al. 2013). Consistently, Colla and colleagues have presented evidence that α-Synucleinopathy in this model is coincident with induction of ER stress and leads to abnormal UPR signal, thus leading to cell death pathway activation in vivo. Additionally, authors show in another paper that a fraction of monomers and aggregates of α-Synuclein locates inside the ER, where it is found in association with ER chaperones (Colla et al. 2012a, b; Lee et al. 2005). Remarkably, treating mutant A53T mice with salubrinal, an inhibitor of eIF2α phosphatase, attenuated the onset of motor dysfunction, although this treatment was not sufficient to protect dopaminergic neuronal death following adeno-associated (AAV) distribution of α-Synuclein mutant A53T in rats (Colla et al. 2012a, b). Other UPR mediators that were shown to be upregulated in the brain of α-Synuclein transgenic mice also include BiP, XBP1, CHOP, and ATF4 (Bellucci et al. 2011; Colla et al. 2012a, b; Belal et al. 2012). Interestingly, using a cell model overexpressing wild-type human α-Synuclein, Jiang and colleagues demonstrated that ER stress induction after tunicamycin treatment was sufficient to increase the α-Synuclein aggregation, thus indicating a causal relation in disease pathogenesis and also sustaining the view that UPR dysfunction may act as a positive feedback for aggregation and cellular death (Jiang et al. 2010). Using a screening system in yeast, a direct explanation for the occurrence of ER stress in PD was provided. Susan Lindquist's laboratory demonstrated that α-Synuclein directly targets and inhibits vesicular trafficking from ER to Golgi apparatus by interacting with RAB1 (Cooper et al. 2006). Overexpression of RAB1 protected against α-Synuclein toxicity on a fly model and also attenuated motor deficits in a rat model (Coune et al. 2011). Remarkably, human dopaminergic neurons generated from induced pluripotent stem (iPS) cells from Parkinson patients harbouring α-Synuclein mutations validated the importance of ER stress in the disease process (Chung et al. 2013). Finally, different reports confirmed that α-Synuclein disrupts COPII ER–Golgi transport, a pathway crucial for activation and signalling by ATF6 (Credle et al. 2015). Creddle and collaborators have shown that α-Synuclein inhibited processing of ATF6 both in a direct and indirect manner thus decreasing ERAD function, a common feature of PMDs (Credle et al. 2015).

The pathogenesis of LRRK2 has been also linked to ER stress as it partially localizes in the ER in dopaminergic neurons of PD patients (Vitte et al. 2010). Using *C. elegans*, it was shown that LRRK2 confers protection to dopaminergic neurons against 6-OHDA treatment or human α-Synuclein expression, a phenomena associated with increased expression of the ER chaperone BiP (Yuan et al. 2011). Additionally, *C. elegans* lacking LRRK2 homolog are more susceptible to ER stress (Sämann et al. 2009). Other important gene linked to familial PD that has been reported to alter ER function is E3 ubiquitin ligase Parkin/PARK2, related to the ubiquitin proteasome system (UPS) and to ERAD (Mercado et al. 2013). Parkin was shown to be transcriptionally regulated by ATF4, thus suggesting that it acts as an ER stress-inducible protein that mediate cytoprotective mechanisms (Bouman et al. 2011) and its subcellular distribution is altered following ER stress (Ledesma et al. 2002). Of interest, Parkin-associated endothelin-receptor like receptor (Pael-R), a target for Parkin-dependent degradation by the proteasome, was shown to induce UPR triggering thus leading to neuronal death, a condition that was worsened by ER chaperone dysfunction (Kitao et al. 2007).

Genetic manipulation of UPR mediators was shown to regulate dopaminergic neuronal death in PD models. Unexpectedly, genetic ablation of XBP1 throughout brain development increased neuronal resistance following 6-OHDA treatment, which was accompanied by the up-regulation of several UPR effectors in the substantia nigra, while there was no activation of pro-apoptotic UPR markers such as CHOP (Valdés et al. 2014). The concept of *hormesis* was proposed where the occurrence of mild ER stress may engage a protective program (Matus et al. 2012; Hetz and Mollereau 2014). In agreement with this, treatment with non-lethal doses of tunicamycin in flies and mice models of PD provided protection against neurodegeneration (Fouillet et al. 2012). Gene therapy to deliver XBP1 into the *substantia nigra* of adult animals protected against neurodegeneration induced by PD-inducing neurotoxins (Valdés et al. 2014; Sado et al. 2009). This observation was also confirmed in a model where XBP1s is delivered to neuronal stem cells transferred to animals treated with rotenone (Si et al. 2012). ATF6 deficient animals exhibit increased ubiquitin positive inclusions and exacerbated dopaminergic neuron loss following MPTP treatment (Egawa et al. 2011; Hashida et al. 2012).

Overexpression of BiP using a gene therapy strategy also proved to be neuroprotective for dopaminergic neurons after overexpression of human α-Synuclein in rats (Gorbatyuk et al. 2012). Finally, genetic ablation of pro-apoptotic CHOP/GADD153 has also been shown to be beneficial for dopaminergic neuronal survival following 6-OHDA but not after MPTP treatment, suggesting that the contribution of the UPR in distinct PD models may be highly influenced by the chosen toxin for dopaminergic degeneration (Silva et al. 2005).

Taken together, ER stress has a pivotal role in the regulation of dopaminergic neuron survival and physiology thus representing an interesting target for disease intervention (Mercado et al. 2016), (please refer to Table 1). However, caution must be taken with interpretation as different models for PD may elicit different UPR components recruitment and the final output of manipulating the UPR may lead to contrasting results in a context dependent manner.

5 Amyotrophic Lateral Sclerosis

ALS is an adult-onset fatal neurodegenerative disease marked by the progressive degeneration of upper and lower motor neurons, in brainstem, cortex and spinal cord leading to paralysis and muscle atrophy, standing as the most common motoneuron disease (Maharjan and Saxena 2016). The etiology of sporadic ALS, which account for nearly 90% of cases is not well defined but inherited genetic defects in distinct genes have been linked to familial cases of ALS (2015). Approximately 20% of familial cases, and 1–2% of all ALS cases, are caused by mutations in the gene encoding superoxide dismutase-1 (SOD1) (Dion et al. 2009) and now nearly 140 different ALS-linked SOD1 mutations have been identified (Sreedharan 2013). Mutated genes implicated in familial ALS also include TAR DNA-binding protein 43 (TDP-43), FUS, ubiquilin-2 and C9ORF72; which are all linked to alterations to mRNA metabolism and proteostasis (Maharjan and Saxena 2016; Sreedharan and Brown 2013).

The involvement of ER stress in ALS is well defined and validated by several groups using different disease models. Importantly, ER stress has been proposed as one of the earliest molecular defects underlying the differential neuronal vulnerability observed (Rozas et al. 2016). Both ALS patients and mutant SOD1 mice present alterations in ER morphology as patients exhibits fragmentation of the rough ER, irregular distension of the rER cisternae and a detachment of ribosomes in degenerating anterior horn cells (Lautenschlaeger et al. 2012; Deitch et al. 2014). Accordingly, ER stress triggering was shown in post-mortem tissue by the presence of UPR markers BiP, calnexin and PDI in co-localization with mutant SOD1 (Wate et al. 2005; Kikuchi et al. 2006; Atkin et al. 2006). Moreover, increased expression of XBP1s and ATF4 has been shown in human postmortem spinal cord tissue from sALS cases (Hetz 2009). Remarkably, using three distinct mouse models for ALS, Saxena and collaborators reported that ER stress specifically occurs in vulnerable motoneurons before any denervation is observed, which was followed by selective axonal degeneration (Saxena et al. 2009). In the same line, proteomic screenings of spinal cord from mutant SOD1 mice revealed significant up-regulation of chaperone PDI and ERp57 (Atkin et al. 2006). Moreover, PDI was found in co-localization with TDP-43 and SOD1 in swollen neurites and neuronal cytoplasmic inclusion of patients with ALS (Honjo et al. 2011) and single nucleotide polymorphisms (SNPs) in intronic regions of the PDIA1 gene were shown to act as risk factors for ALS (Kwok et al. 2013). Of interest, our group recently identified mutations both in PDIA1 and ERp57 as risk factors to develop ALS (Gonzalez-Perez et al. 2015).

Studies in animal models indicated that PDI mutations triggers motor defects associated with a disruption of motoneuron connectivity (Woehlbier et al. 2016). Similarly, calreticulin decreases in ALS mice, which accelerates early-stage muscle weakness and muscle denervation (Bernard-Marissal et al. 2015). Additionally, mutant mice for BiP develop spontaneous motor disease during aging, associated with selective motoneuron degeneration and aggregation of wild-type endogenous SOD1 (Wang et al. 2010). Finally, expression of the BiP co-factor SIL1 was recently reported to underlay in part the differential motoneuron vulnerability in ALS and gene therapy to deliver SIL1 to the nervous system has outstanding protective effects against experimental ALS (Filézac de L'Etang et al. 2015). Thus, alterations on the ER folding capacity may be part of the etiology of the disease.

In addition to increased BIP and PDI expression (Tobisawa et al. 2003), mutant SOD1 expression triggers chronic PERK signaling (Atkin et al. 2006; Saxena et al. 2009; Nagata et al. 2007) and the presence of pro-apoptotic UPR downstream mediator CHOP/GADD153 was confirmed both in spinal cords of sporadic ALS patients and ALS transgenic mice (Ito et al. 2009). A ribosome profiling analysis *in vivo* of motoneurons revealed that chronic ER stress is a major pathological signature in this ALS model (Sun et al. 2015). Similarly, gene expression analysis of ALS brain tissue from patients carrying C9orf72 mutations shows major alteration in ER stress-related genes (Prudencio et al. 2015). In agreement with this, expression of RAN peptides derived from C9ORF72 mutation in cell culture also triggers abnormal levels of ER stress (Zhang et al. 2014).

eIF2α Phosphorylation is also upregulated by TDP-43 overexpression in flies and its pharmacological inhibition was sufficient to attenuate its toxicity (Kim et al. 2014); additionally, targeting one copy of PERK accelerated the disease process (Wang et al. 2011). Based on all this evidence, therapeutic strategies to target eIF2α phosphorylation using pharmacologic approaches have been tested. Accordingly, treatment with salubrinal, a small molecule that inhibits eIF2α dephosphorylation (Boyce 2005), delayed experimental ALS progression (Saxena et al. 2009). Another study in contrast, indicated that guanabenz treatment (a specific inhibitor of the ER stress inducible eIF2α phosphatase that inhibits a negative feedback loop of eIF2α phosphorylation) (Tsaytler 2011) accelerated ALS pathogenesis (Vieira et al. 2015), whereas others observed protection following guanabenz treatment (Wang et al. 2014; Jiang et al. 2014). Remarkably, another small molecule termed Sephin-1, a derivate of guanabenz that similarly inhibits the negative feedback signaling for eIF2α phosphorylation (Crunkhorn 2015), showed almost full protection in mutant SOD1 mice (Das et al. 2015). Our group presented evidence indicating that ATF4 expression modulates ALS pathogenesis (Matus et al. 2013). Deletion of the ATF4 gene was sufficient to increase lifespan in mutant SOD1 mice, reducing expression of pro-apoptotic genes BIM and CHOP. Unexpectedly, ATF4 deletion increased SOD1 aggregation both in vivo and in vitro possibly due to increased oxidative stress, thus confirming its importance in ALS physiopathology (Matus et al. 2013). Interestingly, CHOP increased expression was shown both in neurons and astrocytes, oligodendrocytes and microglia, supporting the theory that UPR glial modulation might also regulate ALS physiopathology (Suzuki and Matsuoka 2012).

The contribution of XBP1 to ALS has also been tested. Using a mutant SOD1 transgenic mouse with deletion of XBP1 specifically in the nervous system, our group has unexpectedly found that XBP1 ablation could decrease the severity of experimental ALS thus increasing lifespan (Hetz et al. 2009). This phenotype correlated with increased macroautophagy machinery activation in motoneurons in the absence of XBP1. Whether direct manipulation of IRE1-XBP1s axis may prove efficient into ALS pathogenesis ablation is still an open question in literature (Rozas et al. 2016). Finally, a novel autosomal-dominant ALS-causative gene was identified in 2004 (Nishimura et al. 2004) which encodes the vesicle-associated membrane protein-associated protein B (VAPB). Interestingly, this protein was shown to interact with ATF6 via its cytosolic domain and its malfunction could disrupt its activity (Gkogkas et al. 2008). The authors suggest that such malfunction may contribute to the pathological mechanisms of degenerative motor neuron disease. Similarly, transgenic mice for ALS-linked mutant VAPB show signs of ER stress (Suzuki et al. 2009). Taken together, the ALS field has witness great advances in experimental strategies with therapeutic potential based on UPR for the future (please refer to Table 1).

6 Huntington's Disease

HD is an autosomal-dominant, progressive neurodegenerative disease with symptoms that include chorea and dystonia, incoordination, cognitive decline, and behavioural difficulties (Walker 2007). HD is caused by a mutation in the first exon of the Huntingtin (HTT) gene, which contains a tract of glutamine residues (polyQ repeats) that can vary in length amongst individuals resulting in a mutated protein HTT with an expanded CAG repeat (Jiang et al. 2016). Longer polyglutamine (polyQ) expansions may exert a toxic gain of function (Walker 2007) although the exact mechanisms leading to neuronal cell death mediated by these expansions are still controversial (Kalathur et al. 2015). Accordingly, polyQ expansions in different proteins are directly related to the pathology of at least nine different neurodegenerative disorders, including HD; one of each represented by a different subset of vulnerable neuronal populations (Orr and Zoghbi 2007). Expression of mutant HTT production and aggregate formation in cytosol and nucleus are believed to overcome the proteostasis machinery, thus altering neuronal physiology (Jiang 2016). Mutant HTT may disrupt different cellular pathways related to protein turnover (Bennett et al. 2007), transport (Gunawardena and Goldstein 2005) and the UPS (Bence et al. 2001; Bennett et al. 2007). Different groups have reported the induction of ER stress in cells expressing poly(Q) peptides resembling the mutations observed in HTT (Nishitoh et al. 2002; Urano et al. 2000; Kouroku et al. 2002). One event particularly disturbed by poly(Q) expanded repeats is ERAD, leading to ER stress (Nishitoh et al. 2008; Duennwald and Lindquist 2008).

UPR downstream targets BiP, HERP and CHOP were shown to be upregulated in brain samples form HD patients (Carnemolla et al. 2009). Similar results were also observed in an animal model for HD (Cho et al. 2009). A recent study performed a bioinformatic analysis that assembled different sets of genes associated with the UPR and examined whether the included genes show differential expression in HD models and patients. They have identified a network of genes that provide a potential link between UPR and HD and its relation to neurodegeneration (Kalathur et al. 2015). Our group have provided data regarding XBP1 modulation in HD pathogenesis and its relation with autophagy. We demonstrated that ablating XBP1 expression in the full-length mutant HTT transgenic mice is sufficient to reduce neuronal loss in the striatum in addition to improve motor performance. These protective effects were in line with decreased HTT accumulation mediated by the induction of autophagy, possibly involving the upregulation of Forkhead box O1 (FoxO1) (Vidal et al. 2012). In contrast, ATF4 deficiency did not affect mutant HTT aggregation (Vidal et al. 2012). As autophagy is suggested to be the preferential degradation pathway for misfolded protein aggregates, including those formed as part of mutant HTT (Jiang et al. 2016) and is proposed to fail in the disease process (Martinez-Vicente et al. 2010), XBP1 may stand as a central mediator of HTT clearance in HD pathogenesis. Accordingly, delivering AAV-XBP1s to the striatum of adult mice overexpressing a mutant HTT reduced its accumulation (Zuleta et al. 2012). Overall, different methodological approaches have indicated that the alterations in the secretory pathway observed in HD may be linked to perturbations at the level of ERAD/protein quality control mechanisms, ER/Golgi trafficking, endocytosis, vesicular trafficking, ER calcium homeostasis and autophagy/lysosomal-mediated protein degradation, thus resulting on ER stress as a common feature (Vidal et al. 2011).

In agreement with the results obtained with XBP1 deficient animals, cell-based functional screening using a mutant HTT aggregation assay identified IRE1 as a potential inducer of its aggregation (Lee et al. 2012). Other studies suggest that PERK signalling may also influence mutant HTT biology. Kouroku and collaborators have shown that eIF2α phosphorylation is a crucial step for autophagy induction mediated by HTT aggregates (Kouroku et al. 2007). Still, the studies linking ER stress with HD are poor and more studies are needed to properly address the relevance of the UPR to HD pathogenesis (Table 1).

7 Prion-Related Disorders

PrDs, also known as transmissible spongiform encephalopathies, are a group of diseases characterized by rapid neurological dysfunction that may include dementia, ataxia and psychiatric disturbances. Its etiology is divided into infectious (derived from the exposure to material contaminated with infectious prions), sporadic (spontaneous origin) or familial (inherited in an autosomal dominant manner) PrDs. Human familial PrDs include some forms of Creutzfeldt-Jacob disease (CJD), Gertmann Straussler-Sheinker syndrome and fatal familial insomnia (Prusiner and Scott 1997). The main molecular event in the pathogenesis of prion diseases is the conversion of the normal cellular prion protein (termed PrP^C) into the pathological form denoted PrP^{Sc} (for scrapie associated PrP) (Prusiner 1998). The possible involvement of ER stress in PrDs was initially highlighted by the finding of chaperones BiP, Grp94 and Grp58/ERp57 in the cortex of patients affected with variant CJD and sporadic CJD (Hetz et al. 2003). Accordingly, ERp57 was reported as the major hit of a proteomic screening in the cerebellum of humans patients affected with sporadic CJD (Yoo et al. 2002). Mice infected with scrapie prions presents a profile of expression in the hippocampus revealing that most genes affected by prion infection are related with mediators of ER stress (Brown et al. 2005). However, other studies in CJD have not found signs of ER stress (Unterberger et al. 2006). Interestingly, mice infected with scrapie prions show markers of ER stress during the pre-symptomatic phase of the disease following PrP^{Sc} accumulation, including high levels of the disulfide isomerase Grp58/ERp57 expression and also BiP and Grp94 increased expression during the symptomatic phase (Hetz et al. 2005). Remarkably, in the same model, neuronal loss is prominent only at the late stage of the disease, which correlates with increased pro-caspase-12 processing and decreased ERp57 expression (Hetz et al. 2003) although knock outs for caspase-12 shows no alteration in disease progression (Steele et al. 2007). ERp57/Grp58 has a neuroprotective role in PrDs as demonstrated in cellular and animal models (Hetz et al. 2005) and ERp57 is also relevant for the synthesis and folding of PrP (Torres et al. 2015). Exposure of cells to purified PrP^{Sc} extracted from the brain of scrapie-infected mice results in ER stress. Remarkably, targeting of the anti-apoptotic protein BCL-2 to the ER membrane could decrease PrP^{Sc} toxicity and cells infected with prions were more susceptible to ER stress-induced apoptosis (Hetz et al. 2003; Apodaca et al. 2006; reviewed in Hetz and Soto 2006). At the molecular level PrP^{Sc} conversion may affect ER calcium homeostasis (Mukherjee et al. 2010; Herms et al. 2000), triggering ER stress (Torres et al. 2010), which may culminate in cellular death (Ferreiro et al. 2008).

Genetic manipulation of the UPR has provided conflicting results. Targeting XBP1 in the brain did not affect prion replication and its pathogenesis in vivo (Hetz et al. 2008). In contrast, Moreno and colleagues reported that persistent protein translational shutdown due to eIF2α phosphorylation triggers synaptic dysfunction and neuronal death in Prion-infected animals (Moreno et al. 2012). Gene therapy to deliver eIF2α phosphatase provided neuroprotection in this model, whereas salubrinal treatment exacerbated the progression of the disease (Moreno et al. 2012). Furthermore, oral treatment with a specific inhibitor of PERK both at preclinical and symptomatic stages of Prion-infected mice attenuated disease progression (Moreno et al. 2013). Similarly, treatment of animals with ISRIB, a small compound that blocks the consequences of eIF2α phosphorylation protected against PrD (Halliday et al. 2015). Of note, PERK inhibition did not attenuate PrP misfolding, suggesting that its protection was specifically linked to synaptic function improvements. More studies are still needed to define how PrP^{Sc} replication triggers ER stress and why its toxicity is only affected by PERK signaling (please refer to Table 1).

8 Closing Remarks

The current state of the field indicates that understanding the relative impact of the UPR to PMDs is difficult to predict and requires systematic studies. Depending on the disease context and the signalling branch analysed, the UPR may have contrasting and even opposing effects. The identification of new drugs that directly interfere with UPR signalling in addition to the possibility of brain in vivo manipulation of ER stress-associated genes in different animal models will provide information regarding the involvement of the UPR in PMD progression (Maly and Papa 2014; Hetz et al. 2013). Table 1 summarizes a collection of the most recent findings concerning UPR manipulation by pharmacological or genetic approaches and its outcome in disease progression of distinct PMDs models in vivo. Notably, UPR manipulation in various disease models might present contrasting results depending on the disease context and the specific signalling module studied. Additionally, since the UPR has a major role in the physiology of many organs like pancreas, liver, and the immune system, serious side effects are predicted of the long-term administration of UPR-targeting drugs (Dufey et al. 2014; Cornejo et al. 2013). Gene therapy is emerging as an interesting strategy to target the UPR locally in specific brain areas (Valenzuela et al. 2016). These tools will likely prove sufficient to clarify the causal relationship between ER proteostasis malfunction and the occurrence of PMDs. Importantly, available data suggests that ER stress may not only affect protein aggregation, but it may be also the direct cause of synaptic dysfunction as demonstrated in AD and PrDs. Since recent findings uncovered a novel role of XBP1s in enhancing learning and memory-related processes through the regulation of BDNF (Martínez et al. 2016). Thus, XBP1s-based gene therapy may actually serve as a dual therapy to target proteostasis alteration, protein aggregation and synaptic function. eIF2α Phosphorylation and ATF4 have been also reported to influence cognitive process at the level of synaptic transmission and neuronal plasticity (Pasini et al. 2015; Costa-Mattioli et al. 2009). The possible impact of the UPR in neuro-inflammation and glial activation remains to be determined. In addition, interesting links between the UPR and energy control and global proteostasis control are also available that could be explored in the future in the context of neurodegenerative diseases.

Acknowledgements Supported by Millennium Institute No. P09-015-F, FONDAP program 15150012, CONICYT-Brazil 441921/2016-7, ALS Therapy Alliance 2014-F-059, Muscular Dystrophy Association 382453, Michael J. Fox Foundation for Parkinson's Research—Target Validation grant No 9277, FONDECYT No. 1140549, Office of Naval Research-Global (ONR-G) N62909-16-1-2003, FONDEF D11E1007, U.S. Air Force Office of Scientific Research FA9550-16-1-0384, FONDEF ID16I10223, and ALSRP Therapeutic Idea Award AL150111 (CH). FC is a postdoctoral fellow funded by FONDAP program 15150012.

References

Abisambra JF, Jinwal UK, Blair LJ et al (2013) Tau accumulation activates the unfolded protein response by impairing endoplasmic reticulum-associated degradation. J Neurosci Off J Soc Neurosci 33(22):9498–9507
Acosta-Alvear D, Zhou Y, Blais A et al (2007) XBP1 controls diverse cell type-and condition-specific transcriptional regulatory networks. Mol Cell 27(1):53–66
Alzheimer's Association (2015) Alzheimer's Dementia: J Alzheimer's Assoc 11(3):332–384
Apodaca J, Kim I, Rao H (2006) Cellular tolerance of prion protein PrP in yeast involves proteolysis and the unfolded protein response. Biochem Biophys Res Commun 347(1):319–326
Atkin JD, Farg MA, Turner BJ et al (2006) Induction of the unfolded protein response in familial amyotrophic lateral sclerosis and association of protein-disulfide isomerase with superoxide dismutase 1. J Biol Chem 281(40):30152–30165
Balch WE, Morimoto RI, Dillin A et al (2008) Adapting proteostasis for disease intervention. Science 319(5865):916–919
Baleriola J, Walker CA, Jean YY et al (2014) Axonally synthesized ATF4 transmits a neurodegenerative signal across brain regions. Cell 158(5):1159–1172
Ballard C, Gauthier S, Corbett A et al (2011) Alzheimer's disease. Lancet 377(9770):1019–1031
Belal C, Ameli NJ, El Kommos A et al (2012) The homocysteine-inducible endoplasmic reticulum (ER) stress protein Herp counteracts mutant α-synuclein-induced ER stress via the homeostatic regulation of ER-resident calcium release channel proteins. Hum Mol Genet 21(5):963–977
Bellucci A, Navarria L, Zaltieri M et al (2011) Induction of the unfolded protein response by α-synuclein in experimental models of Parkinson's disease. J Neurochem 116(4):588–605
Bence NF, Sampat RM, Kopito RR (2001) Impairment of the ubiquitin-proteasome system by protein aggregation. Science 292(5521):1552–1555
Bennett EJ, Shaler TA, Woodman B et al (2007) Global changes to the ubiquitin system in Huntington's disease. Nature 448(7154):704–708
Berezovska O, Lleo A, Herl LD, Frosch MP et al (2005) Familial Alzheimer's disease presenilin 1 mutations cause alterations in the conformation of presenilin and interactions with amyloid precursor protein. J Neurosci Off J Soc Neurosci 25(11):3009–3017
Bernard-Marissal N, Sunyach C, Marissal T, Raoul C et al (2015) Calreticulin levels determine onset of early muscle denervation by fast motoneurons of ALS model mice. Neurobiol Dis 73:130–136
Betarbet R, Sherer TB, MacKenzie G et al (2000) Chronic systemic pesticide exposure reproduces features of Parkinson's disease. Nat Neurosci 3(12):1301–1306

Bouman L, Schlierf A, Lutz AK et al (2011) Parkin is transcriptionally regulated by ATF4: evidence for an interconnection between mitochondrial stress and ER stress. Cell Death Differ 18(5):769–782

Boyce M et al (2005) A selective inhibitor of eIF2α dephosphorylation protects cells from ER stress. Sci 307(5711):935–939

Brandt R, Hundelt M, Shahani N (2005) Tau alteration and neuronal degeneration in tauopathies: mechanisms and models. Biochem Biophys Acta 1739(2–3):331–354

Brown AR, Rebus S, McKimmie CS et al (2005) Gene expression profiling of the preclinical scrapie-infected hippocampus. Biochem Biophys Res Commun 334(1):86–95

Brown RH (1998) SOD1 aggregates in ALS: cause, correlate or consequence? Nat Med 4(12):1362–1364

Carnemolla A, Fossale E, Agostoni E et al (2009) Rrs1 is involved in endoplasmic reticulum stress response in Huntington disease. J Biol Chem 284(27):18167–18173

Casas-Tinto S, Zhang Y, Sanchez-Garcia J et al (2011) The ER stress factor XBP1s prevents amyloid-beta neurotoxicity. Hum Mol Genet 20(11):2144–2160

Castillo-Carranza DL, Zhang Y, Guerrero-Munoz MJ et al (2012) Differential activation of the ER stress factor XBP1 by oligomeric assemblies. Neurochem Res 37(8):1707–1717

Caughey B, Lansbury PT (2003) Protofibrils, pores, fibrils, and neurodegeneration: separating the responsible protein aggregates from the innocent bystanders. Annu Rev Neurosci 26:267–298

Cavedo E, Lista S, Khachaturian Z et al (2014) The road ahead to cure Alzheimer's disease: development of biological markers and neuroimaging methods for prevention trials across all stages and target populations. J Prev Alzheimer's Dis 1(3):181–202

Cho KJ, Lee BI, Cheon SY et al (2009) Inhibition of apoptosis signal-regulating kinase 1 reduces endoplasmic reticulum stress and nuclear huntingtin fragments in a mouse model of Huntington disease. Neuroscience 163(4):1128–1134

Chung CY, Khurana V, Auluck PK et al (2013) Identification and rescue of α-synuclein toxicity in Parkinson patient–derived neurons. Sci 342(6161):983–987

Cisse M, Duplan E, Lorivel T et al (2016) The transcription factor XBP1s restores hippocampal synaptic plasticity and memory by control of the Kalirin-7 pathway in Alzheimer model. Mol Psychiatry

Colla E, Jensen PH, Pletnikova O, Troncoso JC et al (2012a) Accumulation of toxic α-synuclein oligomer within endoplasmic reticulum occurs in α-synucleinopathy in vivo. J Neurosci: Off J Soc Neurosci 32(10):3301–3305

Colla E, Coune P, Liu Y et al (2012b) Endoplasmic reticulum stress is important for the manifestations of α-synucleinopathy in vivo. J Neurosci: Off J Soc Neurosci 32(10):3306–3320

Conn KJ, Gao W, McKee A et al (2004) Identification of the protein disulfide isomerase family member PDIp in experimental Parkinson's disease and Lewy body pathology. Brain Res 1022(1–2):164–172

Cooper AA, Gitler AD, Cashikar A et al (2006) Alpha-synuclein blocks ER-Golgi traffic and Rab1 rescues neuron loss in Parkinson's models. Sci (New York, N.Y.) 313(5785):324–328

Cornejo VH, Hetz C (2013) Seminars in immunopathology. In: The unfolded protein response in Alzheimers disease, pp 277–292

Cornejo VH, Pihán P, Vidal RL et al (2013) Role of the unfolded protein response in organ physiology: lessons from mouse models. IUBMB Life 65(12):962–975

Costa-Mattioli M, Sossin WS, Klann E et al (2009) Translational control of long-lasting synaptic plasticity and memory. Neuron 61(1):10–26

Coune PG, Bensadoun J-C, Aebischer P et al (2011) Rab1A over-expression prevents Golgi apparatus fragmentation and partially corrects motor deficits in an alpha-synuclein based rat model of Parkinson's disease. J Parkinson's Dis 1(4):373–387

Credle JJ, Forcelli PA, Delannoy M et al (2015) α-Synuclein-mediated inhibition of ATF6 processing into COPII vesicles disrupts UPR signaling in Parkinson's disease. Neurobiol Dis 76:112–125

Crunkhorn S (2015) Neurodegenerative disease: phosphatase inhibitor prevents protein-misfolding diseases. Nat Rev Drug Discov 14(6):386

Culmsee C, Landshamer S (2006) Molecular insights into mechanisms of the cell death program: role in the progression of neurodegenerative disorders. Curr Alzheimer Res 3(4):269–283

Das I, Krzyzosiak A, Schneider K, Wrabetz L et al (2015) Preventing proteostasis diseases by selective inhibition of a phosphatase regulatory subunit. Science (New York, N.Y.) 348 (6231):239–242

Deitch JS, Alexander GM, Bensinger A et al (2014) Phenotype of transgenic mice carrying a very low copy number of the mutant human G93A superoxide dismutase-1 gene associated with amyotrophic lateral sclerosis. PLoS ONE 9(6):e99879

Di Domenico F, Head E, Butterfield DA et al (2014) Oxidative stress and proteostasis network: culprit and casualty of Alzheimers-like neurodegeneration. Adv Geriatrics

Dion PA, Daoud H, Rouleau GA (2009) Genetics of motor neuron disorders: new insights into pathogenic mechanisms. Nat Rev Genet 10(11):769–782

Dovey HF, John V, Anderson JP et al (2001) Functional gamma-secretase inhibitors reduce beta-amyloid peptide levels in brain. J Neurochem 76(1):173–181

Duennwald ML, Lindquist S (2008) Impaired ERAD and ER stress are early and specific events in polyglutamine toxicity. Genes Dev 22(23):3308–3319

Dufey E, Sepúlveda D, Rojas-Rivera D et al (2014) Cellular mechanisms of endoplasmic reticulum stress signaling in health and disease. 1. An overview. Am J Physiol Cell Physiol 307(7): C582–C594

Duran-Aniotz C, Martínez G, Hetz C (2014) Memory loss in Alzheimer's disease: are the alterations in the UPR network involved in the cognitive impairment? Front Aging Neurosci 6:8

Duran-Aniotz C et al (2017) IRE1 signaling exacerbates Alzheimer's disease pathogenesis. Acta Neuropathol 1–18

Duvoisin RC (1995) Recent advances in the genetics of Parkinson's disease. Adv Neurol 69:33–40

Egawa N, Yamamoto K, Inoue H et al (2011) The endoplasmic reticulum stress sensor, ATF6α, protects against neurotoxin-induced dopaminergic neuronal death. J Biol Chem 286(10): 7947–7957

Endres K, Reinhardt S (2013) ER-stress in Alzheimer's disease: turning the scale? Am J Neurodegener Dis 2(4):247–265

Fawcett EM, Hoyt JM, Johnson JK et al (2015) Hypoxia disrupts proteostasis in Caenorhabditis elegans. Aging Cell 14(1):92–101

Ferreiro E, Oliveira CR, Pereira CM (2008) The release of calcium from the endoplasmic reticulum induced by amyloid-beta and prion peptides activates the mitochondrial apoptotic pathway. Neurobiol Dis 30(3):331–342

Filézac de L'Etang A, Maharjan N, Braña C et al (2015) Marinesco-Sjögren syndrome protein SIL1 regulates motor neuron subtype-selective ER stress in ALS. Nat Neurosci 18(2):227–238

Fouillet A, Levet C, Virgone A, Robin M et al (2012) ER stress inhibits neuronal death by promoting autophagy. Autophagy 8(6):915–926

Freeman OJ, Mallucci GR (2016) The UPR and synaptic dysfunction in neurodegeneration. Brain Res

Gkogkas C, Middleton S, Kremer AM et al (2008) VAPB interacts with and modulates the activity of ATF6. Hum Mol Genet 17(11):1517–1526

Glenner GG, Wong CW (2012) Alzheimer's disease: initial report of the purification and characterization of a novel cerebrovascular amyloid protein 1984. Biochem Biophys Res Commun 425(3):534–539

Gonzalez-Perez P, Woehlbier U, Chian RJ et al (2015) Identification of rare protein disulfide isomerase gene variants in amyotrophic lateral sclerosis patients. Gene 566(2):158–165

Gorbatyuk MS, Shabashvili A, Chen W et al (2012) Glucose regulated protein 78 diminishes α-synuclein neurotoxicity in a rat model of Parkinson disease. Mol Ther 20(7):1327–1337

Gunawardena S, Goldstein LS (2005) Polyglutamine diseases and transport problems: deadly traffic jams on neuronal highways. Arch Neurol 62(1):46–51

Haass C (2004) Take five–BACE and the gamma-secretase quartet conduct Alzheimer's amyloid beta-peptide generation. EMBO J 23(3):483–488

Halliday M, Radford H, Sekine Y et al (2015) Partial restoration of protein synthesis rates by the small molecule ISRIB prevents neurodegeneration without pancreatic toxicity. Cell Death Dis 6:e1672

Hamos JE, Oblas B, Pulaski-Salo D et al (1991) Expression of heat shock proteins in Alzheimer's disease. Neurol 41(3):345

Hashida K, Kitao Y, Sudo H et al (2012) ATF6alpha promotes astroglial activation and neuronal survival in a chronic mouse model of Parkinsons disease. PLoS ONE 7(10):e47950

Henstridge CM, Pickett E, Spires-Jones TL (2016) Synaptic pathology: a shared mechanism in neurological disease. Ageing Res Rev 28:72–84

Herms JW, Korte S, Gall S et al (2000) Altered intracellular calcium homeostasis in cerebellar granule cells of prion protein-deficient mice. J Neurochem 75(4):1487–1492

Hetz C (2012) The unfolded protein response: controlling cell fate decisions under ER stress and beyond. Nat Rev Mol Cell Biol 13(2):89–102

Hetz CA, Soto C (2006) Stressing out the ER: a role of the unfolded protein response in prion-related disorders. Curr Mol Med 6(1):37–43

Hetz Flores C, Mollereau B (2014) Disturbance of endoplasmic reticulum proteostasis in neurodegenerative diseases. Nat Rev Neurosci 15(4):233

Hetz C, Russelakis-Carneiro M, Maundrell K, Castilla J et al (2003) Caspase-12 and endoplasmic reticulum stress mediate neurotoxicity of pathological prion protein. EMBO J 22(20): 5435–5445

Hetz C, Russelakis-Carneiro M, Wälchli S et al (2005a) The disulfide isomerase Grp58 is a protective factor against prion neurotoxicity. J Neurosci 25(11):2793–2802

Hetz C, Lee A-H, Gonzalez-Romero D, Thielen P et al (2008) Unfolded protein response transcription factor XBP-1 does not influence prion replication or pathogenesis. Proc Natl Acad Sci 105(2):757–762

Hetz C, Thielen P, Matus S et al (2009) XBP-1 deficiency in the nervous system protects against amyotrophic lateral sclerosis by increasing autophagy. Genes Dev 23(19):2294–2306

Hetz C, Chevet E, Harding HP (2013) Targeting the unfolded protein response in disease. Nat Rev Drug Discov 12(9):703–719

Hetz C, Russelakis-Carneiro M, Walchli S et al (2005) The disulfide isomerase Grp58 is a protective factor against prion neurotoxicity. J Neurosci 25:2793–2802

Hippius H, Neundörfer G (2003) The discovery of Alzheimer's disease. Dialogues Clin Neurosci 5 (1):101–108

Ho YS, Yang X, Lau JC et al (2012) Endoplasmic reticulum stress induces tau pathology and forms a vicious cycle: implication in Alzheimer's disease pathogenesis. J Alzheimer's Dis: JAD 28(4):839–854

Honjo Y, Ito H, Horibe T, Takahashi R et al (2010) Protein disulfide isomerase-immunopositive inclusions in patients with Alzheimer disease. Brain Res 1349:90–96

Honjo Y, Kaneko S, Ito H et al (2011) Protein disulfide isomerase-immunopositive inclusions in patients with amyotrophic lateral sclerosis. Amyotroph Later Scler Off Publ World Fed Neurol Res Gr Mot Neuron Dis 12(6):444–450

Hoozemans JJM, Veerhuis R, Van Haastert ES et al (2005) The unfolded protein response is activated in Alzheimers disease. Acta Neuropathol 110(2):165–172

Hoozemans JJM, Van Haastert ES, Eikelenboom P et al (2007) Activation of the unfolded protein response in Parkinsons disease. Biochem Biophys Res Commun 354(3):707–711

Hoozemans JJ, van Haastert ES, Nijholt DA et al (2009) The unfolded protein response is activated in pretangle neurons in Alzheimer's disease hippocampus. Am J Pathol 174(4): 1241–1251

Hoozemans JJ, van Haastert ES, Nijholt DA et al (2012) Activation of the unfolded protein response is an early event in Alzheimer's and Parkinson's disease. Neuro-Degener Dis 10(1–4):212–215

Hu B-R, Janelidze S, Ginsberg MD et al (2001) Protein aggregation after focal brain ischemia and reperfusion. J Cereb Blood Flow Metab 21(7):865–875

Ito Y, Yamada M, Tanaka H et al (2009) Involvement of CHOP, an ER-stress apoptotic mediator, in both human sporadic ALS and ALS model mice. Neurobiol Dis 36(3):470–476

Jankovic J, Aguilar LG (2008) Current approaches to the treatment of Parkinsons disease. Neuropsychiatry Dis Treat 4(4):743–757

Jiang Y, Chadwick SR, Lajoie P (2016) Endoplasmic reticulum stress: the cause and solution to Huntington's disease? Brain Res

Jiang HQ, Ren M, Jiang HZ et al (2014) Guanabenz delays the onset of disease symptoms, extends lifespan, improves motor performance and attenuates motor neuron loss in the SOD1 G93A mouse model of amyotrophic lateral sclerosis. Neuroscience 277:132–138

Jiang P, Gan M, Ebrahim AS et al (2010) ER stress response plays an important role in aggregation of α-synuclein. Mol Neurodegener 5:56

Kalathur RK, Giner-Lamia J, Machado S et al (2015) The unfolded protein response and its potential role in Huntington's disease elucidated by a systems biology approach. F1000Res 4:103

Katayama T, Imaizumi K, Manabe T et al (2004) Induction of neuronal death by ER stress in Alzheimer's disease. J Chem Neuroanat 28(1–2):67–78

Katayama T, Imaizumi K, Sato N et al (1999) Presenilin-1 mutations downregulate the signalling pathway of the unfolded-protein response. Nat Cell Biol 1(8):479–485

Kaushik S, Cuervo AM (2015) Proteostasis and aging. Nat Med 21(12):1406–1415

Kern A, Behl C (2009) The unsolved relationship of brain aging and late-onset Alzheimer disease. Biochem Biophys Acta 1790(10):1124–1132

Kikuchi H, Almer G, Yamashita S et al (2006) Spinal cord endoplasmic reticulum stress associated with a microsomal accumulation of mutant superoxide dismutase-1 in an ALS model. Proc Natl Acad Sci 103(15):6025–6030

Kim HJ, Raphael AR, LaDow ES et al (2014) Therapeutic modulation of eIF2α phosphorylation rescues TDP-43 toxicity in amyotrophic lateral sclerosis disease models. Nat Genet 46(2): 152–160

Kitao Y, Imai Y, Ozawa K et al (2007) Pael receptor induces death of dopaminergic neurons in the substantia nigra via endoplasmic reticulum stress and dopamine toxicity, which is enhanced under condition of parkin inactivation. Hum Mol Genet 16(1):50–60

Kouroku Y, Fujita E, Jimbo A et al (2002) Polyglutamine aggregates stimulate ER stress signals and caspase-12 activation. Hum Mol Genet 11(13):1505–1515

Kouroku Y, Fujita E, Tanida I et al (2007) ER stress (PERK/eIF2alpha phosphorylation) mediates the polyglutamine-induced LC3 conversion, an essential step for autophagy formation. Cell Death Differ 14(2):230–239

Kwok CT, Morris AG, Frampton J et al (2013) Association studies indicate that protein disulfide isomerase is a risk factor in amyotrophic lateral sclerosis. Free Radic Biol Med 58:81–86

Labbadia J, Morimoto RI (2014) Proteostasis and longevity: when does aging really begin? F1000Prime Rep 6:7

De Lau LM, Breteler MM (2006) Epidemiology of Parkinson's disease. Lancet Neurol 5(6): 525–535

Lautenschlaeger J, Prell T, Grosskreutz J (2012) Endoplasmic reticulum stress and the ER mitochondria calcium cycle in amyotrophic lateral sclerosis. Amyotroph Later Scler 13 (2):166–177

Ledesma MD, Galvan C, Hellias B et al (2002) Astrocytic but not neuronal increased expression and redistribution of parkin during unfolded protein stress. J Neurochem 83(6):1431–1440

Lee H-J, Patel S, Lee S-J (2005) Intravesicular localization and exocytosis of α-synuclein and its aggregates. J Neurosci 25(25):6016–6024

Lee DY, Lee KS, Lee HJ et al (2010a) Activation of PERK signaling attenuates Abeta-mediated ER stress. PLoS ONE 5(5):e10489

Lee JH, Won SM et al (2010b) Induction of the unfolded protein response and cell death pathway in Alzheimer's disease, but not in aged Tg2576 mice. Exp Mol Med 42(5):386–394

Lee H, Noh JY, Oh Y et al (2012) IRE1 plays an essential role in ER stress-mediated aggregation of mutant huntingtin via the inhibition of autophagy flux. Hum Mol Genet 21(1):101–114

Liu SY, Wang W, Cai ZY et al (2013) Polymorphism -116C/G of human X-box-binding protein 1 promoter is associated with risk of Alzheimer's disease. CNS Neurosci Ther 19(4):229–234

Loewen CA, Feany MB (2010) The unfolded protein response protects from tau neurotoxicity in vivo. PloS One 5(9)

Lourenco MV, Ferreira ST, De Felice FG (2015) Neuronal stress signaling and eIF2α phosphorylation as molecular links between Alzheimer's disease and diabetes. Prog Neurobiol 129:37–57

Luo Y, Bolon B, Kahn S et al (2001) Mice deficient in BACE1, the Alzheimer's beta-secretase, have normal phenotype and abolished beta-amyloid generation. Nat Neurosci 4(3):231–232

Ma T, Trinh MA, Wexler AJ et al (2013) Suppression of eIF2α kinases alleviates Alzheimer's disease-related plasticity and memory deficits. Nat Neurosci 16(9):1299–1305

Maekawa S, Leigh PN, King A et al (2009) TDP-43 is consistently co-localized with ubiquitinated inclusions in sporadic and Guam amyotrophic lateral sclerosis but not in familial amyotrophic lateral sclerosis with and without SOD1 mutations. Neuropathology 29(6):672–683

Maharjan N, Saxena S (2016) ER strikes again: proteostasis dysfunction in ALS. EMBO J 35 (8):798–800

Maly DJ, Papa FR (2014) Druggable sensors of the unfolded protein response. Nat Chem Biol 10 (11):892–901

Martínez G, Vidal RL, Mardones P et al (2016) Regulation of memory formation by the transcription factor XBP1. Cell Rep 14(6):1382–1394

Martinez-Vicente M, Talloczy Z, Wong E et al (2010) Cargo recognition failure is responsible for inefficient autophagy in Huntington's disease. Nat Neurosci 13(5):567–576

Masters CL, Multhaup G, Simms G et al (1985) Neuronal origin of a cerebral amyloid: neurofibrillary tangles of Alzheimer's disease contain the same protein as the amyloid of plaque cores and blood vessels. EMBO J 4(11):2757

Matus S, Castillo K, Hetz C (2012) Hormesis: protecting neurons against cellular stress in Parkinson disease. Autophagy 8(6):997–1001

Matus S, Lopez E, Valenzuela V, Nassif M et al (2013) Functional contribution of the transcription factor ATF4 to the pathogenesis of amyotrophic lateral sclerosis. PLoS ONE 8(7):e66672

Mays CE, Soto C (2016) The stress of prion disease. Brain Res

McGeer PL, McGeer EG (2013) The amyloid cascade-inflammatory hypothesis of Alzheimer disease: implications for therapy. Acta Neuropathol 126(4):479–497

Mercado G, Valdés P, Hetz C (2013) An ERcentric view of Parkinson's disease. Trends Mol Med 19(3):165–175

Mercado G, Castillo V, Soto P et al (2016) ER stress and Parkinson's disease: pathological inputs that converge into the secretory pathway. Brain Res 1648(Pt B):626–632

Mitsuda T, Hayakawa Y, Itoh M et al (2007) ATF4 regulates γ-secretase activity during amino acid imbalance. Biochem Biophys Res Commun 352(3):722–727

Moreno JA, Radford H, Peretti D et al (2012) Sustained translational repression by eIF2α-P mediates prion neurodegeneration. Nat 485(7399):507–511

Moreno JA, Halliday M, Molloy C et al (2013) Oral treatment targeting the unfolded protein response prevents neurodegeneration and clinical disease in prion-infected mice. Sci Transl Med 5(206):206ra138

Mukherjee A, Morales-Scheihing D, Gonzalez-Romero D et al (2010) Calcineurin inhibition at the clinical phase of prion disease reduces neurodegeneration, improves behavioral alterations and increases animal survival. PLoS Pathog 6(10):e1001138

Nagata T, Ilieva H, Murakami T et al (2007) Increased ER stress during motor neuron degeneration in a transgenic mouse model of amyotrophic lateral sclerosis. Neurol Res 29 (8):767–771

Nishimura AL, Mitne-Neto M, Silva HC, Richieri-Costa A et al (2004) A mutation in the vesicle-trafficking protein VAPB causes late-onset spinal muscular atrophy and amyotrophic lateral sclerosis. Am J Hum Genet 75(5):822–831

Nishitoh H, Matsuzawa A, Tobiume K et al (2002) ASK1 is essential for endoplasmic reticulum stress-induced neuronal cell death triggered by expanded polyglutamine repeats. Genes Dev 16(11):1345–1355

Nishitoh H, Kadowaki H, Nagai A et al (2008) ALS-linked mutant SOD1 induces ER stress-and ASK1-dependent motor neuron death by targeting Derlin-1. Genes Dev 22(11):1451–1464

Oakley H, Cole SL, Logan S, Maus E et al (2006) Intraneuronal beta-amyloid aggregates, neurodegeneration, and neuron loss in transgenic mice with five familial Alzheimer's disease mutations: potential factors in amyloid plaque formation. J Neurosci Off J Soc Neurosci 26(40):10129–10140

Oakes SA, Papa FR (2015) The role of endoplasmic reticulum stress in human pathology. Annu Rev Pathol 10:173–194

Orr HT, Zoghbi HY (2007) Trinucleotide repeat disorders. Annu Rev Neurosci 30:575–621

Page G, Rioux Bilan A, Ingrand S et al (2006) Activated double-stranded RNA-dependent protein kinase and neuronal death in models of Alzheimer's disease. Neurosci 139(4):1343–1354

Pasini S, Corona C, Liu J et al (2015) Specific downregulation of hippocampal ATF4 reveals a necessary role in synaptic plasticity and memory. Cell Rep 11(2):183–191

Powers ET, Balch WE (2013) Diversity in the origins of proteostasis networks–a driver for protein function in evolution. Nat Rev Mol Cell Biol 14(4):237–248

Prudencio M, Belzil VV, Batra R, Ross CA et al (2015) Distinct brain transcriptome profiles in C9orf72-associated and sporadic ALS. Nat Neurosci 18(8):1175–1182

Prusiner SB (1998) Prions. Proc Natl Acad Sci 95(23):13363–13383

Prusiner SB, Scott MR (1997) Genetics of prions. Annu Rev Genet 31:139–175

Reitz C (2012) Alzheimer's disease and the amyloid cascade hypothesis: a critical review. Int J Alzheimers Dis 2012

Rozas P, Bargsted L, Martínez F et al (2016) The ER proteostasis network in ALS: determining the differential motoneuron vulnerability. Neurosci Lett

Ross CA, Poirier MA (2005) What is the role of protein aggregation in neurodegeneration? Nat Rev Mol Cell Biol 6(11):891–898

Ryu EJ, Harding HP, Angelastro JM et al (2002) Endoplasmic reticulum stress and the unfolded protein response in cellular models of Parkinson's disease. J Neurosci 22(24):10690–10698

Sado M, Yamasaki Y, Iwanaga T et al (2009) Protective effect against Parkinson's disease-related insults through the activation of XBP1. Brain Res 1257:16–24

Salminen A, Kauppinen A, Suuronen T et al (2009) ER stress in Alzheimer's disease: a novel neuronal trigger for inflammation and Alzheimer's pathology. J Neuroinflamm 6:41

Sato N, Urano F, Yoon Leem J et al (2000) Upregulation of BiP and CHOP by the unfolded-protein response is independent of presenilin expression. Nat Cell Biol 2(12):863–870

Saxena S, Caroni P (2011) Selective neuronal vulnerability in neurodegenerative diseases: from stressor thresholds to degeneration. Neuron 71(1):35–48

Saxena S, Cabuy E, Caroni P (2009) A role for motoneuron subtype-selective ER stress in disease manifestations of FALS mice. Nat Neurosci 12(5):627–636

Sämann J, Hegermann J, von Gromoff E et al (2009) Caenorhabditis elegans LRK-1 and PINK-1 act antagonistically in stress response and neurite outgrowth. J Biol Chem 284(24):16482–16491

Scheff SW, Ansari MA, Mufson EJ (2016) Oxidative stress and hippocampal synaptic protein levels in elderly cognitively intact individuals with Alzheimer's disease pathology. Neurobiol Aging 42:1–12

Scheper W, Hoozemans JJ (2015) The unfolded protein response in neurodegenerative diseases: a neuropathological perspective. Acta Neuropathol 130(3):315–331

Si L, Xu T, Wang F, Liu Q (2012) X-box-binding protein 1-modified neural stem cells for treatment of Parkinson's disease. Neural Regener Res 7(10):736–740

Silva RM, Ries V, Oo TF, Yarygina O et al (2005) CHOP/GADD153 is a mediator of apoptotic death in substantia nigra dopamine neurons in an in vivo neurotoxin model of parkinsonism. J Neurochem 95(4):974–986

Slodzinski H, Moran LB, Michael GJ et al (2009) Homocysteine-induced endoplasmic reticulum protein (herp) is up-regulated in parkinsonian substantia nigra and present in the core of Lewy bodies. Clin Neuropathol 28(5):333–343

Smith HL, Mallucci GR (2016) The unfolded protein response: mechanisms and therapy of neurodegeneration. Brain J Neurol 139(Pt 8):2113–2121

Soto C (2003) Unfolding the role of protein misfolding in neurodegenerative diseases. Nat Rev Neurosci 4(1):49–60

Sreedharan J, Brown RH (2013) Amyotrophic lateral sclerosis: problems and prospects. Ann Neurol 74(3):309–316

Steele AD, Hetz C, Yi CH et al (2007) Prion pathogenesis is independent of caspase-12. Prion 1 (4):243–247

De Strooper B, Karran E (2016) The cellular phase of Alzheimers disease. Cell 164(4):603–615

Sun S, Sun Y, Ling SC et al (2015) Translational profiling identifies a cascade of damage initiated in motor neurons and spreading to glia in mutant SOD1-mediated ALS. Proc Natl Acad Sci USA 112(50):E6993–E7002

Suzuki H, Matsuoka M (2012) TDP-43 toxicity is mediated by the unfolded protein response-unrelated induction of C/EBP homologous protein expression. J Neurosci Res 90 (3):641–647

Suzuki H, Kanekura K, Levine TP et al (2009) ALS-linked P56S-VAPB, an aggregated loss-of-function mutant of VAPB, predisposes motor neurons to ER stress-related death by inducing aggregation of co-expressed wild-type VAPB. J Neurochem 108(4):973–985

Tobisawa S, Hozumi Y, Arawaka S et al (2003) Mutant SOD1 linked to familial amyotrophic lateral sclerosis, but not wild-type SOD1, induces ER stress in COS7 cells and transgenic mice. Biochem Biophys Res Commun 303(2):496–503

Tompkins MM, Hill WD (1997) Contribution of somal Lewy bodies to neuronal death. Brain Res 775(1–2):24–29

Torres M, Castillo K, Armisén R et al (2010) Prion protein misfolding affects calcium homeostasis and sensitizes cells to endoplasmic reticulum stress. PLoS ONE 5(12):e15658

Torres M, Medinas DB, Matamala JM et al (2015) The protein-disulfide isomerase ERp57 regulates the steady-state levels of the prion protein. J Biol Chem 290(39):23631–23645

Tsaytler P et al (2011) Selective inhibition of a regulatory subunit of protein phosphatase 1 restores proteostasis. Sci 332(6025):91–94

Unterberger U, Höftberger R, Gelpi E et al (2006) Endoplasmic reticulum stress features are prominent in Alzheimer disease but not in prion diseases in vivo. J Neuropathol Exp Neurol 65 (4):348–357

Urano F, Wang X, Bertolotti A et al (2000) Coupling of stress in the ER to activation of JNK protein kinases by transmembrane protein kinase IRE1. Sci 287(5453):664–666

Valdés P, Mercado G, Vidal RL et al (2014) Control of dopaminergic neuron survival by the unfolded protein response transcription factor XBP1. Proc Natl Acad Sci USA 111(18):6804–6809

Valenzuela V, Martínez G, Duran-Aniotz C et al (2016) Gene therapy to target ER stress in brain diseases. Brain Res

van der Harg JM, Nölle A, Zwart R et al (2014) The unfolded protein response mediates reversible tau phosphorylation induced by metabolic stress. Cell Death Dis 5:e1393

Varma D, Sen D (2015) Role of the unfolded protein response in the pathogenesis of Parkinson's disease. Acta Neurobiol Exp 75(1):1–26

Vassar R (2009) Phosphorylation of the translation initiation factor eIF2alpha increases BACE1 levels and promotes amyloidogenesis. Alzheimer's Dementia 5(4):P81–P82

Vidal R, Caballero B, Couve A, Hetz C (2011) Converging pathways in the occurrence of endoplasmic reticulum (ER) stress in Huntington's disease. Curr Mol Med 11(1):1–12

Vidal RL, Figueroa A, Court FA et al (2012) Targeting the UPR transcription factor XBP1 protects against Huntington's disease through the regulation of FoxO1 and autophagy. Hum Mol Genet 21(10):2245–2262

Vieira FG, Ping Q, Moreno AJ et al (2015) Guanabenz treatment accelerates disease in a mutant SOD1 mouse model of ALS. PLoS ONE 10(8):e0135570

Vitte J, Traver S, De Paula M et al (2010) Leucine-rich repeat kinase 2 is associated with the endoplasmic reticulum in dopaminergic neurons and accumulates in the core of Lewy bodies in Parkinson disease. J Neuropathol Exp Neurol 69(9):959–972

Vossel KA, Zhang K, Brodbeck J et al (2010) Tau reduction prevents Abeta-induced defects in axonal transport. Sci (New York, N.Y.) 330(6001):198

Walker FO (2007) Huntington's disease. Lancet 369(9557):218–228

Walker L, McAleese KE, Thomas AJ et al (2015) Neuropathologically mixed Alzheimer's and Lewy body disease: burden of pathological protein aggregates differs between clinical phenotypes. Acta Neuropathol 129(5):729–748

Walter P, Ron D (2011) The unfolded protein response: from stress pathway to homeostatic regulation. Sci 334(6059):1081–1086

Wang M, Kaufman RJ (2016) Protein misfolding in the endoplasmic reticulum as a conduit to human disease. Nat 529(7586):326–335

Wang L, Popko B, Roos RP (2011) The unfolded protein response in familial amyotrophic lateral sclerosis. Hum Mol Genet 20(5):1008–1015

Wang M, Ye R, Barron E et al (2010) Essential role of the unfolded protein response regulator GRP78/BiP in protection from neuronal apoptosis. Cell Death Differ 17(3):488–498

Wang L, Popko B, Tixier E et al (2014) Guanabenz, which enhances the unfolded protein response, ameliorates mutant SOD1-induced amyotrophic lateral sclerosis. Neurobiol Dis 71:317–324

Wate R, Ito H, Zhang JH et al (2005) Expression of an endoplasmic reticulum-resident chaperone, glucose-regulated stress protein 78, in the spinal cord of a mouse model of amyotrophic lateral sclerosis. Acta Neuropathol 110(6):557–562

Woehlbier U, Colombo A, Saaranen MJ et al (2016) ALS-linked protein disulfide isomerase variants cause motor dysfunction. EMBO J 35(8):845–865

Yoo BC, Krapfenbauer K, Cairns N et al (2002) Overexpressed protein disulfide isomerase in brains of patients with sporadic Creutzfeldt-Jakob disease. Neurosci Lett 334(3):196–200

Yoon SO, Park DJ, Ryu JC et al (2012) JNK3 perpetuates metabolic stress induced by Aβ peptides. Neuron 75(5):824–837

Yuan Y, Cao P, Smith MA et al (2011) Dysregulated LRRK2 signaling in response to endoplasmic reticulum stress leads to dopaminergic neuron degeneration in *C. elegans*. PLoS ONE 6(8): e22354

Zhang YJ, Jansen-West K, Xu YF et al (2014) Aggregation-prone c9FTD/ALS poly(GA) RAN-translated proteins cause neurotoxicity by inducing ER stress. Acta Neuropathol 128 (4):505

Zuleta A, Vidal RL, Armentano D et al (2012) AAV-mediated delivery of the transcription factor XBP1s into the striatum reduces mutant Huntingtin aggregation in a mouse model of Huntingtons disease. Biochem Biophys Res Commun 420(3):558–563

Driving Cancer Tumorigenesis and Metastasis Through UPR Signaling

Alexandra Papaioannou and Eric Chevet

Abstract In the tumor microenvironment, cancer cells encounter both external and internal factors that can lead to the accumulation of improperly folded proteins in the Endoplasmic Reticulum (ER) lumen, thus causing ER stress. When this happens, an adaptive mechanism named the Unfolded Protein Response (UPR) is triggered to help the cell cope with this change and restore protein homeostasis in the ER. Sequentially, one would expect that the activation of the three UPR branches, driven namely by IRE1, PERK, and ATF6, are crucial for the adaptation of cancer cells to the changing environment and thus for their survival and further propagation. Indeed, in the last few years, an increasing amount of studies has shown the implication of UPR signaling in different aspects of carcinogenesis and tumor progression. Features such as sustaining proliferation and resistance to cell death, genomic instability, altered metabolism, increased inflammation and tumor-immune infiltration, invasion and metastasis, and angiogenesis, defined as "the hallmarks of cancer", can be regulated by the UPR machinery. At the same time, new potential therapeutic interventions applicable to different kinds of cancers are being revealed. In order to describe the emerging role of UPR in cancer biology, these are the points that will be discussed in this chapter.

List of Abbreviations

ADAM17	ADAM metallopeptidase domain 17
AKT	Protein kinase B (PKB)
ALL	Acute lymphoblastic leukemia
APC	Antigen-presenting cells
APY29	N2-1H-Benzimidazol-6-yl-N4-(5-cyclopropyl-1H-pyrazol-3-yl)-2,4-pyrimidinediamine

---

A. Papaioannou · E. Chevet (✉)
Inserm U1242 «Chemistry, Oncogenesis, Stress and Signaling»,
University of Rennes 1, Rennes, France
e-mail: eric.chevet@inserm.fr

A. Papaioannou
Centre de Lutte contre le Cancer Eugène Marquis,
Avenue de la bataille Flandres Dunkerque, 35000 Rennes, France

ASK1	Apoptosis signal-regulating kinase 1
ATF3	Activating transcription factor 3
ATF4	Activating transcription factor 4
ATF6	Activating transcription factor 6
ATF6f	ATF6 cytoplasmic domain
ATG5	Autophagy protein 5
ATM	Ataxia-telangiectasia mutated (Ser/Thr protein kinase)
BCR	Breakpoint cluster region
BiP	Binding immunoglobin protein or GRP78
CAC	Colitis-associated cancer
CAF	Cancer-associated fibroblasts
CCL2	Chemokine (C-C motif) ligand 2 or MCP-1
CD8/28/40/80/86	Cluster of differentiation 8/28/40/80/86
CHOP	C/EBP-homologous protein
CNS	Central nervous system
CreP	Protein phosphatase 1 regulatory subunit 15B (PPP1R15B)
CRT	Calreticulin
CXCL3/8/10/14	C-X-C motif chemokine ligand 3/8/10/14
DC	Dendritic cells
E2F	Family of transcription factors (TF) in higher eukaryotes
EC	Endothelial cells
ECM	Extracellular matrix
EGF	Epidermal growth factor
EGFR	Epidermal growth factor receptor
EGR-1	Early growth response protein 1
eIF2B	Guanine nucleotide exchange factor for the eukaryotic initiation factor 2
eIF2α	Eukaryotic initiation factor 2α
EMT	Epithelial-to-mesenchymal transition
EPR	Epiregulin
ER	Endoplasmic reticulum
ER$^+$	Estrogen receptor positive
ErbB1	Synonym of EGFR
ERO1α	Endoplasmic reticulum oxidoreductase 1 alpha
FAK	Focal adhesion kinase
FGF	Fibroblast growth factor
FGF2	Fibroblast growth factor 2
FOXO	Forkhead box O
FOXP3	Forkhead box P3
GADD34	Growth arrest and DNA damage-inducible protein GADD34/Protein phosphatase 1 regulatory subunit 15A (PP1R15A)
GBM	Glioblastoma multiforme
GLUT1	Glucose transporter 1

gp96/GRP94	Glycoprotein 96/94 kDa glucose-regulated protein
GRP78	78 kDa glucose-regulated protein
HDAC	Histone deacetylase
HIF1α	Hypoxia-inducible factor 1-alpha
IL1/1β/2/6/8/10/17/23	Interleukin 1/1β/2/6/8/10/17/23
IRE1α	Endoribonuclease inositol-requiring enzyme 1 alpha
ISC	Intestinal stem cells
ISRIB	Integrated stress response inhibitor
JNK	c-Jun N-terminal kinase
Keap1	Kelch-like ECH-associated protein 1
KIRA	Kinase-inhibiting RNase attenuators
LAMP3	Lysosomal-associated membrane protein 3
LC3B	Microtubule-associated proteins 1A/1B-light chain 3B
MAF	MAF BZIP Transcription Factor
MAFB	MAF BZIP transcription factor B
MCP-1	Monocyte chemoattractant protein-1 or CCL2
MDSC	Myeloid-derived suppressor cells
MHC I/II	Major histocompatibility complex I/II
MIP-1α/MIP-1β	Macrophage inflammatory protein-1α/β
miRNA	Micro RNA
MMP	Matrix metalloproteinase
mRNA	Messenger RNA
mTOR	Mechanistic target of rapamycin
mTORC1	Mammalian target of rapamycin complex 1
MΦ	Macrophages
NFκB p65/RelA	Nuclear factor kappa-light-chain-enhancer of activated B cells p65 subunit
NRF2	Nuclear factor (erythroid-derived 2)-like 2
p53	Tumor protein p53
p97/VCP	Valosin-containing protein
PDI	Protein disulfide isomerase
PDIA5	Protein disulfide isomerase family A member 5
PERK	Protein kinase RNA-like endoplasmic reticulum kinase
PLCγ	Phospholipase C gamma
PP1c	Protein phosphatase 1 catalytic subunit
PTM	Post-translational modification
RhoA	Ras homolog gene family, member A
ROS	Reactive oxygen species
rRNA	Ribosomal RNA
RtcB	TRNA-splicing ligase RtcB homolog/C22orf28/HSPC117
RUVBL2	RuvB-like 2
S1P	Site-1 protease
S1PR1	Sphingosine-1-phosphate receptor 1
S2P	Site-2 protease

SCCA1	Squamous cell carcinoma antigen 1
SNAIL	Zinc finger protein SNAI1
SPARC	Secreted protein acidic and rich in cysteine
SRC	Tyrosine-protein kinase Src
STAT3/6	Signal transducer and activator of transcription-3/-6
TAC	Transit-amplifying cells
TAM	Tumor-associated macrophages
TDAG51	T-cell death-associated gene 51 protein
Tg	Thapsigargin
TGF-β	Transforming growth factor beta
TGF-β1	Transforming growth factor beta 1
THBS1	Thrombospondin 1
TNBC	Triple negative breast cancer
TNFα	Tumor necrosis factor alpha
TRAF2	TNF receptor-associated factor 2
Tun	Tunicamycin
Twist	Twist-related protein 1 (TWIST1)
UPR	Unfolded protein response
VEGF-A	Vascular endothelial growth factor A
VEGF	Vascular endothelial growth factor
VEGFR2	Vascular endothelial growth factor receptor 2
WNT	Wingless-related integration site
WNT11	WNT family member 11
XBP1	X-box binding protein 1
XBP1s	Spliced form (active) of X-box binding protein 1

Contents

1	ER Stress Signaling and Its Alterations in Cancer	163
2	Regulation of the Hallmarks of Cancer by the UPR	165
	2.1 Sustaining Cancer Cell Proliferation and Promoting Tumor Dormancy	165
	2.2 Cellular Energetics	168
	2.3 Tumoral UPR Controls the Immune System	169
	2.4 Driving Invasion and Metastasis	171
	2.5 Inducing Angiogenesis	173
	2.6 Genome Instability—Mutations of the UPR Machinery	175
3	Targeting UPR in Cancer	177
4	Conclusion	179
References		179

1 ER Stress Signaling and Its Alterations in Cancer

The endoplasmic reticulum (ER) is a cellular organelle responsible for the synthesis and proper folding of secretory and transmembrane proteins in the eukaryotic cell. Different cellular stresses can occur either in physiological or pathological conditions, during which the demand for protein production (synthesis and folding) can overwhelm cell capacity (Kozutsumi et al. 1988). When this demand outweighs the functional capacity of the ER, improperly folded proteins accumulate in the ER lumen that in turn impact on amino acid, lipid, and sugar metabolism. However, cells have evolved an adaptive mechanism that helps them respond efficiently to ER stress and that is named the Unfolded Protein Response (UPR) (Cox and Walter 1996). The UPR is orchestrated by three distinct ER-resident transmembrane proteins, PERK (protein kinase RNA-like endoplasmic reticulum kinase) (Shi et al. 1998), IRE1α (endoribonuclease inositol-requiring enzyme 1 alpha; referred to as IRE1 hereafter) (Mori et al. 1993), and ATF6 (activating transcription factor 6) (Haze et al. 1999) (Fig. 1). Under non-ER stress conditions, binding immunoglobin protein (BiP or GRP78) binds to all of them thus keeping them in an inactive state, while upon ER stress BiP dissociates from these sensors and binds preferably to the misfolded proteins present in the ER lumen. This results in the oligomerization and trans-autophosphorylation of PERK and IRE1 (Bertolotti et al. 2000), and the revelation of an ER export motif in ATF6 (Shen et al. 2002). Sequentially, PERK phosphorylates eIF2α in turn inhibiting global translation (Ron and Walter 2007), but activating translation of ATF4 (Blais et al. 2004). ATF4 then increases, among others, CHOP expression (Zinszner et al. 1998), which then activates GADD34 and PP1c leading to the dephosphorylation of eIF2α (Novoa et al. 2001), thus creating a negative feedback loop for the latter. Active PERK also phosphorylates NRF2 controlling in this way the antioxidant response pathway (Del Vecchio et al. 2014; Cullinan et al. 2003; Cullinan and Diehl 2004). IRE1 activation is followed by three distinct signaling cascades. It phosphorylates and activates TRAF2, through its kinase domain, leading in turn to ASK1-JNK activation (Urano et al. 2000). In addition, through its RNase domain IRE1 leads to the splicing of *XBP1* mRNA (Calfon et al. 2002; Lee et al. 2002) leading to the generation of the transcription factor XBP1s that stimulates the expression of UPR target genes. The ligation step after the 26nt intron excision (in human) by IRE1 RNase is mediated by the tRNA ligase RtcB (Jurkin et al. 2014; Kosmaczewski et al. 2014; Lu et al. 2014; Ray et al. 2014). The endoribonuclease activity of IRE1 is also responsible for the degradation of certain mRNAs (Hollien et al. 2009; Hollien and Weissman 2006), rRNAs (Iwawaki et al. 2001), and miRNAs (Lerner et al. 2012; Upton et al. 2012) named as Regulated IRE1-Dependent Decay (RIDD) (Maurel et al. 2014). Each of these cascades has been documented to rely partly on the degree of IRE1 oligomerization (Tam et al. 2014; Bouchecareilh et al. 2011). Finally, ATF6 translocates to the Golgi apparatus where its cytoplasmic tail is cleaved by S1P and S2P (Shen and Prywes 2004; Ye et al. 2000) forming ATF6f which then moves to the nucleus and acts as a transcription factor guiding the expression of ER stress genes (Yamamoto et al. 2007; Yoshida et al. 2000; Adachi et al. 2008). ATF6f has also been reported to form heterodimers with XBP1s, thus resulting in different expression patterns than each transcription factor does individually (Shoulders et al. 2013). Overall, the three UPR signaling pathways converge to the upregulation of genes involved in the clearance of the protein overload in the ER lumen and the survival of the cell. However, in cases where the ER stress cannot be cleared due to its high intensity and/or duration the UPR leads to cellular death by favoring the expression of genes involved in apoptotic pathways.

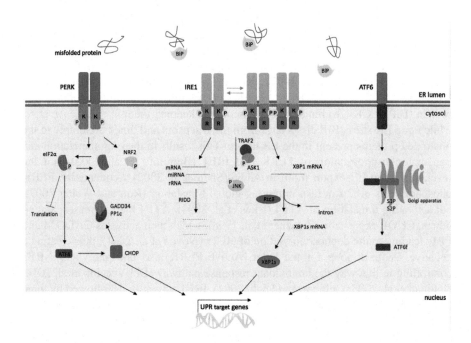

Fig. 1 The Unfolded Protein Response machinery. It consists of three ER-transmembrane proteins: PERK, IRE1α, and ATF6. Under ER stress conditions BiP dissociates from the three sensors and binds to the unfolded proteins accumulated in the ER lumen, thus activating the three ER stress sensors. PERK then phosphorylates eIF2α halting global translation, but activating translation of ATF4. ATF4 then increases, among others, CHOP expression, in turn activating GADD34 and PP1c leading to the dephosphorylation of eIF2α. Active PERK also phosphorylates NRF2 of the antioxidant response pathway. IRE1 activation results in degradation (RIDD) of various RNA molecules, including mRNAs, rRNAs, and miRNAs, phosphorylation and subsequent activation of the ASK1-JNK signaling cascade, and splicing of the *XBP1* mRNA, mediated by the RtcB ligase, generating the transcription factor XBP1s, with each one of these outcomes be dependent on the IRE1 oligomerization status. ATF6 upon its activation translocates to the Golgi apparatus where it is cleaved by S1P and S2P forming the transcription factor ATF6f that eventually moves to the nucleus. In this way, the three UPR arms converge to the upregulation of genes involved either in the clearance of the protein overload in the ER lumen and the survival of the cell or in cell death when the ER stress is so intense in duration and magnitude that cannot be relieved. The signaling pathways are described from left to right

In the tumor microenvironment, various factors such as hypoxia, nutrient deprivation, acidosis, and reactive oxygen species (ROS) production can lead to the accumulation of improperly folded proteins in the ER, therefore causing ER stress and activating the UPR that acts as an adaptive-to-stress mechanism and thus preserves protein homeostasis. In addition to these, other factors coming from the cells per se such as oncogenic activation, genomic instability and imbalance, aneuploidy, and increased mutation rate can cause ER stress and subsequently trigger the activation of the adaptive mechanisms of the UPR machinery to cope with the stress (Urra et al. 2016). The occurrence of ER stress in the tumor environment is evident by the upregulation of UPR markers in different types of tumor. In addition to this, UPR activation in the cells that are part of the tumor surroundings may help toward its faster development.

2 Regulation of the Hallmarks of Cancer by the UPR

2.1 Sustaining Cancer Cell Proliferation and Promoting Tumor Dormancy

Even though UPR activation is perceived as an adaptive mechanism for cancer cells to survive and propagate, it may differ from phase to phase during cancer progression. For instance, in two different models of oncogene-induced malignancies, a Ras-transformed melanoma (Denoyelle et al. 2006) and a Ret-induced fibroblast transformation (Huber et al. 2013), increased activation of the UPR was found to act as a barrier to the initial malignant cell transformation. In contrast to these, in a human squamous carcinoma model UPR has been found to promote tumor cell dormancy (Schewe and Aguirre-Ghiso 2008), thus revealing the dependency of UPR action on the distinct cancerous context and the tumor stage. The UPR is substantially involved in cancers of the gastrointestinal tract, as expected by its crucial role in the physiologic functions of this tract. Activation of only the PERK/eIF2α pathway is enough to trigger the loss of stemness in intestinal stem cells (ISC) (Heijmans et al. 2013), thus regulating differentiation and suggesting a role for this UPR branch in cancer initiation. In oxidative stress conditions, PERK elicits its prosurvival signals, as it leads to the translocation of NRF2 from its cytoplasmic complex with Keap1 to the nucleus, facilitating the expression of genes involved in detoxification and elimination of ROS (Cullinan et al. 2003; Nguyen et al. 2009). However, it has been reported that hypoxic tumors with PERK deficiency have a smaller size and the animal survival is increased, attributed to eIF2α phosphorylation and protein translation control (Bi et al. 2005). In addition to this, in a model of human pancreatic tumor xenograft, inhibition of PERK was shown to negatively affect the mice tumor growth (Axten et al. 2012; Atkins et al. 2013). In different cancers PERK plays its role through controlling additional pathways, including the regulation of redox status of the cell (Cullinan et al. 2003;

Del Vecchio et al. 2014), metabolism (Zhang et al. 2013), and lipid biosynthesis [reviewed in (Pytel et al. 2016)]. Moreover, downstream of PERK during ER stress conditions ATF3 is activated, in turn downregulating early growth response protein 1 (EGR-1) expression, a transcription factor of pro-tumorigenic chemokines, impacting on tumor survival (Park et al. 2014). On the other hand, it has been shown in glioblastoma multiforme (GBM) that PERK is involved in ceramide production, critical for Ca^{2+} induction and ROS formation, ultimately stimulating cell autophagy and death (Yacoub et al. 2010).

A second UPR branch, IRE1 signaling pathway has also been implicated in cancer development. A high XBP1s expression is associated with poor patient prognosis in glioma (Pluquet et al. 2013), triple negative breast cancer (TNBC) (Chen et al. 2014), and pre-B acute lymphoblastic leukemia (ALL) (Kharabi Masouleh et al. 2014). In a GBM model IRE1 was found crucial for tumor growth, apart from invasion and angiogenesis (Auf et al. 2010; Dejeans et al. 2012). Indeed, in GBM cells IRE1-mediated JNK activation induces epiregulin (EPR), a ligand for the EGFR receptor ErbB1, thus taking part in an autocrine proliferation loop through ErbB1 (Auf et al. 2013). In addition to this, XBP1s overactivation has been reported to contribute to the generation of multiple myeloma, observed upon XBP1s ectopic overexpression in mice (Carrasco et al. 2007). In this context, malignant transformation is accompanied by a gene expression signature comprising cyclin D1 and D2, MAF and MAFB, and IL6-dependent signaling, which is directly linked to cancer progression. In line with this, IRE1 has been found to regulate cyclin A1 expression in an XBP1s-dependent manner, thus influencing cell cycle progression and proliferation (Thorpe and Schwarze 2010). In the setting of multiple myeloma, inhibition of IRE1 endoribonuclease activity compensates for the pathogenic role of IRE1 and XBP1s (Mimura et al. 2012; Papandreou et al. 2011; Ri et al. 2012). Initially, in vitro studies showed that the proteasome inhibitor bortezomib might inhibit *XBP1* mRNA splicing, putting a halt in its prosurvival functions (Lee et al. 2003). Later on, several compounds that target and block specifically *XBP1* mRNA splicing displayed important antitumor effects in preclinical models of multiple myeloma [reviewed in (Hetz et al. 2013)]. Moreover, pharmacological inhibition of IRE1 resulted in efficient killing of pre-B lymphoblastic leukemia cells (Kharabi Masouleh et al. 2014; Tang et al. 2014). Loss of XBP1 in these cells negatively affected their survival, by limiting B-cell receptor (BCR) signaling capability and increasing of sphingosine-1-phosphate receptor 1 (S1PR1) on the cell surface. In breast cancer, JNK activation downstream of IRE1 leads to increased cell growth due to elevated cell division attributed to the expression of E2F and cyclin D proteins (Shen et al. 2008). Moreover, XBP1s may regulate the survival of ER^+ breast cancer cells controlling the NFκB p65/RelA expression (Hu et al. 2014). In the same setting, overexpression of XBP1s, by modulating apoptotic and cell cycle genes, can cause anti-estrogen resistance (Gomez et al. 2007), but also estrogen-induced tumor growth (Sengupta et al. 2010). In TNBC, *XBP1* mRNA splicing was observed to promote cell malignant propagation, unexpectedly mediated by the formation of a transcriptional complex with HIF1α (Chen et al. 2014). The spectrum of the genes regulated by XBP1s included, apart from the central proangiogenic factor, vascular endothelial growth factor (VEGF), genes related to cell proliferation, differentiation, and survival, as well as cytoskeletal remodeling. In TNBC patients, this gene expression signature was positively correlated with HIF1α expression and poor prognosis (Chen et al. 2014). Finally, a link between inflammatory bowel disease and tumorigenesis was illustrated upon XBP1 deficiency in epithelial cells that led to elevated levels of colitis-associated cancer (Niederreiter et al. 2013). What accounted for this was an increase in transit-amplifying cells (TAC) and intestinal stem cells (ISC) proliferation along with an improper function in Paneth cells. Of note, the hyperproliferation of ISC was correlated with increased WNT11 expression in Paneth cells, as well as with activation of an ER stress-dependent interleukin/STAT3 pathway. IRE1 was therefore presented as a mediator of ER stress-induced ISC expansion, possibly through its RIDD activity (Niederreiter et al. 2013). Taking a step further to the crosstalk of different cell types within the tumor microenvironment, it has to be noted that tumor-associated macrophages (TAM), resembling M2-like macrophages, facilitate tumor growth and survival, and generally reshape the tumor surroundings, in part through UPR activation [reviewed in (Allavena and Mantovani 2012; Galdiero et al. 2013; Hambardzumyan et al. 2016)]. For instance, EGF (epidermal growth factor) and other growth factors including members of the FGF (fibroblast growth factor) family secreted from TAM can directly lead to growth of the malignant cell population (Allavena et al. 2008). In addition to TAM, cancer-associated fibroblasts (CAF) contribute to tumor development. Indeed, it has been reported that cancer cells can induce oxidative stress in the adjacent CAF, which could then propagate in other cells of the tumor area (Martinez-Outschoorn et al. 2010). Despite the absence of a direct UPR implication in this report, the occurrence of an ER stress response under hypoxic conditions could be causal for the crosstalk between tumor cells and CAF, ultimately promoting the malignant progression of the cancer. The ATF6 arm was also revealed to be important for glioma development according to a genome-wide CRISPR-Cas9 screen (Toledo et al. 2015). In agreement with this, the transcription factor ATF6f activates NFκB and mTOR causing in turn the dephosphorylation of AKT, three factors that are crucial for cancer cell survival and proliferation (Nakajima et al. 2011).

Another feature that cancer cells can acquire and become able to adapt to the changing tumor environment is cell dormancy. During this process, the cells are arrested in the G0/G1 phase of the cell cycle, thus entering into a quiescent state (Aguirre-Ghiso 2007). In this way, the cells can avoid the detrimental environmental conditions and reactivate (start proliferating again and be metabolically active) when optimal conditions are reobtained. Dormant cells can be observed during the early stages of tumor growth, in distant micrometastases as well as during treatments (chemotherapy or radiotherapy) (Paez et al. 2012). In the last case, cell dormancy is the main cause for disease recurrence. Lately, a link between UPR and cell dormancy has been reported, mainly evidenced by ATF6. Indeed, ATF6 expression is increased in recurrent tumors (Ginos et al. 2004), correlating with poor prognosis of colon cancer (Lin et al. 2007), and it is found expressed mostly in metastatic lesions compared to the respective primary tumor (Ramaswamy et al. 2001). Additionally, in a human squamous carcinoma model ATF6 was constitutively active in dormant cells, whereas its silencing in this model reduced cell survival and tumor growth (Schewe and Aguirre-Ghiso 2008). In glioblastoma, ATF6 was also shown to play a role in terms of tumor growth and resistance to radiotherapy, through upregulating GRP78 and Notch-1 proteins (Dadey et al. 2016). In general, ATF6 can modulate the expression of certain proteins related to hepatocarcinogenesis (Arai et al. 2006) and increased chemo-resistance of cancer cells (Higa et al. 2014). Regarding the rest of the UPR constituents, cyclin A1 and D1 are respectively controlled by IRE1 (Thorpe and Schwarze 2010) and PERK (Brewer and Diehl 2000). Due to this negative regulation, PERK induces cell cycle arrest in G1, probably related to cancer cell dormancy (Brewer and Diehl 2000). Finally, PERK activation and eIF2α phosphorylation are involved in drug resistance of dormant carcinoma cells (Ranganathan et al. 2006).

2.2 Cellular Energetics

Cancer cells require nutrients in a greater degree than the tumor microenvironment is able to provide, due to inadequate vascularization, therefore hypoxia and nutrient limitation are generated. However, malignant cells are able to rewire their metabolism by inducing adaptive responses including the UPR and thus are able to support sustained growth (Ma and Hendershot 2004; DeBerardinis et al. 2008). This is also illustrated by the effectiveness of proteostasis modulators against tumor cells, where the adaptive capacity of the UPR is exceeded (Hetz et al. 2013). In addition, it is well-established that the three UPR branches modulate protein synthesis, globally through eIF2α (Harding et al. 1999), and control the expression of ER proteins involved in folding or degradation (Walter and Ron 2011). Beyond adjusting proteostasis, the UPR can increase angiogenesis and activate biosynthetic pathways, thus restoring the metabolic wiring in the tumor. A well-documented example is the VEGF upregulation by the PERK-ATF4 pathway that stimulates angiogenesis (Blais et al. 2006). This can extend from tumor cells to tumor-associated macrophages and even dendritic cells that along with other secreted factors signal to endothelial cells, ultimately promoting angiogenesis (Galdiero et al. 2013; Allavena et al. 2008; Mahadevan et al. 2011, 2012) (for more details see Sect. 2.5). The biosynthetic pathways that are regulated by the UPR can be either anabolic or catabolic ones. For instance, IRE1 through its RIDD activity leads to the degradation of certain mRNAs. Additionally, XBP1s is part of a cellular response that eventually leads to UDP-galactose-4-epimerase transcription, thus providing UDP-Gal substrates for protein glycosylation, compensating for the need of protein folding and PTM (post-translational modification) in tumor cells (Deng et al. 2013). As referred above (Sect. 2.1), XBP1s promotes tumorigenicity by assembling a transcriptional complex with HIF1α, resulting in the upregulation of glycolytic proteins such as glucose transporter 1 (GLUT1) (Chen et al. 2014). XBP1s also regulates the expression of factors involved in the hexosamine biosynthetic pathway (Wang et al. 2014) and along with PERK can control the functionality of FOXO transcription factors (Zhang et al. 2013). Specifically, XBP1s negatively regulates the FOXO1 levels, thereby affecting energy control and glucose metabolism (Zhou et al. 2011), further showing the promotion of glucose uptake by tumor cells through this pathway.

PERK also modulates amino acid and lipid metabolism, as well as autophagy pathways through its downstream target ATF4 (Rzymski et al. 2009). Autophagy is actually a means of yielding additional energy supply and is utilized by cancer cells upon ER stress due to starvation. The PERK/eIF2α/ATF4 signaling pathway induces autophagy through LC3B and ATG5, under hypoxic conditions (Rouschop et al. 2010), increasing the tolerance of tumor cells to unfavorable environmental conditions and promoting their growth (Bi et al. 2005). Moreover, IRE1 signaling to JNK, through its binding to TRAF2, is required for autophagy induction after ER stress (Ogata et al. 2006). Additionally, XBP1 deficiency has been documented to stimulate autophagy pathways, resulting in the cell adaptation and resistance to environmental stresses. The suggested adaptive processes included the clearance of proteotoxic aggregates, restoring the protein folding demand-capacity balance, and the supply of nutrients, through catabolic processes, thus compensating for nutrient deprivation (Hetz et al. 2009).

2.3 Tumoral UPR Controls the Immune System

In the context of different tumor types, UPR has been observed to instrument the immune responses resulting in the avoidance of tumor destruction. A characteristic example is gastrointestinal cancers, where IRE1 seems to play a major role in regulating surrounding immune component cells. The gastrointestinal tract (specifically, the epithelial cells) is the main tissue where the two IRE1 paralogs, IRE1α and IRE1β are expressed (Bertolotti et al. 2001). IRE1β is important for mucin production from airway epithelial cells (Martino et al. 2013) and goblet cells (Tsuru et al. 2013). IRE1β deficiency has been associated with increased ER stress and inflammation of the intestine, as well as lack of resistance to chemically induced colitis (Bertolotti et al. 2001). Moreover, XBP1 deficiency or expression of *XBP1* variants was associated with Crohn's disease and ulcerative colitis (Kaser et al. 2008). A link between inflammatory bowel disease and tumorigenesis was also illustrated upon XBP1 deficiency in epithelial cells that leaded to elevated levels of colitis-associated cancer (CAC) (Niederreiter et al. 2013). In addition, *XBP1* deletion in Paneth cells revealed that the IRE1α-TNFα/NFκB pathway plays a central role in the generation of inflammation after ER stress (Adolph et al. 2014). Further to this, the activation of autophagy pathways through the PERK/eIF2α/ATF4 signaling axis caused a restraint in IRE1 activation and ER stress-induced intestinal inflammation (Adolph et al. 2014).

In the tumor mass except for cancer cells, one can also witness the presence of cells of the immune system, endothelial cells, and cancer-associated fibroblasts (CAF). Among the immune system components, tumor-associated macrophages (TAM), dendritic cells (DC), and T cells are some of the common tumor infiltrates. TAM have an M2-like phenotype (Mantovani et al. 2002), meaning that they elicit a tumor-supportive function, and do not secrete cytokines that can stimulate the immune response against the tumor. In this way, the presence of TAM in the tumor microenvironment is correlated with poor patient prognosis (Komohara et al. 2008; Bingle et al. 2002). Indeed, it has been found that TAM coming from postoperative tissue specimens of glioma patients do not express the costimulatory molecules CD86, CD80, and CD40 that are critical for T cell activation and cannot yield pro-inflammatory cytokines such as TNFα, IL1, and IL6, accounting for their suppressive phenotype (Hussain et al. 2006). The UPR is involved in this tumor-immune system crosstalk as shown through an increasing number of studies. For instance, activation of IRE1/XBP1s, PERK, and CHOP in macrophages (MΦ) promotes production of pro-tumorigenic IL6 and TNFα (Martinon et al. 2010b; Chen et al. 2009), while CHOP in DC controls IL23 production (Goodall et al. 2010). In addition, absence of the ER-resident chaperone gp96/GRP94, an ER-resident chaperon, in MΦ leads to the reduction of IL17A, IL17F, IL23, and TNFα levels in tumor mouse models (Morales et al. 2014). Importantly, conditioned media from ER-stressed tumor cells has been reported to trigger the global ER stress response in recipient MΦ and DC, as shown by the transcriptional upregulation of GRP78, XBP1s, and CHOP (Mahadevan et al. 2011, 2012). In these cells the production of certain cytokines was also upregulated, including the pro-inflammatory cytokines IL6, IL23, and TNFα, as well as the chemokines MIP-1α and MIP-1β (macrophage inflammatory protein-1α and β), and CCL2/MCP-1 (monocyte chemoattractant protein-1) (Mahadevan et al. 2011, 2012). CCL2 in turn has been correlated with further TAM accumulation at the tumor site (Balkwill and Mantovani 2001) and along with other chemokines contributes to tumor progression, promotion of inflammation, and angiogenesis (Ueno et al. 2000). Cytokines also produced by TAM such as IL6 and TGF-β/IL10 can act in an immunosuppressive way towards DC by blocking their maturation, subsequently interfering with T cell activation (Grohmann et al. 2003; Allavena et al. 2008).

The UPR machinery has also been involved in the perturbation of antigen presentation process by affecting the antigen-presenting cells (APC). Upon ER stress, presentation of high-affinity peptides by MHC I and II is reduced (Wheeler et al. 2008; Granados et al. 2009). Additionally, UPR signals transmitted from cancer cells to DC disables their antigen cross-presentation function and subsequent activation of CD8[+] T cells (Mahadevan et al. 2012), and stimulates arginase, possessing a T cell suppressive role (Norian et al. 2009). Moreover, concerning the population of CD8[+] T cells, an increase in IL2, IL10, and FOXP3 and a decrease in CD28 was indicative of a suppressive phenotype (Mahadevan et al. 2012). Due to these cancer-cell-derived signals, DC have a constitutively active IRE1/XBP1s pathway, independent of an ER stress occurrence (Osorio et al. 2014). XBP1s signaling is important for the development and survival of DC (Iwakoshi et al. 2007), and for their ability to present antigens. Loss of XBP1 in DC resulted in dysregulated ER homeostasis, while IRE1α-mediated RIDD accounted for phenotypic and functional defects in the dendritic cells (Osorio et al. 2014). However, in a model of primary and metastatic ovarian carcinoma, DC were shown to exhibit high levels of spliced *XBP1* mRNA, that led to accelerated progression of the cancer (Cubillos-Ruiz et al. 2015). In that model, DC-specific abrogation of *XBP1* restored their immuno-stimulatory function and prolonged host survival by inducing a protective antitumor-immune response (Cubillos-Ruiz et al. 2015). The nature of the ER stress in DC was associated with an increased ROS production and the generation of lipid peroxidation byproducts. This may account for the initiation and maintenance of pro-tumorigenic ER stress in infiltrating DC. Another report also indicated that in myeloid-derived suppressor cells (MDSC), CHOP is upregulated by tumor-induced ROS and peroxynitrite, and in turn modulates the immune inhibitory activity, inducing anti-tumor responses (Thevenot et al. 2014). All this evidence supports a new function of the UPR in modulating the inflammatory environment of tumors, thus fine tuning immune responses in cancer.

2.4 Driving Invasion and Metastasis

Cancer cells are able to spread to other tissues where they can start forming another tumor. Epithelial cells are transformed to mesenchymal (EMT) thus being able to migrate, invade their surrounding environment and home to the new site of the organism due to the development of stem cell features (Thiery et al. 2009; Nieto and Cano 2012; Baum et al. 2008). EMT and UPR mutually lead to the activation of one another. UPR signaling can result in the acquisition of an EMT-like phenotype; it affects the organization of the polarized epithelial monolayer and the formation of actin stress fibers, as well as downregulates epithelial markers including E-cadherin and upregulates mesenchymal markers such as vimentin. First reported in thyroid cells, tunicamycin (Tun) or thapsigargin (Tg) triggered SRC signaling, leaded to their dedifferentiation and induced an EMT-like phenotype (Ulianich et al. 2008). The SRC and β-catenin/WNT pathways were also activated in alveolar epithelial cells by the UPR after Tun administration or overexpression of a variant protein (Tanjore et al. 2011). In renal proximal tubular epithelial cells, only Tg but not Tun induced an EMT, owing to the fact that Tg is an ER stressor that alters Ca^{2+} fluxes between the ER lumen and the cytosol and causes TDAG51 and TGF-β1 upregulation (Carlisle et al. 2012). In mammary epithelial cells overexpression of SCCA1, a Ser/Cys protease inhibitor, stimulates chronic UPR, in turn activating NFκB leading to IL6 production, that induces EMT-like phenotypes (Sheshadri et al. 2014). At the same time, during EMT, malignant cells present changes in their secretory phenotype mainly regarding extracellular matrix (ECM) protein secretion that can stimulate ER stress (Kirk et al. 2009; Fox et al. 2010). In colorectal carcinoma cells, HIF1α stabilization or serum starvation induces EMT and subsequent activation of the UPR (Zeindl-Eberhart et al. 2014). In mammary epithelial cells, overexpression of the transcription factor Twist stimulated EMT and was correlated with constitutive activation of PERK, whereas inhibition of PERK activity attenuated cells' ability to migrate and form tumor spheres (Feng et al. 2014). In addition, PERK signaling in dedifferentiated cells upon epithelial-to-mesenchymal transition leads to constitutive activation of NRF2, a master regulator of cellular response to oxidative damage, enabling these cells to express antioxidant enzymes and drug efflux pumps, thereby becoming chemoresistant (Alam et al. 1999, 2000; Maines 1988). The increased VEGF-A expression, resulting from PERK activation, is related to an elevated potential of medulloblastoma cell migration, mediated by VEGFR2 (Jamison et al. 2015). Of note, both PERK and ATF6 arms of the UPR induce a member of disintegrins and metalloproteases family, ADAM17, influencing cancer initiation and progression (Rzymski et al. 2012). Another factor downstream of PERK, ATF4, mediates the activation of LAMP3 (lysosomal-associated membrane protein 3), which orchestrates the metastasis process in hypoxic breast cancer cells (Mujcic et al. 2013; Nagelkerke et al. 2013). In esophageal squamous carcinoma, ATF4 expression is increased promoting metastasis and invasion, possibly through the modulation of matrix metalloproteinases (Zhu et al. 2014). In addition, downstream of PERK, ATF3 downregulates the EGR-1 expression, transcriptionally altering the expression of pro-tumorigenic chemokines, thus affecting cancer survival and metastasis (Park et al. 2014). Moreover, in glioblastoma IRE1 activity affects the infiltrating properties of glioma cells in the brain concerning cell adhesion, migration and invasion (Auf et al. 2010; Dejeans et al. 2012), in addition to controlling the production of certain pro-inflammatory chemokines like IL6, IL8, and CXCL3 (Auf et al. 2010; Pluquet et al. 2013). It this cancer context, IRE1 signaling leads to the inhibition of migration through *SPARC* mRNA degradation via its RIDD activity, and through the following RhoA and FAK signaling attenuation, as it impacts on the glioma cells–ECM interaction (Auf et al. 2010). On the contrary, in breast cancer XBP1s upregulation is correlated with the tumor progression and matched metastatic tumors have been found with overexpressed XBP1s, acting via SNAIL signaling toward promoting EMT (Li et al. 2014). Specifically, in TNBC XBP1s expression leads to an effective lung metastasis, being related to invasion of tumor cells (Chen et al. 2014). Importantly, UPR activation also in tumor-associated macrophages (TAM) is shown to remodel ECM thus promoting metastasis [reviewed in (Allavena and Mantovani 2012; Galdiero et al. 2013; Hambardzumyan et al. 2016)]. Indeed, MMP (matrix metalloproteinase) production from TAM was dependent on TNFα, stimulated by IRE1 and PERK arms of the UPR (Martinon et al. 2010a; Chen et al. 2009), and was found to increase tumor cell invasiveness (Pollard 2004). In addition, STAT3/STAT6 signaling pathways stimulated cathepsin secretion from MΦ, through IRE1 activation, contributing to tumor invasion (Yan et al. 2016). Together these studies highlight the need for clarifying in what way UPR is involved in invasion and metastasis in different cancerous environments. So far, the outcome appears to originate from both a fine balance between the three UPR branches and the inner attributes of the distinct cancer type.

2.5 Inducing Angiogenesis

Nutrient deprivation, hypoxia, and a resultant reduction in energy (ATP) production become present in solid tumors due to decreased vascularization and thus insufficient perfusion. However, these factors themselves lead to the formation of new blood vessels to restore the nutrient supply as well as the oxygen and energy levels (Jain 2003). A strong link between the UPR, especially IRE1 and PERK, and angiogenesis has been reported in various cancers including breast, brain, lung cancer, and head and neck squamous cell carcinoma (Wang et al. 2012b; Drogat et al. 2007). Indeed, the transcription factor ATF4 downstream of PERK binds to the promoter of VEGF regulating its expression, found increased in tumor cells that undergo ER stress (Wang et al. 2012b; Pereira et al. 2010; Binet and Sapieha 2015). In addition to VEGF, the expression of other proangiogenic factors such as FGF2, IL6, IL8, and IL1β is increased (Wang et al. 2012b; Pereira et al. 2010), while expression of certain antiangiogenic factors like THBS1, CXCL14, and CXCL10 is decreased after ATF4 activation independently of HIF1α (Wang et al. 2012b). In a GBM model under ER stress, due to lack of oxygen or glucose, IRE1 activation led to upregulation of proangiogenic factors including VEGF-A, IL6, IL8, and IL1β and downregulation of antiangiogenic factors such as THBS1 and decorin (Auf et al. 2010; Drogat et al. 2007), contributing to angiogenesis and tumor growth also in vivo (Auf et al. 2010; Drogat et al. 2007). In the same model IRE1 had the same effect through the cleavage of the circadian mRNA *PERIOD1* (Pluquet et al. 2013). In a model of TNBC, cancer cell growth was dependent on XBP1s and its regulation of HIF1α transcriptional targets (mentioned in Sect. 2.1), one of which was VEGF, thus impacting on angiogenesis of breast cancer (Chen et al. 2014). Importantly, UPR involvement in angiogenesis does not only become evident in tumor cells, but also in endothelial cells (EC). In EC XBP1s possibly through physical interaction again with HIF1α (Chen et al. 2014), and through regulation of pro-inflammatory chemokines (Auf et al. 2010), plays a crucial role in angiogenesis. Reciprocally, VEGF upon binding to its receptor activates the UPR in EC unconventionally through PLCγ and mTORC1, even when ER stress is not evident, promoting survival of EC (Karali et al. 2014) and inducing angiogenesis in vivo (Karali et al. 2014). Moreover, increased production of VEGF, TNFα, FGF, IL6, IL8/CXCL8, and CCL2/MCP-1 from TAM results in EC recruitment to the tumor and neoangiogenesis (Allavena et al. 2008; Galdiero et al. 2013). Importantly, the evident production of such factors by dendritic cells as well could have the same impact on endothelial cell accumulation at the tumor site (Mahadevan et al. 2011, 2012). In this way, the evidence about the "angiogenic switch" induction also by myeloid cells [reviewed extensively in (Allavena and Mantovani 2012; Galdiero et al. 2013; Hambardzumyan et al. 2016)] further supports the interrelationship between the constituents of the tumor microenvironment, that as a whole lead to the propagation of the tumor.

In trying to define how this complex control of the tumor microenvironment spreads from cell-to-cell, an interesting concept arose (Fig. 2). In this, the ER stress

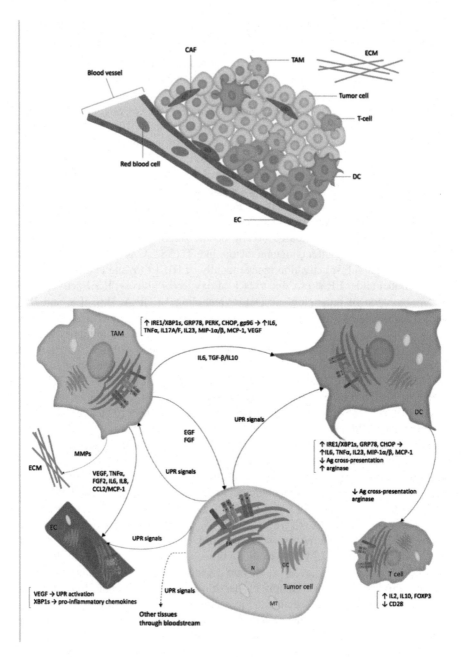

response can be transmitted from tumor cells to myeloid cells, including tumor-associated MΦ and DC, as well as endothelial cells, so the existence of a "transmissible ER stress-inducing factor" was proposed (Mahadevan et al. 2011). This indirect UPR activation by unknown factors in the recipient cells elicits in turn

Fig. 2 **The cell crosstalk within the tumor microenvironment mediated by the UPR activation and transmission**. *Top* An overview of the cellular tumor microenvironment. *Bottom* A zoom-in to the various interplays between the tumor microenvironment cell components. The cell-intrinsic and -extrinsic stresses elicited in the tumor are able to activate the ER stress response in cancer cells. Yet unknown secreted factors can transmit this response from tumor to stromal cells, including tumor-associated macrophages (TAM), dendritic cells (DC), T cells (indirectly), cancer-associated fibroblasts (CAF), and endothelial cells (EC). The transmission of the so-called "UPR signals" to the macrophages and DC leads to the activation of the IRE1 and PERK/CHOP pathways, and chaperones' upregulation leading to increased production of pro-inflammatory/-tumorigenic cytokines and chemokines. In response to this stimulus, (described clockwise) TAM promote immunosuppression of DC through IL6 and TGF-β/IL10, which in turn reduce their Ag cross-presentation activity and secrete arginase. Therefore, DC are not able to activate T cells, which present increased IL2, IL10, and FOXP3 and reduced CD28. TAM additionally promote growth of tumor cells through the secretion of growth factors (e.g., EGF, FGF), angiogenesis through signaling of VEGF, TNFα, FGF2, IL6, IL8, and CCL2/MCP-1 to endothelial cells, and metastasis and invasion by reshaping the ECM (e.g., TNFα-dependent MMP secretion). UPR signals are also transmitted from tumor cells to EC, increasing VEGF production and activating the UPR in these cells, with XBP1s to increase the secretion of pro-inflammatory chemokines, thus amplifying angiogenesis in the tumor area. Based on observations outside of the cancer system spectrum, one could additionally visualize the transmission of UPR signals from tumor cells to distant tissues through the bloodstream, making tumor metastasis more potent. *ECM* extracellular matrix, *ER* endoplasmic reticulum, *GC* Golgi complex, *MT* mitochondrion, *N* nucleus

signals that favor tumor development and spreading. The above-mentioned studies fall in line with this idea, introducing the concept of cell non-autonomous activation of UPR next to the classical cell autonomous one. Indeed, current advances in the field support this view in a range of experimental systems (Mardones et al. 2015), still to be determined in cancer, and take a step further by setting the phenomenon at an organismal level. The majority of the data comes from *C. elegans* models, where undefined molecules are reported to transmit UPR activation from the central nervous system (CNS) to peripheral tissues, controlling in this way systemic proteostasis (Mardones et al. 2015; Taylor and Dillin 2013). If this is applied in a cancerous context, one could hypothesize that UPR signals emanated from tumor cells could travel through the bloodstream reaching distant tissues, transmitting an ER stress to the resident cells. This in turn could well contribute to the generation of a metastatic niche that would favor the arrival and the further malignant development of cancer cells at the new site. Overall, the reciprocal communication between tumor and surrounding cells now comes to the forefront, further suggesting the targeting of stromal cells in the tumor as a promising therapeutic strategy.

2.6 Genome Instability—Mutations of the UPR Machinery

2.6.1 Genome Instability

Genomic instability referring to high mutation number, chromosomal rearrangements, and aneuploidy is prevalent in human cancers. Components of the UPR machinery have been reported to control genomic imbalance and at the same time alterations in genomic integrity activate the UPR, as reported in different experimental systems (Yamamori et al. 2013; Epple et al. 2013; Hsu et al. 2009; Al-Rawashdeh et al. 2010; Feng et al. 2011), which means that a disturbance in their physiological reciprocal regulation could enhance malignant progression. Indeed, polyploidy like the generation of tetraploid cells is accompanied by basal levels of ER stress. In this setting, the ER chaperone calreticulin (CRT) is exposed at the cell surface contributing to immunogenic cell death. Since CRT exposure pathway relies partly on PERK signaling, this could be assigned to the function of PERK specifically in mitochondrial associated membranes (MAM), where it regulates intercellular Ca^{2+} fluxes and ROS production (Zitvogel et al. 2013). Indeed, PERK inactivation triggers genomic instability, related to the occurrence of oxidative DNA damage and uncontrolled ROS production (Bobrovnikova-Marjon et al. 2010). In a similar way, Ire1p deficiency in yeast causes an increase in chromosomal loss, thus negatively affecting the genomic integrity (Henry et al. 2010). Although this has not been validated in mammals, XBP1s appears to control a cluster of DNA repair genes (Acosta-Alvear et al. 2007). In general, there is the suggestion that ER stress generates oxidative stress resulting in DNA damage, possibly through a deregulation of the PDI-ERO1α cycle that controls the redox state in the ER (Farooqi et al. 2015). ER stress is also linked with important components of the DNA damage response. As an example, cells lacking ATM or p53 develop spontaneous ER stress (He et al. 2009; Dioufa et al. 2012; Duplan et al. 2013), while p53-deficient mice exhibit constitutive ER stress (Dioufa et al. 2009).

In parallel, aneuploidy and gene copy number variations can be translated to alterations in global proteostasis and subsequent proteotoxicity (Oromendia et al. 2012). Proteostasis is also controlled at an epigenetic level. Histone deacetylase (HDAC) inhibitors synergize with ER stress in cancer cells (Yang et al. 2013). Proteostasis control by HDAC is also related to p97/VCP and RUVBL2 proteins (Boyault et al. 2006; Marza et al. 2015), implicated in misfolded protein disaggregation and degradation, and genome integrity. In a reciprocal way, ER stress impacts on chromatin remodeling by affecting the post-translation modifications of histones (e.g., methylation, acetylation) [reviewed in (Dicks et al. 2015)].

2.6.2 Cancer-Associated Somatic Mutations in UPR Genes

Recently, genome-wide sequencing studies have reported somatic mutations in UPR sensor-encoding genes in different cancers, supporting in a greater extent its contribution to tumorigenic processes (Greenman et al. 2007). IRE1 mutations have been found in glioblastoma (Parsons et al. 2008) and hepatocellular carcinoma (Guichard et al. 2012), being one of the kinases mostly mutated in cancer (Greenman et al. 2007). Interestingly, although some of these mutations have a functional kinase and endoribonuclease activity, the ability to induce apoptosis in culture cells is lost (Ghosh et al. 2014b; Xue et al. 2011). An increasing number of mutations in all three UPR sensors are now reported in the COSMIC, cBIOportal, and IntOGen databases (Gonzalez-Perez et al. 2013; Cerami et al. 2012; Forbes et al. 2009; Chevet et al. 2015). In the IRE1 gene (ERN1), mostly silent mutations as well as in frame deletions and insertions are found; in the PERK gene (EIF2AK3) missense mutations are mainly observed, while in the ATF6 gene (ATF6) nonsense mutations prevail (Chevet et al. 2015). The mutation occurrence of each branch is also specific to the tissue bearing the tumor. In that sense, PERK gene mutations are more prevalent in bone cancers, ATF6 mutations in genital cancers and IRE1 in cancers of the nervous system, falling in agreement with previous functional observations in a GBM setting (Pluquet et al. 2013). Additionally, ATF6 along with IRE1 mutations are highly observed in cancers of the gastrointestinal tract, while ATF6 along with PERK are frequently mutated in urologic and lung cancers. These observations reveal the distinct mutational profile that each UPR arm has and the specific prevalence of them depending on the cancer type. Although the functionality of these mutations in a biological context (gain/loss of function and impact on the tumor characteristics) has not been fully reported, one can tell the importance of this distinct mutational spectrum for the selective targeting of each UPR branch in cancer.

3 Targeting UPR in Cancer

Given the major implication of the UPR machinery in the modulation of carcinogenesis and tumor progression, a great number of pharmacological compounds have been and are currently being developed. These therapeutic molecules aim either to the reduction of the adaptive capability of the tumor cells or to the elevation of ER stress to a degree that the cancer cells are not able to alleviate. Some of them are reported to act along with other chemotherapeutic agents in the context of combination therapies, through cell sensitization (Atkins et al. 2013; Higa et al. 2014; Gallagher and Walter 2016; Gallagher et al. 2016; Plate et al. 2016; Cross et al. 2012; Volkmann et al. 2011; Mimura et al. 2012; Sanches et al. 2014; Lhomond et al. 2015; Ghosh et al. 2014a; Papandreou et al. 2011; Wang et al. 2012a; Sidrauski et al. 2013; Boyce et al. 2005; Tsaytler et al. 2011; Das et al. 2015). First, the compounds 16F16, ceapins, and compounds 147 and 263 have been so far reported to control ATF6 signaling. Treatment with the PDI inhibitor 16F16 as well as genetic inhibition of PDIA5 showed that PDIA5 allows the export of ATF6 from the ER under stress conditions and that both confer sensitization of leukemia cells to imatinib treatment (Higa et al. 2014). Ceapins, a class of pyrazole amides, were also shown to abrogate the export of ATF6 from the ER under stress conditions (Gallagher and Walter 2016; Gallagher et al. 2016). More recently two small molecules, compounds 147 and 263, were reported to specifically target ATF6 (Plate et al. 2016). Regarding IRE1 pharmacological modulation, most of the identified compounds target its RNase activity including 4μ8C, 3-methoxy-6-bromosalicylaldehyde, MKC analogs, and STF-083010 (Obacz et al. 2017). Both 4μ8C (Cross et al. 2012; Hetz et al. 2013) and 3-methoxy-6-bromosalicylaldehyde (Volkmann et al. 2011) by binding in the catalytic site of the protein in a reversible manner halt the cleavage of *XBP1* mRNA and RIDD. The compounds STF-083010 and, among the MKC analogs, MKC-3946 are reported to reduce tumor growth in a mouse model of multiple myeloma (Papandreou et al. 2011; Mimura et al. 2012), with the latter to additionally show synergistic effects with bortezomib, a proteasome inhibitor (Mimura et al. 2012). In a compound screening toyocamycin, trierixin, and quinotrierixin were identified as RNase inhibitors (Ri et al. 2012; Kawamura et al. 2008; Futamura et al. 2007). Regarding toyocamycin, it was shown not to affect IRE1 phosphorylation and the other UPR branches, and to increase the effect of bortezomib on the apoptosis of multiple myeloma cells (Ri et al. 2012). The other category of IRE1 regulators consists of compounds targeting its kinase domain by binding to the ATP-binding pocket. In a cellular context, a type II ATP kinase inhibitor named compound 3 inhibits first of all the kinase activity of IRE1, as well as its oligomerization and its RNase activity (Wang et al. 2012a). In contrast to this, the broad type I ATP-competitive kinase inhibitors APY29 and sunitinib stimulate the RNase activity of IRE1 in vitro (Ali et al. 2011). Another class of ATP-competitive inhibitors called KIRA (Kinase-Inhibiting RNase Attenuators) has been lately identified to break the IRE1 oligomers, thus allosterically inhibiting its RNase activity. Among them, KIRA6 targets IRE1 in vivo leading to cell survival under ER stress (Ghosh et al. 2014a). Importantly, it was found that peptides derived from IRE1 can alter the oligomeric status of the protein under ER stress conditions and uncouple the cleavage of *XBP1* mRNA from RIDD in vitro and in vivo (Bouchecareilh et al. 2011). Concerning PERK pharmacological modulation, compound 38 (GSK2606414) blocks the protein's phosphorylation reducing pancreatic tumor growth in mice (Axten et al. 2012), and GSK2656157 affects cellular metabolism and angiogenesis in tumors in vivo (Atkins et al. 2013). Additional molecules target PERK signaling by regulation of downstream factors. For instance, ISRIB controls eIF2α phosphorylation through the exchange factor eIF2B (Sidrauski et al. 2013, 2015a, b). In the same line, the molecules salubrinal (Boyce et al. 2005), guanabenz (Tsaytler et al. 2011; Hetz et al. 2013) under ER stress and, more recently discovered, sephin 1 (Das et al. 2015) block the PP1c–GADD34 complex that dephosphorylates eIF2α, causing a delay in translation. Salubrinal also blocks the PP1c–CreP complex with the same cellular effect but in a constitutive manner (Boyce et al. 2005). Of note, the resultant prolonged eIF2α phosphorylation leads to an increase in expression of pro-apoptotic genes, despite the general halt in translation (Hetz et al. 2013). Importantly and even more recently trazodone, an anti-depressant, and dibenzoylmethane were discovered to act similar to ISRIB downstream of eIF2α phosphorylation and proved to be neuroprotective in mice, strongly representing potential treatments for various neurodegenerative disorders and which could possibly be applicable to cancers (Halliday et al. 2017).

Based on the various cellular constituents within the tumor microenvironment, a new perspective is given to the UPR targeting in cancer. Due to the apparent UPR activation in the immune infiltrate as well as the rest of the tumor stromal cells, the different pharmacological modulators could be administered not strictly to the tumor cells per se, but also, or alternatively, to the tumor infiltrate or stroma.

4 Conclusion

According to the evidence highlighted in this chapter, we have now a well-documented role of the UPR adaptive mechanisms in cancer. Its activation in both the tumor cells and the stroma leads to the adaptation of the developing malignancy in a changing environment. It is involved in the malignant cell proliferation, ECM reshaping, cell migration, invasion and metastasis, angiogenesis, in the rewiring of cell metabolism, and even in genomic instability with a parallel increase in UPR gene mutations. That is why some unmet therapeutic alternatives are being revealed and tested currently, including the cancer drug administration beyond the tumor cells per se.

Acknowledgements This work was funded by grants from the Institut National du Cancer (INCa) and EU H2020 MSCA ITN-675448 (TRAINERS) to EC.

References

Acosta-Alvear D, Zhou Y, Blais A, Tsikitis M, Lents NH, Arias C, Lennon CJ, Kluger Y, Dynlacht BD (2007) XBP1 controls diverse cell type- and condition-specific transcriptional regulatory networks. Mol Cell 27(1):53–66

Adachi Y, Yamamoto K, Okada T, Yoshida H, Harada A, Mori K (2008) ATF6 is a transcription factor specializing in the regulation of quality control proteins in the endoplasmic reticulum. Cell Struct Funct 33(1):75–89 JST.JSTAGE/csf/07044 [pii]

Adolph TE, Tomczak MF, Niederreiter L, Ko HJ, Bock J, Martinez-Naves E, Glickman JN, Tschurtschenthaler M, Hartwig J, Hosomi S, Flak MB, Cusick JL, Kohno K, Iwawaki T, Billmann-Born S, Raine T, Bharti R, Lucius R, Kweon MN, Marciniak SJ, Choi A, Hagen SJ, Schreiber S, Rosenstiel P, Kaser A, Blumberg RS (2014) Paneth cells as a site of origin for intestinal inflammation. Nature 503(7475):272–276. doi:10.1038/nature12599 [pii]

Aguirre-Ghiso JA (2007) Models, mechanisms and clinical evidence for cancer dormancy. Nat Rev Cancer 7(11):834–846. doi:10.1038/nrc2256

Al-Rawashdeh FY, Scriven P, Cameron IC, Vergani PV, Wyld L (2010) Unfolded protein response activation contributes to chemoresistance in hepatocellular carcinoma. Eur J Gastroenterol Hepatol 22(9):1099–1105. doi:10.1097/MEG.0b013e3283378405

Alam J, Stewart D, Touchard C, Boinapally S, Choi AM, Cook JL (1999) Nrf2, a Cap'n'Collar transcription factor, regulates induction of the heme oxygenase-1 gene. J Biol Chem 274(37):26071–26078

Alam J, Wicks C, Stewart D, Gong P, Touchard C, Otterbein S, Choi AM, Burow ME, Tou J (2000) Mechanism of heme oxygenase-1 gene activation by cadmium in MCF-7 mammary epithelial cells. Role of p38 kinase and Nrf2 transcription factor. J Biol Chem 275(36):27694–27702. doi:10.1074/jbc.M004729200 [pii]

Ali MM, Bagratuni T, Davenport EL, Nowak PR, Silva-Santisteban MC, Hardcastle A, McAndrews C, Rowlands MG, Morgan GJ, Aherne W, Collins I, Davies FE, Pearl LH (2011) Structure of the Ire1 autophosphorylation complex and implications for the unfolded protein response. EMBO J 30(5):894–905. doi:10.1038/emboj.2011.18

Allavena P, Mantovani A (2012) Immunology in the clinic review series; focus on cancer: tumour-associated macrophages: undisputed stars of the inflammatory tumour microenvironment. Clin Exp Immunol 167(2):195–205. doi:10.1111/j.1365-2249.2011.04515.x

Allavena P, Sica A, Garlanda C, Mantovani A (2008) The Yin-Yang of tumor-associated macrophages in neoplastic progression and immune surveillance. Immunol Rev 222:155–161. doi:10.1111/j.1600-065X.2008.00607.x

Arai M, Kondoh N, Imazeki N, Hada A, Hatsuse K, Kimura F, Matsubara O, Mori K, Wakatsuki T, Yamamoto M (2006) Transformation-associated gene regulation by ATF6alpha during hepatocarcinogenesis. FEBS Lett 580(1):184–190. doi:10.1016/j.febslet.2005.11.072 S0014-5793(05)01449-3 [pii]

Atkins C, Liu Q, Minthorn E, Zhang SY, Figueroa DJ, Moss K, Stanley TB, Sanders B, Goetz A, Gaul N, Choudhry AE, Alsaid H, Jucker BM, Axten JM, Kumar R (2013) Characterization of a novel PERK kinase inhibitor with antitumor and antiangiogenic activity. Cancer Res 73 (6):1993–2002. doi:10.1158/0008-5472.CAN-12-3109

Auf G, Jabouille A, Delugin M, Guerit S, Pineau R, North S, Platonova N, Maitre M, Favereaux A, Vajkoczy P, Seno M, Bikfalvi A, Minchenko D, Minchenko O, Moenner M (2013) High epiregulin expression in human U87 glioma cells relies on IRE1alpha and promotes autocrine growth through EGF receptor. BMC Cancer 13:597. doi:10.1186/1471-2407-13-597 [pii]

Auf G, Jabouille A, Guerit S, Pineau R, Delugin M, Bouchecareilh M, Magnin N, Favereaux A, Maitre M, Gaiser T, von Deimling A, Czabanka M, Vajkoczy P, Chevet E, Bikfalvi A, Moenner M (2010) Inositol-requiring enzyme 1alpha is a key regulator of angiogenesis and invasion in malignant glioma. Proc Natl Acad Sci USA 107(35):15553–15558. doi:10.1073/pnas.0914072107

Axten JM, Medina JR, Feng Y, Shu A, Romeril SP, Grant SW, Li WH, Heerding DA, Minthorn E, Mencken T, Atkins C, Liu Q, Rabindran S, Kumar R, Hong X, Goetz A, Stanley T, Taylor JD, Sigethy SD, Tomberlin GH, Hassell AM, Kahler KM, Shewchuk LM, Gampe RT (2012) Discovery of 7-methyl-5-(1-{[3-(trifluoromethyl)phenyl]acetyl}-2,3-dihydro-1H-indol-5-yl)-7H-p yrrolo[2,3-d]pyrimidin-4-amine (GSK2606414), a potent and selective first-in-class inhibitor of protein kinase R (PKR)-like endoplasmic reticulum kinase (PERK). J Med Chem 55(16):7193–7207. doi:10.1021/jm300713s

Balkwill F, Mantovani A (2001) Inflammation and cancer: back to Virchow? Lancet 357 (9255):539–545. doi:10.1016/S0140-6736(00)04046-0

Baum B, Settleman J, Quinlan MP (2008) Transitions between epithelial and mesenchymal states in development and disease. Semin Cell Dev Biol 19(3):294–308. doi:10.1016/j.semcdb.2008. 02.001 S1084-9521(08)00022-0 [pii]

Bertolotti A, Wang X, Novoa I, Jungreis R, Schlessinger K, Cho JH, West AB, Ron D (2001) Increased sensitivity to dextran sodium sulfate colitis in IRE1beta-deficient mice. J Clin Invest 107(5):585–593

Bertolotti A, Zhang Y, Hendershot LM, Harding HP, Ron D (2000) Dynamic interaction of BiP and ER stress transducers in the unfolded-protein response. Nat Cell Biol 2(6):326–332. doi:10.1038/35014014

Bi M, Naczki C, Koritzinsky M, Fels D, Blais J, Hu N, Harding H, Novoa I, Varia M, Raleigh J, Scheuner D, Kaufman RJ, Bell J, Ron D, Wouters BG, Koumenis C (2005) ER stress-regulated translation increases tolerance to extreme hypoxia and promotes tumor growth. EMBO J 24 (19):3470–3481. doi:10.1038/sj.emboj.7600777 [pii]

Binet F, Sapieha P (2015) ER stress and angiogenesis. Cell Metab 22(4):560–575. doi:10.1016/j. cmet.2015.07.010

Bingle L, Brown NJ, Lewis CE (2002) The role of tumour-associated macrophages in tumour progression: implications for new anticancer therapies. J Pathol 196(3):254–265. doi:10.1002/path.1027

Blais JD, Addison CL, Edge R, Falls T, Zhao H, Wary K, Koumenis C, Harding HP, Ron D, Holcik M, Bell JC (2006) Perk-dependent translational regulation promotes tumor cell adaptation and angiogenesis in response to hypoxic stress. Mol Cell Biol 26(24):9517–9532. doi:10.1128/MCB.01145-06 [pii]

Blais JD, Filipenko V, Bi M, Harding HP, Ron D, Koumenis C, Wouters BG, Bell JC (2004) Activating transcription factor 4 is translationally regulated by hypoxic stress. Mol Cell Biol 24 (17):7469–7482. doi:10.1128/MCB.24.17.7469-7482.2004 [pii]

Bobrovnikova-Marjon E, Grigoriadou C, Pytel D, Zhang F, Ye J, Koumenis C, Cavener D, Diehl JA (2010) PERK promotes cancer cell proliferation and tumor growth by limiting oxidative DNA damage. Oncogene 29(27):3881–3895. doi:10.1038/onc.2010.153 [pii]

Bouchecareilh M, Higa A, Fribourg S, Moenner M, Chevet E (2011) Peptides derived from the bifunctional kinase/RNase enzyme IRE1{alpha} modulate IRE1{alpha} activity and protect cells from endoplasmic reticulum stress. FASEB J 25(9):3115–3129. doi:10.1096/fj.11-182931 [pii]

Boyault C, Gilquin B, Zhang Y, Rybin V, Garman E, Meyer-Klaucke W, Matthias P, Muller CW, Khochbin S (2006) HDAC6-p97/VCP controlled polyubiquitin chain turnover. EMBO J 25 (14):3357–3366. doi:10.1038/sj.emboj.7601210

Boyce M, Bryant KF, Jousse C, Long K, Harding HP, Scheuner D, Kaufman RJ, Ma D, Coen DM, Ron D, Yuan J (2005) A selective inhibitor of eIF2alpha dephosphorylation protects cells from ER stress. Science 307(5711):935–939. doi:10.1126/science.1101902 307/5711/935 [pii]

Brewer JW, Diehl JA (2000) PERK mediates cell-cycle exit during the mammalian unfolded protein response. Proc Natl Acad Sci USA 97(23):12625–12630. doi:10.1073/pnas.220247197 [pii]

Calfon M, Zeng H, Urano F, Till JH, Hubbard SR, Harding HP, Clark SG, Ron D (2002) IRE1 couples endoplasmic reticulum load to secretory capacity by processing the XBP-1 mRNA. Nature 415(6867):92–96. doi:10.1038/415092a [pii]

Carlisle RE, Heffernan A, Brimble E, Liu L, Jerome D, Collins CA, Mohammed-Ali Z, Margetts PJ, Austin RC, Dickhout JG (2012) TDAG51 mediates epithelial-to-mesenchymal transition in human proximal tubular epithelium. Am J Physiol Renal Physiol 303(3):F467–481. doi:10.1152/ajprenal.00481.2011 [pii]

Carrasco DR, Sukhdeo K, Protopopova M, Sinha R, Enos M, Carrasco DE, Zheng M, Mani M, Henderson J, Pinkus GS, Munshi N, Horner J, Ivanova EV, Protopopov A, Anderson KC, Tonon G, DePinho RA (2007) The differentiation and stress response factor XBP-1 drives multiple myeloma pathogenesis. Cancer Cell 11(4):349–360

Cerami E, Gao J, Dogrusoz U, Gross BE, Sumer SO, Aksoy BA, Jacobsen A, Byrne CJ, Heuer ML, Larsson E, Antipin Y, Reva B, Goldberg AP, Sander C, Schultz N (2012) The cBio cancer genomics portal: an open platform for exploring multidimensional cancer genomics data. Cancer Discov 2(5):401–404. doi:10.1158/2159-8290.CD-12-0095

Chen L, Jarujaron S, Wu X, Sun L, Zha W, Liang G, Wang X, Gurley EC, Studer EJ, Hylemon PB, Pandak WM Jr, Zhang L, Wang G, Li X, Dent P, Zhou H (2009) HIV protease inhibitor lopinavir-induced TNF-alpha and IL-6 expression is coupled to the unfolded protein response and ERK signaling pathways in macrophages. Biochem Pharmacol 78(1):70–77. doi:10.1016/j.bcp.2009.03.022

Chen X, Iliopoulos D, Zhang Q, Tang Q, Greenblatt MB, Hatziapostolou M, Lim E, Tam WL, Ni M, Chen Y, Mai J, Shen H, Hu DZ, Adoro S, Hu B, Song M, Tan C, Landis MD, Ferrari M, Shin SJ, Brown M, Chang JC, Liu XS, Glimcher LH (2014) XBP1 promotes triple-negative breast cancer by controlling the HIF1alpha pathway. Nature 508(7494):103–107. doi:10.1038/nature13119

Chevet E, Hetz C, Samali A (2015) Endoplasmic reticulum stress-activated cell reprogramming in oncogenesis. Cancer Discov 5(6):586–597. doi:10.1158/2159-8290.CD-14-1490

Cox JS, Walter P (1996) A novel mechanism for regulating activity of a transcription factor that controls the unfolded protein response. Cell 87(3):391–404

Cross BC, Bond PJ, Sadowski PG, Jha BK, Zak J, Goodman JM, Silverman RH, Neubert TA, Baxendale IR, Ron D, Harding HP (2012) The molecular basis for selective inhibition of unconventional mRNA splicing by an IRE1-binding small molecule. Proc Natl Acad Sci USA 109(15):E869–878. doi:10.1073/pnas.1115623109

Cubillos-Ruiz JR, Silberman PC, Rutkowski MR, Chopra S, Perales-Puchalt A, Song M, Zhang S, Bettigole SE, Gupta D, Holcomb K, Ellenson LH, Caputo T, Lee AH, Conejo-Garcia JR,

Glimcher LH (2015) ER stress sensor XBP1 controls anti-tumor immunity by disrupting dendritic cell homeostasis. Cell 161(7):1527–1538. doi:10.1016/j.cell.2015.05.025

Cullinan SB, Diehl JA (2004) PERK-dependent activation of Nrf2 contributes to redox homeostasis and cell survival following endoplasmic reticulum stress. J Biol Chem 279 (19):20108–20117

Cullinan SB, Zhang D, Hannink M, Arvisais E, Kaufman RJ, Diehl JA (2003) Nrf2 is a direct PERK substrate and effector of PERK-dependent cell survival. Mol Cell Biol 23(20):7198–7209

Dadey DY, Kapoor V, Khudanyan A, Urano F, Kim AH, Thotala D, Hallahan DE (2016) The ATF6 pathway of the ER stress response contributes to enhanced viability in glioblastoma. Oncotarget 7(2):2080–2092. doi:10.18632/oncotarget.6712

Das I, Krzyzosiak A, Schneider K, Wrabetz L, D'Antonio M, Barry N, Sigurdardottir A, Bertolotti A (2015) Preventing proteostasis diseases by selective inhibition of a phosphatase regulatory subunit. Science 348(6231):239–242. doi:10.1126/science.aaa4484

DeBerardinis RJ, Lum JJ, Hatzivassiliou G, Thompson CB (2008) The biology of cancer: metabolic reprogramming fuels cell growth and proliferation. Cell Metab 7(1):11–20. doi:10.1016/j.cmet.2007.10.002 S1550-4131(07)00295-1 [pii]

Dejeans N, Pluquet O, Lhomond S, Grise F, Bouchecareilh M, Juin A, Meynard-Cadars M, Bidaud-Meynard A, Gentil C, Moreau V, Saltel F, Chevet E (2012) Autocrine control of glioma cells adhesion and migration through IRE1alpha-mediated cleavage of SPARC mRNA. J Cell Sci 125(Pt 18):4278–4287. doi:10.1242/jcs.099291

Del Vecchio CA, Feng Y, Sokol ES, Tillman EJ, Sanduja S, Reinhardt F, Gupta PB (2014) De-differentiation confers multidrug resistance via noncanonical PERK-Nrf2 signaling. PLoS Biol 12(9):e1001945. doi:10.1371/journal.pbio.1001945 PBIOLOGY-D-14-00485 [pii]

Deng Y, Wang ZV, Tao C, Gao N, Holland WL, Ferdous A, Repa JJ, Liang G, Ye J, Lehrman MA, Hill JA, Horton JD, Scherer PE (2013) The Xbp1s/GalE axis links ER stress to postprandial hepatic metabolism. J Clin Invest 123(1):455–468. doi:10.1172/JCI62819 [pii]

Denoyelle C, Abou-Rjaily G, Bezrookove V, Verhaegen M, Johnson TM, Fullen DR, Pointer JN, Gruber SB, Su LD, Nikiforov MA, Kaufman RJ, Bastian BC, Soengas MS (2006) Anti-oncogenic role of the endoplasmic reticulum differentially activated by mutations in the MAPK pathway. Nat Cell Biol 8(10):1053–1063. doi:10.1038/ncb1471 [pii]

Dicks N, Gutierrez K, Michalak M, Bordignon V, Agellon LB (2015) Endoplasmic reticulum stress, genome damage, and cancer. Front Oncol 5:11. doi:10.3389/fonc.2015.00011

Dioufa N, Chatzistamou I, Farmaki E, Papavassiliou AG, Kiaris H (2009) p53 antagonizes the unfolded protein response and inhibits ground glass hepatocyte development during endoplasmic reticulum stress. Exp Biol Med (Maywood) 237(10):1173–1180. doi:10.1258/ebm.2012.012140 [pii]

Dioufa N, Chatzistamou I, Farmaki E, Papavassiliou AG, Kiaris H (2012) p53 antagonizes the unfolded protein response and inhibits ground glass hepatocyte development during endoplasmic reticulum stress. Exp Biol Med (Maywood) 237(10):1173–1180. doi:10.1258/ebm.2012.012140

Drogat B, Auguste P, Nguyen DT, Bouchecareilh M, Pineau R, Nalbantoglu J, Kaufman RJ, Chevet E, Bikfalvi A, Moenner M (2007) IRE1 signaling is essential for ischemia-induced vascular endothelial growth factor-A expression and contributes to angiogenesis and tumor growth in vivo. Cancer Res 67(14):6700–6707. doi:10.1158/0008-5472.CAN-06-3235 67/14/6700 [pii]

Duplan E, Giaime E, Viotti J, Sevalle J, Corti O, Brice A, Ariga H, Qi L, Checler F, Alves da Costa C (2013) ER-stress-associated functional link between Parkin and DJ-1 via a transcriptional cascade involving the tumor suppressor p53 and the spliced X-box binding protein XBP-1. J Cell Sci 126(Pt 9):2124–2133. doi:10.1242/jcs.127340 [pii]

Epple LM, Dodd RD, Merz AL, Dechkovskaia AM, Herring M, Winston BA, Lencioni AM, Russell RL, Madsen H, Nega M, Dusto NL, White J, Bigner DD, Nicchitta CV, Serkova NJ, Graner MW (2013) Induction of the unfolded protein response drives enhanced metabolism and chemoresistance in glioma cells. PLoS ONE 8(8):e73267. doi:10.1371/journal.pone. 0073267 PONE-D-12-22337 [pii]

Farooqi AA, Li K-T, Fayyaz S, Chang Y-T, Ismail M, Liaw C-C, Yuan S-SF, Tang J-Y, Chang H-W (2015) Anticancer drugs for the modulation of endoplasmic reticulum stress and oxidative stress. Tumor Biology 36(8):5743–5752. doi:10.1007/s13277-015-3797-0

Feng R, Zhai WL, Yang HY, Jin H, Zhang QX (2011) Induction of ER stress protects gastric cancer cells against apoptosis induced by cisplatin and doxorubicin through activation of p38 MAPK. Biochem Biophys Res Commun 406(2):299–304. doi:10.1016/j.bbrc.2011.02.036 S0006-291X(11)00219-1 [pii]

Feng YX, Sokol ES, Del Vecchio CA, Sanduja S, Claessen JH, Proia TA, Jin DX, Reinhardt F, Ploegh HL, Wang Q, Gupta PB (2014) Epithelial-to-mesenchymal transition activates PERK-eIF2alpha and sensitizes cells to endoplasmic reticulum stress. Cancer Discov 4(6):702–715. doi:10.1158/2159-8290.CD-13-0945 [pii]

Forbes SA, Tang G, Bindal N, Bamford S, Dawson E, Cole C, Kok CY, Jia M, Ewing R, Menzies A, Teague JW, Stratton MR, Futreal PA (2009) COSMIC (the catalogue of somatic mutations in cancer): a resource to investigate acquired mutations in human cancer. Nucleic Acids Res 38(Database issue):D652–657. doi: 10.1093/nar/gkp995 [pii]

Fox RM, Hanlon CD, Andrew DJ (2010) The CrebA/Creb3-like transcription factors are major and direct regulators of secretory capacity. J Cell Biol 191(3):479–492. doi:10.1083/jcb. 201004062 [pii]

Futamura Y, Tashiro E, Hironiwa N, Kohno J, Nishio M, Shindo K, Imoto M (2007) Trierixin, a novel Inhibitor of ER stress-induced XBP1 activation from Streptomyces sp. II. structure elucidation. J Antibiot (Tokyo) 60(9):582–585. doi:10.1038/ja.2007.74 JST. JSTAGE/antibiotics/60.582 [pii]

Galdiero MR, Garlanda C, Jaillon S, Marone G, Mantovani A (2013) Tumor associated macrophages and neutrophils in tumor progression. J Cell Physiol 228(7):1404–1412. doi:10. 1002/jcp.24260

Gallagher CM, Garri C, Cain EL, Ang KK, Wilson CG, Chen S, Hearn BR, Jaishankar P, Aranda-Diaz A, Arkin MR, Renslo AR, Walter P (2016) Ceapins are a new class of unfolded protein response inhibitors, selectively targeting the ATF6alpha branch. Elife 5. doi:10.7554/ eLife.11878

Gallagher CM, Walter P (2016) Ceapins inhibit ATF6alpha signaling by selectively preventing transport of ATF6alpha to the Golgi apparatus during ER stress. Elife 5. doi:10.7554/eLife. 11880

Ghosh R, Wang L, Wang ES, Perera BG, Igbaria A, Morita S, Prado K, Thamsen M, Caswell D, Macias H, Weiberth KF, Gliedt MJ, Alavi MV, Hari SB, Mitra AK, Bhhatarai B, Schurer SC, Snapp EL, Gould DB, German MS, Backes BJ, Maly DJ, Oakes SA, Papa FR (2014a) Allosteric inhibition of the IRE1alpha RNase preserves cell viability and function during endoplasmic reticulum stress. Cell. doi:10.1016/j.cell.2014.07.002 S0092-8674(14)00878-2 [pii]

Ghosh R, Wang L, Wang ES, Perera BG, Igbaria A, Morita S, Prado K, Thamsen M, Caswell D, Macias H, Weiberth KF, Gliedt MJ, Alavi MV, Hari SB, Mitra AK, Bhhatarai B, Schurer SC, Snapp EL, Gould DB, German MS, Backes BJ, Maly DJ, Oakes SA, Papa FR (2014b) Allosteric inhibition of the IRE1alpha RNase preserves cell viability and function during endoplasmic reticulum stress. Cell 158(3):534–548. doi:10.1016/j.cell.2014.07.002

Ginos MA, Page GP, Michalowicz BS, Patel KJ, Volker SE, Pambuccian SE, Ondrey FG, Adams GL, Gaffney PM (2004) Identification of a gene expression signature associated with recurrent disease in squamous cell carcinoma of the head and neck. Cancer Res 64(1):55–63

Gomez BP, Riggins RB, Shajahan AN, Klimach U, Wang A, Crawford AC, Zhu Y, Zwart A, Wang M, Clarke R (2007) Human X-box binding protein-1 confers both estrogen independence and antiestrogen resistance in breast cancer cell lines. FASEB J 21(14):4013–4027. doi:10.1096/fj.06-7990com [pii]

Gonzalez-Perez A, Perez-Llamas C, Deu-Pons J, Tamborero D, Schroeder MP, Jene-Sanz A, Santos A, Lopez-Bigas N (2013) IntOGen-mutations identifies cancer drivers across tumor types. Nat Methods 10(11):1081–1082. doi:10.1038/nmeth.2642

Goodall JC, Wu C, Zhang Y, McNeill L, Ellis L, Saudek V, Gaston JS (2010) Endoplasmic reticulum stress-induced transcription factor, CHOP, is crucial for dendritic cell IL-23 expression. Proc Natl Acad Sci USA 107(41):17698–17703. doi:10.1073/pnas.1011736107

Granados DP, Tanguay PL, Hardy MP, Caron E, de Verteuil D, Meloche S, Perreault C (2009) ER stress affects processing of MHC class I-associated peptides. BMC Immunol 10:10. doi:10.1186/1471-2172-10-10

Greenman C, Stephens P, Smith R, Dalgliesh GL, Hunter G, Bignell G, Davies H, Teague J, Butler A, Stevens C, Edkins S, O'Meara S, Vastrik I, Schmidt EE, Avis T, Barthorpe S, Bhamra G, Buck G, Choudhury B, Clements J, Cole J, Dicks E, Forbes S, Gray K, Halliday K, Harrison R, Hills K, Hinton J, Jenkinson A, Jones D, Menzies A, Mironenko T, Perry J, Raine K, Richardson D, Shepherd R, Small A, Tofts C, Varian J, Webb T, West S, Widaa S, Yates A, Cahill DP, Louis DN, Goldstraw P, Nicholson AG, Brasseur F, Looijenga L, Weber BL, Chiew YE, DeFazio A, Greaves MF, Green AR, Campbell P, Birney E, Easton DF, Chenevix-Trench G, Tan MH, Khoo SK, Teh BT, Yuen ST, Leung SY, Wooster R, Futreal PA, Stratton MR (2007) Patterns of somatic mutation in human cancer genomes. Nature 446(7132):153–158. doi:10.1038/nature05610 [pii]

Grohmann U, Fallarino F, Puccetti P (2003) Tolerance, DCs and tryptophan: much ado about IDO. Trends Immunol 24(5):242–248

Guichard C, Amaddeo G, Imbeaud S, Ladeiro Y, Pelletier L, Maad IB, Calderaro J, Bioulac-Sage P, Letexier M, Degos F, Clement B, Balabaud C, Chevet E, Laurent A, Couchy G, Letouze E, Calvo F, Zucman-Rossi J (2012) Integrated analysis of somatic mutations and focal copy-number changes identifies key genes and pathways in hepatocellular carcinoma. Nat Genet 44(6):694–698. doi:10.1038/ng.2256 [pii]

Halliday M, Radford H, Zents KAM, Molloy C, Moreno JA, Verity NC, Smith E, Ortori CA, Barrett DA, Bushell M, Mallucci GR (2017) Repurposed drugs targeting eIF2alpha-P-mediated translational repression prevent neurodegeneration in mice. Brain. doi:10.1093/brain/awx074

Hambardzumyan D, Gutmann DH, Kettenmann H (2016) The role of microglia and macrophages in glioma maintenance and progression. Nat Neurosci 19(1):20–27. doi:10.1038/nn.4185

Harding HP, Zhang Y, Ron D (1999) Protein translation and folding are coupled by an endoplasmic-reticulum-resident kinase. Nature 397(6716):271–274. doi:10.1038/16729

Haze K, Yoshida H, Yanagi H, Yura T, Mori K (1999) Mammalian transcription factor ATF6 is synthesized as a transmembrane protein and activated by proteolysis in response to endoplasmic reticulum stress. Mol Biol Cell 10(11):3787–3799

He L, Kim SO, Kwon O, Jeong SJ, Kim MS, Lee HG, Osada H, Jung M, Ahn JS, Kim BY (2009) ATM blocks tunicamycin-induced endoplasmic reticulum stress. FEBS Lett 583(5):903–908. doi:10.1016/j.febslet.2009.02.002 S0014-5793(09)00094-5 [pii]

Heijmans J, de Jeude JFVL, Koo BK, Rosekrans SL, Wielenga MC, van de Wetering M, Ferrante M, Lee AS, Onderwater JJ, Paton JC, Paton AW, Mommaas AM, Kodach LL, Hardwick JC, Hommes DW, Clevers H, Muncan V, van den Brink GR (2013) ER stress causes rapid loss of intestinal epithelial stemness through activation of the unfolded protein response. Cell Rep 3(4):1128–1139

Henry KA, Blank HM, Hoose SA, Polymenis M (2010) The unfolded protein response is not necessary for the G1/S transition, but it is required for chromosome maintenance in Saccharomyces cerevisiae. PLoS ONE 5(9):e12732. doi:10.1371/journal.pone.0012732

Hetz C, Chevet E, Harding HP (2013) Targeting the unfolded protein response in disease. Nat Rev Drug Discov 12(9):703–719. doi:10.1038/nrd3976 [pii]

Hetz C, Thielen P, Matus S, Nassif M, Court F, Kiffin R, Martinez G, Cuervo AM, Brown RH, Glimcher LH (2009) XBP-1 deficiency in the nervous system protects against amyotrophic lateral sclerosis by increasing autophagy. Genes Dev 23(19):2294–2306

Higa A, Taouji S, Lhomond S, Jensen D, Fernandez-Zapico ME, Simpson JC, Pasquet JM, Schekman R, Chevet E (2014) Endoplasmic reticulum stress-activated transcription factor

ATF6alpha requires the disulfide isomerase PDIA5 to modulate chemoresistance. Mol Cell Biol 34(10):1839–1849. doi:10.1128/MCB.01484-13 [pii]

Hollien J, Lin JH, Li H, Stevens N, Walter P, Weissman JS (2009) Regulated Ire1-dependent decay of messenger RNAs in mammalian cells. J Cell Biol 186(3):323–331

Hollien J, Weissman JS (2006) Decay of endoplasmic reticulum-localized mRNAs during the unfolded protein response. Science 313(5783):104–107

Hsu JL, Chiang PC, Guh JH (2009) Tunicamycin induces resistance to camptothecin and etoposide in human hepatocellular carcinoma cells: role of cell-cycle arrest and GRP78. Naunyn Schmiedebergs Arch Pharmacol 380(5):373–382. doi:10.1007/s00210-009-0453-5

Hu R, Warri A, Jin L, Zwart A, Riggins RB, Clarke R (2014) NFkappaB signaling is required for XBP1 (U and S) mediated effects on antiestrogen responsiveness and cell fate decisions in breast cancer. Mol Cell Biol. doi:10.1128/MCB.00847-14 [pii]

Huber AL, Lebeau J, Guillaumot P, Petrilli V, Malek M, Chilloux J, Fauvet F, Payen L, Kfoury A, Renno T, Chevet E, Manie SN (2013) p58(IPK)-mediated attenuation of the proapoptotic PERK-CHOP pathway allows malignant progression upon low glucose. Mol Cell 49(6):1049–1059

Hussain SF, Yang D, Suki D, Aldape K, Grimm E, Heimberger AB (2006) The role of human glioma-infiltrating microglia/macrophages in mediating antitumor immune responses. Neuro Oncol 8(3):261–279. doi:10.1215/15228517-2006-008

Iwakoshi NN, Pypaert M, Glimcher LH (2007) The transcription factor XBP-1 is essential for the development and survival of dendritic cells. J Exp Med 204(10):2267–2275. doi:10.1084/jem. 20070525

Iwawaki T, Hosoda A, Okuda T, Kamigori Y, Nomura-Furuwatari C, Kimata Y, Tsuru A, Kohno K (2001) Translational control by the ER transmembrane kinase/ribonuclease IRE1 under ER stress. Nat Cell Biol 3:158–164. doi:10.1038/35055065

Jain RK (2003) Molecular regulation of vessel maturation. Nat Med 9(6):685–693. doi:10.1038/nm0603-685

Jamison S, Lin Y, Lin W (2015) Pancreatic endoplasmic reticulum kinase activation promotes medulloblastoma cell migration and invasion through induction of vascular endothelial growth factor A. PLoS ONE 10(3):e0120252. doi:10.1371/journal.pone.0120252

Jurkin J, Henkel T, Nielsen AF, Minnich M, Popow J, Kaufmann T, Heindl K, Hoffmann T, Busslinger M, Martinez J (2014) The mammalian tRNA ligase complex mediates splicing of XBP1 mRNA and controls antibody secretion in plasma cells. EMBO J. doi:10.15252/embj. 201490332 [pii]

Karali E, Bellou S, Stellas D, Klinakis A, Murphy C, Fotsis T (2014) VEGF signals through ATF6 and PERK to promote endothelial cell survival and angiogenesis in the absence of ER stress. Mol Cell 54(4):559–572. doi:10.1016/j.molcel.2014.03.022

Kaser A, Lee A-H, Franke A, Glickman JN, Zeissig S, Tilg H, Nieuwenhuis EES, Higgins DE, Schreiber S, Glimcher LH, Blumberg RS (2008) XBP1 links ER stress to intestinal inflammation and confers genetic risk for human inflammatory bowel disease. Cell 134(5):743–756. doi:10.1016/j.cell.2008.07.021

Kawamura T, Tashiro E, Yamamoto K, Shindo K, Imoto M (2008) SAR study of a novel triene-ansamycin group compound, quinotrierixin, and related compounds, as inhibitors of ER stress-induced XBP1 activation. J Antibiot (Tokyo) 61(5):303–311. doi:10.1038/ja.2008.43 JST.JSTAGE/antibiotics/61.303 [pii]

Kharabi Masouleh B, Geng H, Hurtz C, Chan LN, Logan AC, Chang MS, Huang C, Swaminathan S, Sun H, Paietta E, Melnick AM, Koeffler P, Muschen M (2014) Mechanistic rationale for targeting the unfolded protein response in pre-B acute lymphoblastic leukemia. Proc Natl Acad Sci USA 111(21):E2219–2228. doi:10.1073/pnas.1400958111 [pii]

Kirk SJ, Cliff JM, Thomas JA, Ward TH (2009) Biogenesis of secretory organelles during B cell differentiation. J Leukoc Biol 87(2):245–255. doi:10.1189/jlb.1208774 [pii]

Komohara Y, Ohnishi K, Kuratsu J, Takeya M (2008) Possible involvement of the M2 anti-inflammatory macrophage phenotype in growth of human gliomas. J Pathol 216(1):15–24. doi:10.1002/path.2370

Kosmaczewski SG, Edwards TJ, Han SM, Eckwahl MJ, Meyer BI, Peach S, Hesselberth JR, Wolin SL, Hammarlund M (2014) The RtcB RNA ligase is an essential component of the metazoan unfolded protein response. EMBO Rep. doi:10.15252/embr.201439531 [pii]

Kozutsumi Y, Segal M, Normington K, Gething MJ, Sambrook J (1988) The presence of malfolded proteins in the endoplasmic reticulum signals the induction of glucose-regulated proteins. Nature 332(6163):462–464. doi:10.1038/332462a0

Lee AH, Iwakoshi NN, Anderson KC, Glimcher LH (2003) Proteasome inhibitors disrupt the unfolded protein response in myeloma cells. Proc Natl Acad Sci USA 100(17):9946–9951

Lee K, Tirasophon W, Shen X, Michalak M, Prywes R, Okada T, Yoshida H, Mori K, Kaufman RJ (2002) IRE1-mediated unconventional mRNA splicing and S2P-mediated ATF6 cleavage merge to regulate XBP1 in signaling the unfolded protein response. Genes Dev 16 (4):452–466

Lerner AG, Upton JP, Praveen PV, Ghosh R, Nakagawa Y, Igbaria A, Shen S, Nguyen V, Backes BJ, Heiman M, Heintz N, Greengard P, Hui S, Tang Q, Trusina A, Oakes SA, Papa FR (2012) IRE1alpha induces thioredoxin-interacting protein to activate the NLRP3 inflammasome and promote programmed cell death under irremediable ER stress. Cell Metab 16 (2):250–264. doi:10.1016/j.cmet.2012.07.007 S1550-4131(12)00284-7. [pii]

Lhomond S, Pallares N, Barroso K, Schmit K, Dejeans N, Fazli H, Taouji S, Patterson JB, Chevet E (2015) Adaptation of the secretory pathway in cancer through IRE1 signaling. Methods Mol Biol 1292:177–194. doi:10.1007/978-1-4939-2522-3_13

Li H, Chen X, Gao Y, Wu J, Zeng F, Song F (2014) XBP1 induces snail expression to promote epithelial-to-mesenchymal transition and invasion of breast cancer cells. Cell Signal. doi:10.1016/j.cellsig.2014.09.018 S0898-6568(14)00322-2 [pii]

Lin YH, Friederichs J, Black MA, Mages J, Rosenberg R, Guilford PJ, Phillips V, Thompson-Fawcett M, Kasabov N, Toro T, Merrie AE, van Rij A, Yoon HS, McCall JL, Siewert JR, Holzmann B, Reeve AE (2007) Multiple gene expression classifiers from different array platforms predict poor prognosis of colorectal cancer. Clin Cancer Res Official J Am Assoc Cancer Res 13(2 Pt 1):498–507. doi:10.1158/1078-0432.CCR-05-2734

Lu Y, Liang FX, Wang X (2014) A synthetic biology approach identifies the mammalian UPR RNA ligase RtcB. Mol Cell 55(5):758–770. doi:10.1016/j.molcel.2014.06.032 S1097-2765(14)00566-8 [pii]

Ma Y, Hendershot LM (2004) The role of the unfolded protein response in tumour development: friend or foe? Nat Rev Cancer 4(12):966–977. doi:10.1038/nrc1505 [pii]

Mahadevan NR, Anufreichik V, Rodvold JJ, Chiu KT, Sepulveda H, Zanetti M (2012) Cell-extrinsic effects of tumor ER stress imprint myeloid dendritic cells and impair $CD8^+$ T cell priming. PLoS ONE 7(12):e51845. doi:10.1371/journal.pone.0051845

Mahadevan NR, Rodvold J, Sepulveda H, Rossi S, Drew AF, Zanetti M (2011) Transmission of endoplasmic reticulum stress and pro-inflammation from tumor cells to myeloid cells. Proc Natl Acad Sci USA 108(16):6561–6566

Maines MD (1988) Heme oxygenase: function, multiplicity, regulatory mechanisms, and clinical applications. FASEB J 2(10):2557–2568

Mantovani A, Sozzani S, Locati M, Allavena P, Sica A (2002) Macrophage polarization: tumor-associated macrophages as a paradigm for polarized M2 mononuclear phagocytes. Trends Immunol 23(11):549–555

Mardones P, Martinez G, Hetz C (2015) Control of systemic proteostasis by the nervous system. Trends Cell Biol 25(1):1–10. doi:10.1016/j.tcb.2014.08.001

Martinez-Outschoorn UE, Balliet RM, Rivadeneira DB, Chiavarina B, Pavlides S, Wang C, Whitaker-Menezes D, Daumer KM, Lin Z, Witkiewicz AK, Flomenberg N, Howell A, Pestell RG, Knudsen ES, Sotgia F, Lisanti MP (2010) Oxidative stress in cancer associated fibroblasts drives tumor-stroma co-evolution: a new paradigm for understanding tumor metabolism, the field effect and genomic instability in cancer cells. Cell Cycle 9(16):3256–3276. doi:10.4161/cc.9.16.12553

Martino MB, Jones L, Brighton B, Ehre C, Abdulah L, Davis CW, Ron D, O'Neal WK, Ribeiro CM (2013) The ER stress transducer IRE1beta is required for airway epithelial mucin production. Mucosal Immunol 6(3):639–654. doi:10.1038/mi.2012.105

Martinon F, Chen X, Lee A-H, Glimcher LH (2010a) TLR activation of the transcription factor XBP1 regulates innate immune responses in macrophages. Nat Immunol 11(5):411–418. doi:10.1038/ni.1857

Martinon F, Chen X, Lee AH, Glimcher LH (2010b) TLR activation of the transcription factor XBP1 regulates innate immune responses in macrophages. Nat Immunol 11(5):411–418

Marza E, Taouji S, Barroso K, Raymond AA, Guignard L, Bonneu M, Pallares-Lupon N, Dupuy JW, Fernandez-Zapico ME, Rosenbaum J, Palladino F, Dupuy D, Chevet E (2015) Genome-wide screen identifies a novel p97/CDC-48-dependent pathway regulating ER-stress-induced gene transcription. EMBO Rep 16(3):332–340. doi:10.15252/embr.201439123

Maurel M, Chevet E, Tavernier J, Gerlo S (2014) Getting RIDD of RNA: IRE1 in cell fate regulation. Trends Biochem Sci 39(5):245–254. doi:10.1016/j.tibs.2014.02.008 S0968-0004(14)00033-4 [pii]

Mimura N, Fulciniti M, Gorgun G, Tai YT, Cirstea D, Santo L, Hu Y, Fabre C, Minami J, Ohguchi H, Kiziltepe T, Ikeda H, Kawano Y, French M, Blumenthal M, Tam V, Kertesz NL, Malyankar UM, Hokenson M, Pham T, Zeng Q, Patterson JB, Richardson PG, Munshi NC, Anderson KC (2012) Blockade of XBP1 splicing by inhibition of IRE1alpha is a promising therapeutic option in multiple myeloma. Blood 119(24):5772–5781. doi:10.1182/blood-2011-07-366633

Morales C, Rachidi S, Hong F, Sun S, Ouyang X, Wallace C, Zhang Y, Garret-Mayer E, Wu J, Liu B, Li Z (2014) Immune chaperone gp96 drives the contributions of macrophages to inflammatory colon tumorigenesis. Cancer Res 74(2):446–459. doi:10.1158/0008-5472.CAN-13-1677

Mori K, Ma W, Gething MJ, Sambrook J (1993) A transmembrane protein with a cdc2$^+$/CDC28-related kinase activity is required for signaling from the ER to the nucleus. Cell 74(4):743–756 0092-8674(93)90521-Q [pii]

Mujcic H, Nagelkerke A, Rouschop KM, Chung S, Chaudary N, Span PN, Clarke B, Milosevic M, Sykes J, Hill RP, Koritzinsky M, Wouters BG (2013) Hypoxic activation of the PERK/eIF2alpha arm of the unfolded protein response promotes metastasis through induction of LAMP3. Clin Cancer Res 19(22):6126–6137. doi:10.1158/1078-0432.CCR-13-0526

Nagelkerke A, Bussink J, Mujcic H, Wouters BG, Lehmann S, Sweep FC, Span PN (2013) Hypoxia stimulates migration of breast cancer cells via the PERK/ATF4/LAMP3-arm of the unfolded protein response. Breast Cancer Res BCR 15(1):R2. doi:10.1186/bcr3373

Nakajima S, Hiramatsu N, Hayakawa K, Saito Y, Kato H, Huang T, Yao J, Paton AW, Paton JC, Kitamura M (2011) Selective abrogation of BiP/GRP78 blunts activation of NF-kappaB through the ATF6 branch of the UPR: involvement of C/EBPbeta and mTOR-dependent dephosphorylation of Akt. Mol Cell Biol 31(8):1710–1718. doi:10.1128/MCB.00939-10 [pii]

Nguyen T, Nioi P, Pickett CB (2009) The Nrf2-antioxidant response element signaling pathway and its activation by oxidative stress. J Biol Chem 284(20):13291–13295. doi:10.1074/jbc.R900010200

Niederreiter L, Fritz TM, Adolph TE, Krismer AM, Offner FA, Tschurtschenthaler M, Flak MB, Hosomi S, Tomczak MF, Kaneider NC, Sarcevic E, Kempster SL, Raine T, Esser D, Rosenstiel P, Kohno K, Iwawaki T, Tilg H, Blumberg RS, Kaser A (2013) ER stress transcription factor Xbp1 suppresses intestinal tumorigenesis and directs intestinal stem cells. J Exp Med 210(10):2041–2056. doi:10.1084/jem.20122341 [pii]

Nieto MA, Cano A (2012) The epithelial-mesenchymal transition under control: global programs to regulate epithelial plasticity. Semin Cancer Biol 22(5–6):361–368. doi:10.1016/j.semcancer.2012.05.003 S1044-579X(12)00079-X [pii]

Norian LA, Rodriguez PC, O'Mara LA, Zabaleta J, Ochoa AC, Cella M, Allen PM (2009) Tumor-infiltrating regulatory dendritic cells inhibit CD8$^+$ T cell function via L-arginine metabolism. Cancer Res 69(7):3086–3094. doi:10.1158/0008-5472.CAN-08-2826

Novoa I, Zeng H, Harding HP, Ron D (2001) Feedback inhibition of the unfolded protein response by GADD34-mediated dephosphorylation of eIF2alpha. J Cell Biol 153(5):1011–1022

Obacz J, Avril T, Le Reste PJ, Urra H, Quillien V, Hetz C, Chevet E (2017) Endoplasmic reticulum proteostasis in glioblastoma-From molecular mechanisms to therapeutic perspectives. Sci Signal 10(470). doi:10.1126/scisignal.aal2323

Ogata M, Hino S, Saito A, Morikawa K, Kondo S, Kanemoto S, Murakami T, Taniguchi M, Tanii I, Yoshinaga K, Shiosaka S, Hammarback JA, Urano F, Imaizumi K (2006) Autophagy is activated for cell survival after endoplasmic reticulum stress. Mol Cell Biol 26(24):9220–9231

Oromendia AB, Dodgson SE, Amon A (2012) Aneuploidy causes proteotoxic stress in yeast. Genes Dev 26(24):2696–2708. doi:10.1101/gad.207407.112

Osorio F, Tavernier SJ, Hoffmann E, Saeys Y, Martens L, Vetters J, Delrue I, De Rycke R, Parthoens E, Pouliot P, Iwawaki T, Janssens S, Lambrecht BN (2014) The unfolded-protein-response sensor IRE-1alpha regulates the function of CD8alpha+ dendritic cells. Nat Immunol 15(3):248–257. doi:10.1038/ni.2808

Paez D, Labonte MJ, Bohanes P, Zhang W, Benhanim L, Ning Y, Wakatsuki T, Loupakis F, Lenz HJ (2012) Cancer dormancy: a model of early dissemination and late cancer recurrence. Clin Cancer Res Official J Am Assoc Cancer Res 18(3):645–653. doi:10.1158/1078-0432.CCR-11-2186

Papandreou I, Denko NC, Olson M, Van Melckebeke H, Lust S, Tam A, Solow-Cordero DE, Bouley DM, Offner F, Niwa M, Koong AC (2011) Identification of an Ire1alpha endonuclease specific inhibitor with cytotoxic activity against human multiple myeloma. Blood 117(4):1311–1314. doi:10.1182/blood-2010-08-303099 [pii]

Park SH, Kim J, Do KH, Park J, Oh CG, Choi HJ, Song BG, Lee SJ, Kim YS, Moon Y (2014) Activating transcription factor 3-mediated chemo-intervention with cancer chemokines in a noncanonical pathway under endoplasmic reticulum stress. J Biol Chem 289(39):27118–27133. doi:10.1074/jbc.M114.568717

Parsons DW, Jones S, Zhang X, Lin JC, Leary RJ, Angenendt P, Mankoo P, Carter H, Siu IM, Gallia GL, Olivi A, McLendon R, Rasheed BA, Keir S, Nikolskaya T, Nikolsky Y, Busam DA, Tekleab H, Diaz LA Jr, Hartigan J, Smith DR, Strausberg RL, Marie SK, Shinjo SM, Yan H, Riggins GJ, Bigner DD, Karchin R, Papadopoulos N, Parmigiani G, Vogelstein B, Velculescu VE, Kinzler KW (2008) An integrated genomic analysis of human glioblastoma multiforme. Science 321(5897):1807–1812. doi:10.1126/science.1164382 [pii]

Pereira ER, Liao N, Neale GA, Hendershot LM (2010) Transcriptional and post-transcriptional regulation of proangiogenic factors by the unfolded protein response. PLoS One 5(9). doi:10.1371/journal.pone.0012521

Plate L, Cooley CB, Chen JJ, Paxman RJ, Gallagher CM, Madoux F, Genereux JC, Dobbs W, Garza D, Spicer TP, Scampavia L, Brown SJ, Rosen H, Powers ET, Walter P, Hodder P, Wiseman RL, Kelly JW (2016) Small molecule proteostasis regulators that reprogram the ER to reduce extracellular protein aggregation. Elife 5. doi:10.7554/eLife.15550

Pluquet O, Dejeans N, Bouchecareilh M, Lhomond S, Pineau R, Higa A, Delugin M, Combe C, Loriot S, Cubel G, Dugot-Senant N, Vital A, Loiseau H, Gosline SJ, Taouji S, Hallett M, Sarkaria JN, Anderson K, Wu W, Rodriguez FJ, Rosenbaum J, Saltel F, Fernandez-Zapico ME, Chevet E (2013) Posttranscriptional regulation of PER1 underlies the oncogenic function of IREalpha. Cancer Res 73(15):4732–4743. doi:10.1158/0008-5472.CAN-12-3989 [pii]

Pollard JW (2004) Tumour-educated macrophages promote tumour progression and metastasis. Nat Rev Cancer 4(1):71–78. doi:10.1038/nrc1256

Pytel D, Majsterek I, Diehl JA (2016) Tumor progression and the different faces of the PERK kinase. Oncogene 35(10):1207–1215. doi:10.1038/onc.2015.178

Ramaswamy S, Tamayo P, Rifkin R, Mukherjee S, Yeang CH, Angelo M, Ladd C, Reich M, Latulippe E, Mesirov JP, Poggio T, Gerald W, Loda M, Lander ES, Golub TR (2001) Multiclass cancer diagnosis using tumor gene expression signatures. Proc Natl Acad Sci USA 98(26):15149–15154. doi:10.1073/pnas.211566398

Ranganathan AC, Zhang L, Adam AP, Aguirre-Ghiso JA (2006) Functional coupling of p38-induced up-regulation of BiP and activation of RNA-dependent protein kinase-like endoplasmic reticulum kinase to drug resistance of dormant carcinoma cells. Cancer Res 66 (3):1702–1711. doi:10.1158/0008-5472.CAN-05-3092

Ray A, Zhang S, Rentas C, Caldwell KA, Caldwell GA (2014) RTCB-1 mediates neuroprotection via XBP-1 mRNA splicing in the unfolded protein response pathway. J Neurosci 34 (48):16076–16085. doi:10.1523/JNEUROSCI.1945-14.2014 34/48/16076 [pii]

Ri M, Tashiro E, Oikawa D, Shinjo S, Tokuda M, Yokouchi Y, Narita T, Masaki A, Ito A, Ding J, Kusumoto S, Ishida T, Komatsu H, Shiotsu Y, Ueda R, Iwawaki T, Imoto M, Iida S (2012) Identification of Toyocamycin, an agent cytotoxic for multiple myeloma cells, as a potent inhibitor of ER stress-induced XBP1 mRNA splicing. Blood Cancer J 2(7):e79. doi:10.1038/bcj.2012.26

Ron D, Walter P (2007) Signal integration in the endoplasmic reticulum unfolded protein response. Nat Rev Mol Cell Biol 8(7):519–529

Rouschop KM, van den Beucken T, Dubois L, Niessen H, Bussink J, Savelkouls K, Keulers T, Mujcic H, Landuyt W, Voncken JW, Lambin P, van der Kogel AJ, Koritzinsky M, Wouters BG (2010) The unfolded protein response protects human tumor cells during hypoxia through regulation of the autophagy genes MAP1LC3B and ATG5. J Clin Invest 120(1):127–141. doi:10.1172/JCI40027 [pii]

Rzymski T, Milani M, Singleton DC, Harris AL (2009) Role of ATF4 in regulation of autophagy and resistance to drugs and hypoxia. Cell Cycle 8(23):3838–3847 10086 [pii]

Rzymski T, Petry A, Kracun D, Riess F, Pike L, Harris AL, Gorlach A (2012) The unfolded protein response controls induction and activation of ADAM17/TACE by severe hypoxia and ER stress. Oncogene 31(31):3621–3634. doi:10.1038/onc.2011.522

Sanches M, Duffy NM, Talukdar M, Thevakumaran N, Chiovitti D, Canny MD, Lee K, Kurinov I, Uehling D, Al-awar R, Poda G, Prakesch M, Wilson B, Tam V, Schweitzer C, Toro A, Lucas JL, Vuga D, Lehmann L, Durocher D, Zeng Q, Patterson JB, Sicheri F (2014) Structure and mechanism of action of the hydroxy-aryl-aldehyde class of IRE1 endoribonuclease inhibitors. Nat Commun 5:4202. doi:10.1038/ncomms5202

Schewe DM, Aguirre-Ghiso JA (2008) ATF6alpha-Rheb-mTOR signaling promotes survival of dormant tumor cells in vivo. Proc Natl Acad Sci USA 105(30):10519–10524. doi:10.1073/pnas.0800939105 [pii]

Sengupta S, Sharma CG, Jordan VC (2010) Estrogen regulation of X-box binding protein-1 and its role in estrogen induced growth of breast and endometrial cancer cells. Horm Mol Biol Clin Investig 2(2):235–243. doi:10.1515/HMBCI.2010.025

Shen J, Chen X, Hendershot L, Prywes R (2002) ER stress regulation of ATF6 localization by dissociation of BiP/GRP78 binding and unmasking of Golgi localization signals. Dev Cell 3 (1):99–111 S1534580702002034 [pii]

Shen J, Prywes R (2004) Dependence of site-2 protease cleavage of ATF6 on prior site-1 protease digestion is determined by the size of the luminal domain of ATF6. J Biol Chem 279 (41):43046–43051. doi:10.1074/jbc.M408466200 [pii]

Shen Q, Uray IP, Li Y, Krisko TI, Strecker TE, Kim HT, Brown PH (2008) The AP-1 transcription factor regulates breast cancer cell growth via cyclins and E2F factors. Oncogene 27(3):366–377. doi:10.1038/sj.onc.1210643

Sheshadri N, Catanzaro JM, Bott A, Sun Y, Ullman E, Chen E, Pan JA, Wu S, Crawford HC, Zhang J, Zong WX (2014) SCCA1/SerpinB3 promotes oncogenesis and epithelial-mesenchymal transition via the unfolded protein response and IL-6 signaling. Cancer Res. doi:10.1158/0008-5472.CAN-14-0798 [pii]

Shi Y, Vattem KM, Sood R, An J, Liang J, Stramm L, Wek RC (1998) Identification and characterization of pancreatic eukaryotic initiation factor 2 alpha-subunit kinase, PEK, involved in translational control. Mol Cell Biol 18(12):7499–7509

Shoulders MD, Ryno LM, Genereux JC, Moresco JJ, Tu PG, Wu C, Yates JR 3rd, Su AI, Kelly JW, Wiseman RL (2013) Stress-independent activation of XBP1s and/or ATF6 reveals three functionally diverse ER proteostasis environments. Cell Rep 3(4):1279–1292. doi:10.1016/j.celrep.2013.03.024 S2211-1247(13)00131-9 [pii]

Sidrauski C, Acosta-Alvear D, Khoutorsky A, Vedantham P, Hearn BR, Li H, Gamache K, Gallagher CM, Ang KK, Wilson C, Okreglak V, Ashkenazi A, Hann B, Nader K, Arkin MR, Renslo AR, Sonenberg N, Walter P (2013) Pharmacological brake-release of mRNA translation enhances cognitive memory. Elife 2:e00498. doi:10.7554/eLife.00498 [pii]

Sidrauski C, McGeachy AM, Ingolia NT, Walter P (2015a) The small molecule ISRIB reverses the effects of eIF2alpha phosphorylation on translation and stress granule assembly. Elife 4. doi:10.7554/eLife.05033

Sidrauski C, Tsai JC, Kampmann M, Hearn BR, Vedantham P, Jaishankar P, Sokabe M, Mendez AS, Newton BW, Tang EL, Verschueren E, Johnson JR, Krogan NJ, Fraser CS, Weissman JS, Renslo AR, Walter P (2015b) Pharmacological dimerization and activation of the exchange factor eIF2B antagonizes the integrated stress response. Elife 4:e07314. doi:10.7554/eLife.07314

Tam AB, Koong AC, Niwa M (2014) Ire1 has distinct catalytic mechanisms for XBP1/HAC1 splicing and RIDD. Cell Reports 9:1–9

Tang CH, Ranatunga S, Kriss CL, Cubitt CL, Tao J, Pinilla-Ibarz JA, Del Valle JR, Hu CC (2014) Inhibition of ER stress-associated IRE-1/XBP-1 pathway reduces leukemic cell survival. J Clin Invest 124(6):2585–2598. doi:10.1172/JCI73448 [pii]

Tanjore H, Cheng DS, Degryse AL, Zoz DF, Abdolrasulnia R, Lawson WE, Blackwell TS (2011) Alveolar epithelial cells undergo epithelial-to-mesenchymal transition in response to endoplasmic reticulum stress. J Biol Chem 286(35):30972–30980. doi:10.1074/jbc.M110.181164 [pii]

Taylor RC, Dillin A (2013) XBP-1 is a cell-nonautonomous regulator of stress resistance and longevity. Cell 153(7):1435–1447. doi:10.1016/j.cell.2013.05.042

Thevenot PT, Sierra RA, Raber PL, Al-Khami AA, Trillo-Tinoco J, Zarreii P, Ochoa AC, Cui Y, Del Valle L, Rodriguez PC (2014) The stress-response sensor chop regulates the function and accumulation of myeloid-derived suppressor cells in tumors. Immunity 41(3):389–401. doi:10.1016/j.immuni.2014.08.015

Thiery JP, Acloque H, Huang RY, Nieto MA (2009) Epithelial-mesenchymal transitions in development and disease. Cell 139(5):871–890. doi:10.1016/j.cell.2009.11.007 S0092-8674(09)01419-6 [pii]

Ja Thorpe, Schwarze SR (2010) IRE1alpha controls cyclin A1 expression and promotes cell proliferation through XBP-1. Cell Stress Chaperones 15:497–508. doi:10.1007/s12192-009-0163-4

Toledo CM, Ding Y, Hoellerbauer P, Davis RJ, Basom R, Girard EJ, Lee E, Corrin P, Hart T, Bolouri H, Davison J, Zhang Q, Hardcastle J, Aronow BJ, Plaisier CL, Baliga NS, Moffat J, Lin Q, Li XN, Nam DH, Lee J, Pollard SM, Zhu J, Delrow JJ, Clurman BE, Olson JM, Paddison PJ (2015) Genome-wide CRISPR-Cas9 screens reveal loss of redundancy between PKMYT1 and WEE1 in glioblastoma stem-like cells. Cell Rep 13(11):2425–2439. doi:10.1016/j.celrep.2015.11.021

Tsaytler P, Harding HP, Ron D, Bertolotti A (2011) Selective inhibition of a regulatory subunit of protein phosphatase 1 restores proteostasis. Science 332(6025):91–94. doi:10.1126/science.1201396 [pii]

Tsuru A, Fujimoto N, Takahashi S, Saito M, Nakamura D, Iwano M, Iwawaki T, Kadokura H, Ron D, Kohno K (2013) Negative feedback by IRE1beta optimizes mucin production in goblet cells. Proc Natl Acad Sci USA 110(8):2864–2869

Ueno T, Toi M, Saji H, Muta M, Bando H, Kuroi K, Koike M, Inadera H, Matsushima K (2000) Significance of macrophage chemoattractant protein-1 in macrophage recruitment, angiogenesis, and survival in human breast cancer. Clin Cancer Res 6(8):3282–3289

Ulianich L, Garbi C, Treglia AS, Punzi D, Miele C, Raciti GA, Beguinot F, Consiglio E, Di Jeso B (2008) ER stress is associated with dedifferentiation and an epithelial-to-mesenchymal transition-like phenotype in PC Cl3 thyroid cells. J Cell Sci 121(Pt 4):477–486. doi:10.1242/jcs.017202 [pii]

Upton JP, Wang L, Han D, Wang ES, Huskey NE, Lim L, Truitt M, McManus MT, Ruggero D, Goga A, Papa FR, Oakes SA (2012) IRE1alpha cleaves select microRNAs during ER stress to derepress translation of proapoptotic Caspase-2. Science 338(6108):818–822. doi:10.1126/science.1226191 [pii]

Urano F, Wang X, Bertolotti A, Zhang Y, Chung P, Harding HP, Ron D (2000) Coupling of stress in the ER to activation of JNK protein kinases by transmembrane protein kinase IRE1. Science 287(5453):664–666 8218 [pii]

Urra H, Dufey E, Avril T, Chevet E, Hetz C (2016) Endoplasmic reticulum stress and the hallmarks of cancer. Trends in Cancer 2(5):252–262

Volkmann K, Lucas JL, Vuga D, Wang X, Brumm D, Stiles C, Kriebel D, Der-Sarkissian A, Krishnan K, Schweitzer C, Liu Z, Malyankar UM, Chiovitti D, Canny M, Durocher D, Sicheri F, Patterson JB (2011) Potent and selective inhibitors of the inositol-requiring enzyme 1 endoribonuclease. J Biol Chem 286(14):12743–12755. doi:10.1074/jbc.M110.199737 [pii]

Walter P, Ron D (2011) The unfolded protein response: from stress pathway to homeostatic regulation. Science 334(6059):1081–1086. doi:10.1126/science.1209038 334/6059/1081 [pii]

Wang L, Perera BG, Hari SB, Bhhatarai B, Backes BJ, Seeliger MA, Schurer SC, Oakes SA, Papa FR, Maly DJ (2012a) Divergent allosteric control of the IRE1alpha endoribonuclease using kinase inhibitors. Nat Chem Biol 8(12):982–989. doi:10.1038/nchembio.1094

Wang Y, Alam GN, Ning Y, Visioli F, Dong Z, Nor JE, Polverini PJ (2012b) The unfolded protein response induces the angiogenic switch in human tumor cells through the PERK/ATF4 pathway. Cancer Res 72(20):5396–5406. doi:10.1158/0008-5472.CAN-12-0474

Wang ZV, Deng Y, Gao N, Pedrozo Z, Li DL, Morales CR, Criollo A, Luo X, Tan W, Jiang N, Lehrman MA, Rothermel BA, Lee AH, Lavandero S, Mammen PP, Ferdous A, Gillette TG, Scherer PE, Hill JA (2014) Spliced X-box binding protein 1 couples the unfolded protein response to hexosamine biosynthetic pathway. Cell 156(6):1179–1192. doi:10.1016/j.cell.2014.01.014 S0092-8674(14)00025-7 [pii]

Wheeler MC, Rizzi M, Sasik R, Almanza G, Hardiman G, Zanetti M (2008) KDEL-retained antigen in B lymphocytes induces a proinflammatory response: a possible role for endoplasmic reticulum stress in adaptive T cell immunity. J Immunol 181(1):256–264

Xue Z, He Y, Ye K, Gu Z, Mao Y, Qi L (2011) A conserved structural determinant located at the interdomain region of mammalian inositol-requiring enzyme 1{alpha}. J Biol Chem 286:30859–30866. doi:10.1074/jbc.M111.273714

Yacoub A, Hamed HA, Allegood J, Mitchell C, Spiegel S, Lesniak MS, Ogretmen B, Dash R, Sarkar D, Broaddus WC, Grant S, Curiel DT, Fisher PB, Dent P (2010) PERK-dependent regulation of ceramide synthase 6 and thioredoxin play a key role in mda-7/IL-24-induced killing of primary human glioblastoma multiforme cells. Cancer Res 70(3):1120–1129. doi:10.1158/0008-5472.CAN-09-4043

Yamamori T, Meike S, Nagane M, Yasui H, Inanami O (2013) ER stress suppresses DNA double-strand break repair and sensitizes tumor cells to ionizing radiation by stimulating proteasomal degradation of Rad51. FEBS Lett 587(20):3348–3353. doi:10.1016/j.febslet.2013.08.030 S0014-5793(13)00659-5 [pii]

Yamamoto K, Sato T, Matsui T, Sato M, Okada T, Yoshida H, Harada A, Mori K (2007) Transcriptional induction of mammalian ER quality control proteins is mediated by single or combined action of ATF6alpha and XBP1. Dev Cell 13(3):365–376

Yan D, Wang HW, Bowman RL, Joyce JA (2016) STAT3 and STAT6 signaling pathways synergize to promote cathepsin secretion from macrophages via IRE1alpha activation. Cell Rep 16(11):2914–2927. doi:10.1016/j.celrep.2016.08.035

Yang C, Huntoon K, Ksendzovsky A, Zhuang Z, Lonser RR (2013) Proteostasis modulators prolong missense VHL protein activity and halt tumor progression. Cell Rep 3(1):52–59. doi:10.1016/j.celrep.2012.12.007

Ye J, Rawson RB, Komuro R, Chen X, Dave UP, Prywes R, Brown MS, Goldstein JL (2000) ER stress induces cleavage of membrane-bound ATF6 by the same proteases that process SREBPs. Mol Cell 6(6):1355–1364 S1097-2765(00)00133-7 [pii]

Yoshida H, Okada T, Haze K, Yanagi H, Yura T, Negishi M, Mori K (2000) ATF6 activated by proteolysis binds in the presence of NF-Y (CBF) directly to the cis-acting element responsible for the mammalian unfolded protein response. Mol Cell Biol 20(18):6755–6767

Zeindl-Eberhart E, Brandl L, Liebmann S, Ormanns S, Scheel SK, Brabletz T, Kirchner T, Jung A (2014) Epithelial-mesenchymal transition induces endoplasmic-reticulum-stress response in human colorectal tumor cells. PLoS ONE 9(1):e87386. doi:10.1371/journal.pone.0087386 PONE-D-13-31507 [pii]

Zhang W, Hietakangas V, Wee S, Lim SC, Gunaratne J, Cohen SM (2013) ER stress potentiates insulin resistance through PERK-mediated FOXO phosphorylation. Genes Dev 27(4):441–449. doi:10.1101/gad.201731.112 27/4/441 [pii]

Zhou Y, Lee J, Reno CM, Sun C, Park SW, Chung J, Fisher SJ, White MF, Biddinger SB, Ozcan U (2011) Regulation of glucose homeostasis through a XBP-1-FoxO1 interaction. Nat Med 17(3):356–365. doi:10.1038/nm.2293 nm.2293 [pii]

Zhu H, Chen X, Chen B, Song W, Sun D, Zhao Y (2014) Activating transcription factor 4 promotes esophageal squamous cell carcinoma invasion and metastasis in mice and is associated with poor prognosis in human patients. PLoS ONE 9(7):e103882. doi:10.1371/journal.pone.0103882

Zinszner H, Kuroda M, Wang X, Batchvarova N, Lightfoot RT, Remotti H, Stevens JL, Ron D (1998) CHOP is implicated in programmed cell death in response to impaired function of the endoplasmic reticulum. Genes Dev 12(7):982–995

Zitvogel L, Galluzzi L, Smyth MJ, Kroemer G (2013) Mechanism of action of conventional and targeted anticancer therapies: reinstating immunosurveillance. Immunity 39(1):74–88. doi:10.1016/j.immuni.2013.06.014 S1074-7613(13)00281-1 [pii]

ER Protein Quality Control and the Unfolded Protein Response in the Heart

A. Arrieta, E.A. Blackwood and C.C. Glembotski

Abstract Cardiac myocytes are the cells responsible for the robust ability of the heart to pump blood throughout the circulatory system. Cardiac myocytes grow in response to a variety of physiological and pathological conditions; this growth challenges endoplasmic reticulum-protein quality control (ER-PQC), a major feature of which includes the unfolded protein response (UPR). ER-PQC and the UPR in cardiac myocytes growing under physiological conditions, including normal development, exercise, and pregnancy, are sufficient to support hypertrophic growth of each cardiac myocyte. However, the ER-PQC and UPR are insufficient to respond to the challenge of cardiac myocyte growth under pathological conditions, including myocardial infarction and heart failure. In part, this insufficiency is due to a continual decline in the expression levels of important adaptive UPR components as a function of age and during myocardial pathology. This chapter will discuss the physiological and pathological conditions unique to the heart that involves ER-PQC, and whether the UPR is adaptive or maladaptive under these circumstances.

Contents

1	Introduction...	194
2	ER Protein Quality Control in Cardiac Myocytes..	196
3	Physiological Conditions that Challenge the ER-PQC in Cardiac Myocytes...	198
4	Pathological Conditions that Challenge the ER-PQC in Cardiac Myocytes.....	201
5	Experiment Methods and Models for Studying ER-PQC in Cardiac Pathology...	204

A. Arrieta · E.A. Blackwood · C.C. Glembotski (✉)
San Diego State University Heart Institute and the Department of Biology,
San Diego State University, San Diego, CA 92182, USA
e-mail: cglembotski@mail.sdsu.edu

Current Topics in Microbiology and Immunology (2018) 414:193–214
DOI 10.1007/82_2017_54
© Springer International Publishing AG 2017
Published Online: 13 October 2017

6	Studies of ER-PQC in Cardiac Pathology	205
	6.1 PERK/CHOP	205
	6.2 IRE-1/XBP1	206
	6.3 ATF6	207
7	Conclusions	208
References		208

1 Introduction

In most cells, including those comprising the heart, the rough endoplasmic reticulum (ER) serves as the location for the synthesis of most secreted and membrane proteins (Palade and Siekevitz 1956; Reid and Nicchitta 2012). This accounts for at least 35% of all proteins, emphasizing the importance of ensuring the quality of these proteins. In the heart, a number of pathological conditions including ischemia, ischemia/reperfusion, and hypertrophic myocyte growth can place high demands on the ER protein folding machinery and, in some cases, can cause ER stress (Glembotski 2007; Minamino and Kitakaze 2010) (Fig. 1a), which impairs ER protein folding and leads to activation of the unfolded protein response (UPR). As in other tissue types, the UPR in the heart, which comprises the ER protein quality control (ER-PQC) system, consists of three major branches that are mediated by the ER-transmembrane sensors of ER-misfolded proteins, PERK, IRE-1, and ATF6 (Fig. 1b–d), each of which transduces the signal initiated by SR/ER protein misfolding to downstream events (Glembotski 2007, 2008; Doroudgar and Glembotski 2013b). The ER-PQC is responsible for ensuring that proteins made in the ER are properly folded and correctly modified post-translationally, for example, by glycosylation or disulfide bond formation, before they are permitted to leave the ER on the way to their eventual target destinations (Walter and Ron 2011; Gardner et al. 2013). Proteins that do not pass ER-PQC are not permitted to move on, and instead are translocated back out of the ER lumen and subjected to proteasome-mediated degradation in a process called ER associated degradation or ERAD (Plemper and Wolf 1999; Pisoni and Molinari 2016) (Fig. 1e). Acute activation of PERK, IRE-1, and ATF6 leads to restoration of ER protein folding and cell survival. However, in the event that ER protein folding is not restored, the UPR remains activated, and this chronic activation leads to initiation of cell death. While the precise mechanisms responsible for switching the UPR from adaptive to maladaptive are not well understood, it is apparent that, if activated chronically, each branch of the UPR can lead to the initiation of death pathways. Temporal changes in the UPR gene program are likely to be at least partly responsible for the transition of the UPR from adaptive to maladaptive (Walter and Ron 2011; Hetz et al. 2015; Ryoo 2016). In the heart, it is apparent that prolonged activation of the UPR leads to decreased expression of adaptive UPR gene products, such as ER chaperones and protein disulfide isomerases, and increased expression of pro-death UPR gene products, such as C/EBP homologous protein (CHOP), which can contribute to the switch of the UPR from adaptive to maladaptive (Sano and Reed 2013; Li et al. 2014; Ryoo 2016). Targeted gene deletion studies, which are discussed in the last section of this review, have shown that PERK is adaptive in the heart (Fig. 1b), and is required for optimal cardiac function in models of myocardial disease, while CHOP, which lies downstream of PERK, is maladaptive (Fig. 1f). Additional studies have shown that the UPR-regulated transcription factors XBP1$_s$ and ATF6 serve novel adaptive roles in the heart by inducing genes required for cytosolic protein O-GlcNAcylation (Fig. 1g) and an array of about 400 UPR gene products (Fig. 1h), respectively.

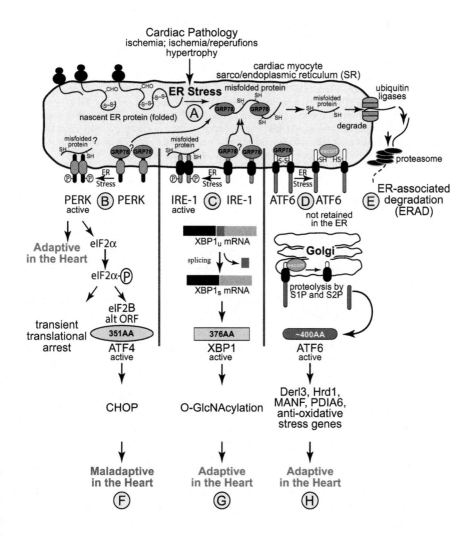

◂**Fig. 1** Effects of cardiac pathology on ER stress and the unfolded protein response in cardiac myocytes: Shown is a diagram of the rough ER with attached ribosomes translating mRNAs that encode ER-luminal proteins. Conditions that impair the folding of nascent ER proteins, which include ischemia, ischemia/reperfusion, and hypertrophy, can result in ER stress (**a**). Under non-stressed conditions, the ER-luminal chaperone, glucose-regulated protein 78 kDa (GRP78) associates with the luminal domains of the 3 proximal effectors of ER stress, PERK (**b**), IRE-1, (**c**) and ATF6 (**d**). Upon ER stress, GRP78 relocates from the luminal domains of these proteins to misfolded proteins and either facilitates their folding, or escorts them to the degradation machinery (**e**). The disassociation of GRP78 from PERK, IRE-1, as well as the binding of misfolded proteins to IRE-1 and, possibly to PERK, allows their oligomerization, which fosters trans-phosphorylation and activation of these effectors. In the case of ATF6, dissociation of GRP78 and the association of escorts, such as thrombospondin 4, facilitate the movement of ATF6 to the Golgi, where it is cleaved by site 1 and site 2 proteases that reside in the Golgi. The resulting N-terminal fragment is liberated from the Golgi, translocates to the nucleus, binds to ER stress response elements in ER stress response (ERSR) genes, and regulates their transcription (**d**). Activated PERK (**b**) phosphorylates eIF2α, which fosters transient global translational repression and the translation of the ATF4 mRNA from an alternate start site to generate active ATF4 using an alternate open reading frame (ORF). While PERK appears to be cardioprotective, the ATF4-mediated induction of CHOP is pro-apoptotic in the heart (**f**). Activated IRE-1 splices the unspliced form of XBP1 mRNA (XBP1$_u$ mRNA) to generate a splice variant form (XBP1$_s$ mRNA) that encodes the active transcription factor, XBP1$_s$ (**c**). Like ATF6, XBP1$_s$ translocates to the nucleus and bind to various types of regulator elements in ERSRs to regulate their expression. One important group of XBP1$_s$ genes in the heart is those that increase cytosolic protein O-GlcNAcylation, which is cardioprotective (**g**). Depending on whether ER stress and UPR gene induction is acute or chronic, as with other tissues, in the heart the results are either adaptive (acute), oriented toward resolution of ER stress and cell survival (**g, h**), or maladaptive (chronic), activating death pathways (**f**)

2 ER Protein Quality Control in Cardiac Myocytes

Heart disease includes both vascular and cardiac diseases. Vascular diseases, such as atherosclerosis, affect blood vessels, and blood, while cardiac diseases affect the heart muscle, or myocardium and impair its ability to efficiently propel blood to organs and tissues. While ER-PQC, ER stress, and the UPR have been studied extensively in the setting of vascular diseases (Zhou and Tabas 2013; Ivanova and Orekhov 2016), until relatively recently, these topics have attracted less attention in studies of the myocardium. Among the possible reasons for this are the lack of appreciation of the importance of the ER as a site of protein synthesis in cardiac myocytes, as well as the relatively ill-defined location and function of the ER protein synthetic machinery in cardiac myocytes. Contributing to these reasons is the complicated structure of the ER in cardiac myocytes, most of which comprises the sarco/endoplasmic reticulum (SR), a specialized form of the ER that surrounds the contractile elements of myocytes, and is recognized primarily for its roles in the regulated release and storage of the calcium that underlies contraction and relaxation, respectively, in a process called contractile calcium handling (Fig. 2a) (Bers 2002, 2008; Sobie and Lederer 2012). In addition to the SR, there is a perinuclear ER that is structurally contiguous with the SR and as in most cells (Franke 1977), in cardiac myocytes, appears to have the molecular machinery necessary for the synthesis of secreted and membrane proteins (Fig. 2b) (Nakayama et al. 2010).

What remains unknown is the extent of overlap between the SR and ER in terms of contractile calcium handling, and secreted and membrane protein synthesis (Glembotski 2012; Millott et al. 2012; Groenendyk et al. 2013). Although there have been few studies that directly address these unknowns, it appears as though both the ER and SR of cardiac myocytes have some of the molecular machinery usually identified with secreted and membrane protein synthesis (Fig. 2c) and quality control, such as the ER chaperone GRP78, consistent with a possible role for the SR in the synthesis of secreted and membrane proteins. It also appears as though the calcium stored within the perinuclear ER of cardiac myocytes may be released into the nucleus (Fig. 2d), perhaps in coordination with contraction and each depolarization event (Ljubojevic and Bers 2015). Moreover, in contrast to SR calcium, ER calcium in cardiac myocytes can also be released in an IP_3-dependent manner, somewhat like other cell types; this IP_3-sensitive calcium pool may facilitate cell surface receptor-regulated events in cardiac myocytes that rely on the local release of calcium that does not drive contraction (Fig. 2e) (Nakayama et al. 2010). Thus, there seems to be some, but not complete overlap in ER and SR functions.

Cardiac myocytes comprise the major mass of the myocardium (Zak 1974; Banerjee et al. 2007), where they are responsible for contraction and the movement of blood throughout the body. Therefore, conditions that impair cardiac myocyte contraction, such as those that affect contractile calcium handling, or the ability of sarcomeres to generate force can be life threatening. The left ventricle (LV) is the largest chamber of the heart and is the main pump that supplies all organs and tissues with oxygenated blood. Because of its dominant role as a pumping chamber,

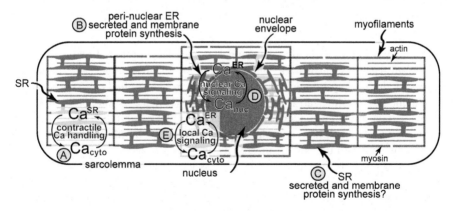

Fig. 2 Sarco/endoplasmic reticulum network in a cardiac myocyte: Shown is a diagram of a cardiac myocyte depicting the relationships between the sarcoplasmic reticulum (SR) involved in contractile calcium handling (**a**), the perinuclear ER involved in secreted and membrane protein synthesis (**b**), the SR that may be involved in secreted and membrane protein synthesis (**c**), the perinuclear ER involved in nuclear calcium signaling (**d**), and the perinuclear ER involved in local cytosolic calcium signaling (**e**). Also shown are the actin and myosin myofilaments that comprise the contractile apparatus of cardiac myocytes

diseases that affect LV function have widespread impact and can be life threatening. Therefore, most studies of the heart focus on the structure and function of the myocytes that comprise the LV, with a particular emphasis on sarcomeric proteins and SR calcium handling, and less focus on the synthesis of secreted and membrane proteins in the ER. This is surprising, considering that many membrane and secreted proteins required for proper cardiac contractility and endocrine/paracrine function are made in the ER (Glembotski 2012). It is interesting to note that atrial myocytes, which are less critical for propelling blood than ventricular myocytes, are distinct from ventricular myocytes, since they are professional secretory cells, responsible for the production of the hemodynamic hormone, atrial natriuretic peptide (de Bold 2011). Relatively recent studies demonstrating that, in addition to ER-targeted proteins, proteins bound to other locations are also translated on ER ribosomes, coupled with the finding that ER ribosome translation is more efficient than free ribosome translation (Reid and Nicchitta 2012, 2015) have spawned new interest in determining roles for ER-PQC and the UPR in cardiac myocytes under physiological and pathological conditions. A number of conditions can arise in cardiac myocytes that can affect ER-PQC, including decreased ER/SR calcium and the generation of reactive oxygen species (ROS), both of which occur during a number of cardiac pathologies (Kranias and Bers 2007; Ward et al. 2014). However, when it comes to the UPR, one often overlooked condition that may challenge the ER-PQC in health and disease is cardiac myocyte growth (Glembotski 2014; Doroudgar et al. 2015b).

3 Physiological Conditions that Challenge the ER-PQC in Cardiac Myocytes

Because of the increased protein synthesis, and thus, protein folding demands during growth, PQC in the ER and elsewhere are of great importance in cardiac myocytes during physiological and pathological growth of the heart (Shimizu and Minamino 2016). The heart is a very plastic organ (Hill and Olson 2008), and there are numerous conditions under which cardiac myocytes exhibit dramatic changes in number and size, beginning with development (Heineke and Molkentin 2006). The embryonic myocardium grows mostly by the rapid division of relatively small cardiac myocytes or hyperplasia. This process is depicted for the LV in (Fig. 3a). However, soon after birth, within 7–14 days in mice, the replicative capacity of cardiac myocytes decreases dramatically; thus, during most of postnatal development, when the heart increases in size to meet the demand for circulating greater volumes of blood, myocardial growth is primarily the result of increased cardiac myocyte size or hypertrophy (Fig. 3b). During both of these growth phases, it is essential that the protein folding machinery in cardiac myocytes, including the ER-PQC system, has the capacity to meet the demands placed on it by increased protein synthesis. Thus, we posited that the UPR contributes significantly to the

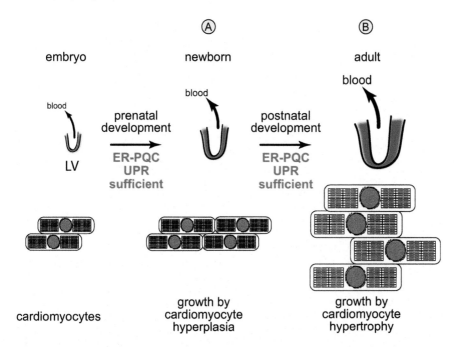

Fig. 3 Growth of the left ventricle during pre- and postnatal development: Cross sections of the mammalian left ventricle (LV) of the heart are shown diagrammatically, with the relative size depicting the changes in LV mass at different stages of development in the newborn (**a**) or the adult (**b**). The *red areas* are representations of the myocardium. The relative sizes of the arrows and blood represent the relative amounts of blood pumped, or ejected by the LV at different stages of development. The cardiac myocyte diagrams below each LV represent the relative changes in number and size during development. Since increases in LV muscle mass by hyperplasic or hypertrophic growth require increases in protein folding machinery, and since there is not ER stress or cell death associated with LV myocyte growth during pre- and postnatal development, endoplasmic reticulum-protein quality control (ER-PQC) and unfolded protein response (UPR) machinery are sufficient to support myocardial growth

PQC machinery that supports hyperplastic and hypertrophic cardiac myocyte growth. Consistent with this hypothesis is the finding that expression of numerous UPR genes is relatively high in the neonatal and juvenile rodent heart, compared to the adult heart, when cardiac myocyte growth has ceased (Doroudgar et al. 2015a; Taylor 2016). Moreover, neonatal and juvenile mouse hearts exhibit a robust adaptive UPR, which is much weaker in the adult mouse heart. It is precisely this decrease of UPR gene expression in the adult heart that poses a potential problem in the event that cardiac myocyte growth is reinitiated, depending on the stimulus for that growth.

There are also physiological conditions under which cardiac myocytes resume growth in the adult, including exercise and pregnancy (Hill and Olson 2008; Shimizu and Minamino 2016). Under these conditions, myocardial growth is adaptive, resulting in increases in the ability of the LV to pump blood to meet the

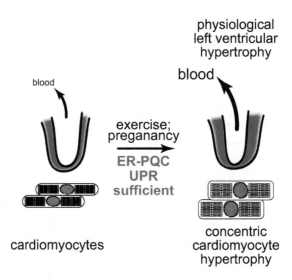

Fig. 4 Growth of the left ventricle during exercise and pregnancy: Cross sections of the LV and cardiomyocytes are depicted, as in Fig. 3. During exercise and pregnancy, the LV grows adaptively by concentric hypertrophy of each cardiomyocyte, i.e., increased diameter but not increased length, while the number of myocytes remains unchanged. Since this growth is adaptive and is not associated with ER stress or myocyte death, the ER-PQC and UPR are sufficient to support the growth

new demands for increased blood flow. During both exercise and pregnancy, myocardial growth is hypertrophic (Fig. 4). In this case, as with myocardial growth during development, myocytes grow in diameter more than in length, a process that is called concentric hypertrophy (van Berlo et al. 2013). Myocardial growth during exercise and pregnancy is reversible and does not lead to a pathological growth of the myocardium (Wasfy and Weiner 2015; Roh et al. 2016). Finally, while it is not well studied, the increased myocyte size during exercise and pregnancy is associated with increased protein synthesis (Catalucci et al. 2008; Li et al. 2012; Chung and Leinwand 2014) and thus would be expected to increase the demands on protein folding machinery in the ER, as well as in the cytosol; however, it is evident that the UPR in the adult heart is sufficient to meet the demands of increased protein synthesis, because under these conditions, there does not appear to be any myocyte death. Therefore, since this growth is limited in extent and time, and since the signal transduction initiated during physiological growth differs significantly from that during pathological growth (Maillct ct al. 2013; van Berlo et al. 2013; Lerchenmuller and Rosenzweig 2014), it appears as though physiological growth of the adult myocardium does not pose a challenge to the ER-PQC machinery. However, the impact on the ER-PQC may be different under conditions of growth during myocardial pathology (Glembotski 2014).

4 Pathological Conditions that Challenge the ER-PQC in Cardiac Myocytes

Diseases that affect the myocardium can be genetic or environmental. For example, mutations (Burke et al. 2016; Wilsbacher and McNally 2016), certain drugs, such as anti-neoplastics (Lee 2015; Molinaro et al. 2015), and pathogens (Fung et al. 2016; Yajima and Knowlton 2009) can affect sarcomere structure and impact contractile function to the point where LV pump action is severely impaired, leading eventually to heart failure and death (Fig. 5a). Although less common than mutations in sarcomeric proteins, mutations in calcium handling proteins and ion channels can lead to arrhythmias as well as contraction deficits that can contribute to sudden death, cardiac arrest, and, in some cases, heart failure (Bezzina et al. 2015; Wagner et al. 2015; Curran and Mohler 2015; Vatta and Ackerman 2010). In many of these cases, expression of UPR components is upregulated, although it is currently unclear as to whether this upregulation is a cause or a consequence of the myocardial defect (Liu and Dudley 2015).

Diseases that affect the myocardium can also develop as a result of injuries, such as myocardial infarction (MI) (Fig. 5b). In many cases, MI is the result of insufficient coronary blood flow to the myocardium, which is sometimes a consequence of atherosclerosis. This insufficient blood flow causes a lack of oxygen and nutrients, i.e., ischemia, that can eventually lead to death of the affected myocardium

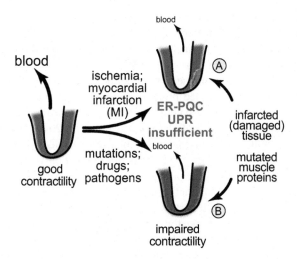

Fig. 5 Effect of ischemia, myocardia infarction, mutations, drugs, and pathogens on cardiac contractility: **a** Myocardial infarction (MI) results in irreparable damage to the myocardium, which decreases LV contractility. Ischemic cardiomyocytes that eventually die from ischemia have insufficient ER-PQC and UPR machinery. **b** Mutations, drugs, mostly anti-neoplastic agents, and pathogens, such as viruses, also impair cardiomyocyte contractility. These insults are associated with an insufficient ER-PQC and UPR, leading to maladaptive ER stress, which contributes to myocyte dropout by apoptosis

(Frangogiannis 2015). Surrounding the infarcted myocardium is the peri-infarct region, or border zone, in which the myocytes remain viable, but are stressed, due to the near lethal ischemia. In border zone myocytes, markers of the UPR are increased, and it is possible that these genes encode proteins that increase the survival of cardiac myocytes in the border zone.

Since the damaged myocardium in adults cannot be regenerated, MI damage is essentially permanent and can set in motion a pathological remodeling of the heart that can lead to serious functional impairment, culminating eventually in life-threatening heart failure (Rosenzweig 2012). Heart failure afflicts about 6 million people (Roger et al. 2012) and is the single most common discharge diagnosis in those over 65 (DeFrances and Podgornik 2006). Moreover, the disease is especially sobering because the limited regenerative capacity of the myocardium, coupled with the current absence of an effective therapy for the disease, leaves heart transplant as the only cure. In the early stages of heart failure, the myocardium grows by concentric hypertrophy by a process called hypertrophic cardiomyopathy, or HCM (Fig. 6a). In addition to MI, hypertension can also initiate HCM (Hoenig et al. 2008). In contrast to physiological hypertrophy, which stops when optimal benefit is obtained, this pathological hypertrophy, which is caused by a concentric growth of cardiac myocytes, does not resolve (van Berlo et al. 2013; Hill and Olson 2008; Frey et al. 2004). Initially, HCM is thought by some to be compensatory, providing sustained, perhaps even improved, LV function. However, as the myocardium remodels, this HCM often transforms into dilated cardiomyopathy (DCM), a condition associated with a decrease in myocyte number and an eccentric hypertrophy associated with the elongation of cardiomyocytes, in the absence of a compensating increase in diameter (Fig. 6b). This chamber dilation, associated with a severe weakening of the LV muscle, leads to decreased LV blood ejection and ultimately, to life-threatening heart failure (HF) (Lazzeroni et al. 2016).

While ER-PQC is compromised in ischemic heart disease and hypertrophic cardiomyopathy, where misfolded proteins accumulate within cardiac myocytes, it is underappreciated that cardiac pathology can also stem from the accumulation of misfolded proteins in the extracellular space (Shi et al. 2010). In cancer patients diagnosed with light chain amyloidosis, clonally expanded plasma cells secrete amyloidogenic light chains that misfold in the circulation (Cooley et al. 2014). These circulating amyloids aggregate into soluble proteotoxic oligomers, forming amyloid fibril deposits predominantly in the heart. This leads to impaired myocyte contractility, myocyte oxidative stress, and ultimately, myocyte death. Patients diagnosed with this disease typically cannot tolerate standard treatments meant to preserve cardiac function during heart failure, including β-adrenergic receptor blockers, angiotensin converting enzyme inhibitors, and digoxin (Meier-Ewert et al. 2011). Another form of cardiac amyloidosis is familial atrial fibrillation, in which the electrical conduction system of the heart, specifically in the atria, is disrupted by deposition of mutant amyloidogenic forms of transthyretin and atrial natriuretic peptide, secreted by the liver and the atria, respectively (Hodgson-Zingman et al. 2008). Patients with dilated cardiomyopathy stemming from cardiac amyloidosis and atrial fibrillation do not tolerate standard arrhythmia treatments. In heart failure

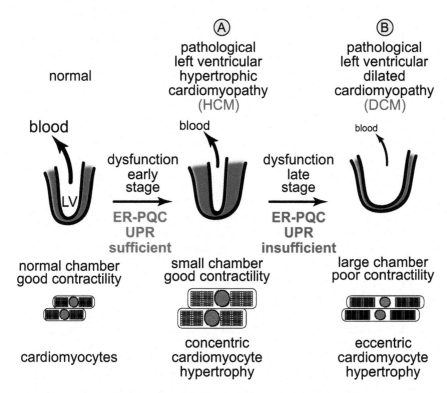

Fig. 6 Development of cardiomyopathy in the diseased heart: Ischemia, MI, mutations, drugs, and pathogens often lead to **a** hypertrophic cardiomyopathy (HCM), a condition in which each LV myocyte grows concentrically, i.e., increased diameter. This hypertrophic growth of the LV does not lead to increased pump function, and in many cases LV pump function is decreased, as the LV wall growth is so extensive that the chamber size is decreased. This hypertrophy is considered pathological because it usually leads to further remodeling and deterioration of LV structure and LV dysfunction in the late stage of pathology, associated with cardiomyocyte apoptosis, as well as an eccentric growth of remaining cardiomyocyte, i.e., increased length, which does not enhance cardiomyocyte contractility. This results in a dilation of the LV so chamber size increases, but since LV muscle mass is decreased, the ability of the LV to pump blood is severely impaired, leading to dilated hypertrophic cardiomyopathy (DCM) and eventually to heart failure (**b**). In the early stage of the disease, the ER-PQC and UPR are sufficient to handle cardiomyocyte growth; however, in the late stage of the disease, maladaptive ER stress ensues, and the ER-PQC and UPR are not sufficient to maintain LV structure and function

patients diagnosed with light chain cardiac amyloidosis, cancer remission coincides with improved cardiac function (Meier-Ewert et al. 2011), suggesting that stemming the production of amyloidogenic proteins may be a therapeutic approach to treating cardiac pathologies resulting from compromised protein quality control in the extracellular space. Interestingly, it has been shown that activation of the UPR increases expression and secretion of a subset of chaperones that prevent aggregation of misfolded proteins in the extracellular space (Genereux et al. 2015). Additionally, it has been shown that stress-independent activation of the UPR can prevent secretion of amyloidogenic light chains while not affecting secretion of non-amyloidogenic light chains (Cooley et al. 2014). Given that the activation of the UPR can also attenuate oxidative stress-mediated myocyte death, the UPR can be therapeutically targeted to treat forms of heart disease where proteotoxicity occurs from outside of the cell.

5 Experiment Methods and Models for Studying ER-PQC in Cardiac Pathology

For the most part, the relationship between cardiac pathologies and the UPR has been examined in several models, including primary cultures of cardiac myocytes and hearts of genetically modified mice. In these model systems, there are several maneuvers that are accepted as mimics of heart disease in humans. For example, cultured cardiac myocytes can be subjected to conditions that simulate those in the ischemic heart, i.e., oxygen and glucose deprivation. But even more clinically relevant is subjecting cultured cardiac myocytes to simulated ischemia followed by reperfusion (I/R), which is meant to mimic the clinically relevant scenario the heart of an MI patient might experience upon reopening of a blocked coronary vessel by angioplasty (Webster et al. 1995). Cardiac myocyte damage and viability are impacted during both ischemia and reperfusion, with reperfusion damage being caused primarily by reactive oxygen species (ROS) (Chen and Zweier 2014). Finally, in some cultured cardiac myocyte models, hypertrophic growth of myocytes can be induced with growth promoters, such as α-adrenergic receptor agonists, which mimic some aspects of pathological growth (O'Connell et al. 2014). However, each of these cultured cell models has limitations. For example, due to technical reasons, hypertrophic myocyte growth is studied in cultured neonatal rat cardiac myocytes, which has limitations because, at this stage of development, cardiac myocytes are much more plastic and able to exhibit growth in ways that do not replicate the growth observed in the adult rodent heart (Louch et al. 2011).

As important as cultured cell models are for examining molecular mechanisms, more relevant to human pathology are mouse models in which heart disease can be studied in vivo. Three surgical mouse models of myocardial heart disease are commonly used to mimic ischemia, ischemia/reperfusion, and pathological hypertrophy; they are myocardial infarction (MI), ischemia/reperfusion (I/R), and transaortic constriction (TAC), respectively (Abarbanell et al. 2010). MI surgery involves permanent ligation of the left anterior descending coronary artery (LAD), which is the artery most commonly blocked in patients suffering from an MI. I/R surgery involves ligating the LAD in the same manner as MI surgery, but for only about 30 minutes, followed by release and subsequent reperfusion. TAC surgery involves a partial restriction of the aorta, which increases the pressure against which the ventricle must work to expel blood, thus mimicking hypertension. Following these surgical procedures, cardiac function can be examined by several ways, but the most widely used noninvasive approach is echocardiography. The effect of each of these surgeries on cardiac structure is generally examined in postmortem histological examination of hearts for myocyte size and number, as well as apoptosis and gene expression.

6 Studies of ER-PQC in Cardiac Pathology

While there have been numerous reports of how various mouse models of heart disease effect changes in UPR gene expression, such reports are of limited impact to our understanding of whether and how the UPR contributes to or is a result of cardiac pathology. Instead, more revealing are mouse models in which specific genes of interest have been overexpressed, knocked down or knockout in the heart, which allows a more reliable interrogation of the effects of gain- or loss-of-function of known master regulators of the UPR on cardiac structure and function in response to models of heart disease. Accordingly, instead of compiling many studies that demonstrate a correlation between heart disease and the UPR, in the following section, a few of the highest impact studies that more clearly establish cause and effect between the UPR and heart disease are summarized. These studies have been categorized by which major branch of the UPR has been examined.

6.1 PERK/CHOP

The major downstream effect of activating the PERK branch of the UPR is eIF2α phosphorylation on serine 51, which attenuates global translation but increases translation of the transcription factor, ATF4 (Fig. 1b). Upon prolonged activation of the UPR, ATF4 increases expression of the pro-apoptotic transcription factor, C/EBP homologous protein (CHOP) (Malhotra and Kaufman 2007; Ron and Walter 2007). Since apoptosis is a major contributor to the decline in cardiac function observed during heart failure and other cardiac pathologies (Foo et al. 2005), and since CHOP expression is increased in experimental models of heart disease (Okada et al. 2004), several studies have focused on the effects of CHOP gene deletion in the mouse heart.

In genetically modified mice in which CHOP has been deleted in all tissues, i.e., global knockout, in the setting of MI, it was shown that compared to WT mice, CHOP KO mice exhibited increased death; however, the mice that survived had increased pathological hypertrophy of the LV, but no significant change in cardiac function or fibrosis (Luo et al. 2015). These results suggested that CHOP deletion does not negatively effect cardiac structure and function during permanent occlusion of the LAD, which is not usually associated with significant increases in apoptosis, leaving unanswered the question of whether CHOP contributes to apoptosis in the diseased heart.

The effect of CHOP on apoptosis in the diseased myocardium was examined in a different study in which the same line of CHOP KO mice was subjected to I/R surgery (Miyazaki et al. 2011), a maneuver known to increase apoptosis in the heart. In that study, it was shown that CHOP deletion decreased infarct size by decreasing apoptosis, as well as decreasing myocardial inflammation. In the same study, a mechanistic examination in mouse hearts and in cultured cardiac myocytes showed that in WT cardiac myocytes, the ROS generated upon simulated reperfusion activated CHOP, which induced apoptosis and inflammation gene expression.

Since CHOP is increased in heart failure, a study designed to examine the effects of CHOP in the hearts of mice subjected to TAC, which induces heart failure, also used the same line of CHOP KO mice (Fu et al. 2010). In that study, it was found that the CHOP KO mice exhibited reduced signs of heart failure after TAC surgery, including reduced hypertrophy, myocyte size, myocyte apoptosis, lung weight, and cardiac fibrosis, compared to WT mice. These results suggested that CHOP contributes to UPR-mediated apoptotic myocyte death and, therefore, is responsible for at least a portion of heart failure in this model of hypertension.

The effects of PERK deletion have also been examined in the mouse heart (Liu et al. 2014). However, since global deletion of PERK causes mouse growth retardation, this study examined the effects of deleting PERK specifically in cardiac myocytes using a conditional gene targeting approach. In this study, it was found that under non-stress conditions, PERK deletion had no effect on mouse heart structure and function. However, in a TAC model of heart failure, compared to WT mice, PERK KO mice exhibited decreased cardiac function, as well as exacerbated LV fibrosis and myocyte apoptosis. These findings indicate that PERK is required to protect the heart from pressure-overload induced heart failure. Taken together, these studies indicate that in the heart, the PERK branch of the UPR moderates the damaging effects of TAC-induced heart failure, so in that regard PERK is in some way protective. While the effects of PERK deletion on MI and I/R have not yet been studied, it is apparent that the apoptotic effector of the PERK branch of the UPR, CHOP, exacerbates the TAC-induced heart failure, as well as I/R damage in the mouse heart.

6.2 IRE-1/XBP1

One of the major targets of IRE-1 is the transcription factor, XBP1. When activated by ER stress, the endonuclease activity of IRE1 promotes the splicing of the mRNA encoding XBP1 from a form encoding an inactive protein, to one encoding an active transcription factor, spliced XBP1, $XBP1_s$ (Groenendyk et al. 2013). In mouse hearts, ischemia activates the UPR and increases the formation of $XBP1_s$, which protects from I/R damage of cardiac myocytes (Thuerauf et al. 2006). Moreover, I/R increases protein O-GlcNAcylation in the hearts of mice subjected to I/R, in vivo, and O-GlcNAcylation decreases I/R damage in mouse hearts (Ngoh et al. 2011). In a study of the effects of XBP1 gain- and loss-of-function in cardiac myocytes of mice, it was shown that XBP1$_s$ protects mouse heart from I/R damage by transcriptionally inducing key genes responsible for the hexosamine biosynthetic pathway required for protein O-GlcNAcylation (Wang et al. 2014). Interestingly, O-GlcNAcylation in the heart is not always protective. For example, in models of Type II diabetes, O-GlcNAcylation has been shown to increase cardiac dysfunction (Dassanayaka and Jones 2014), partly through the O-GlcNAcylation and hyper-activation of CaMKII, which deleteriously affects myocardial contractility (Erickson et al. 2013).

6.3 ATF6

ATF6 is activated, and many of its downstream target genes are induced in a variety of cardiac pathologies, including those induced in mice by MI, I/R, and TAC (Glembotski 2014). Moreover, acute conditional activation of ATF6 in mouse hearts protects the heart from I/R damage (Martindale et al. 2006). It has been known for sometime that in the absence of ER stress, the ER-resident chaperone, GRP78, can bind to, and anchor ATF6 in the lumen of the ER (Shen et al. 2002). Upon ER stress, GRP78 releases its stronghold on ATF6, allowing it to leave the ER and translocate to the Golgi, where it is cleaved by S1 and S2 proteases to generate its transcriptionally active form (Ye et al. 2000). However, a recent study showed that thrombospondin 4, a component of the extracellular matrix that is made and folded in the ER lumen, serves an escort role by binding to, and facilitating the movement of ATF6 from the ER to the Golgi (Lynch et al. 2012). In this study, it was shown that even in the absence of ER stress, ATF6 translocates from ER to Golgi upon overexpression of thrombospondin 4 in the heart, while thrombospondin 4 deletion decreases this translocation. Moreover, thrombospondin 4 gain- and loss-of-function in the heart decreased or increased, respectively, the damage in mouse hearts subjected to MI surgery. Thus, during secretion, thrombospondin 4 functions as an escort for ATF6 relocation from the ER to the Golgi, and then after it is secreted (Fig. 1d escort), thrombospondin 4 plays the second role as an extracellular matrix protein. As a result of this study, it has been hypothesized that thrombospondin 4 competes with GRP78 for binding to ATF6, and that while GRP78 cloaks the Golgi localization sequence on ATF6, by displacing GRP78, thrombospondin 4 reveals this sequence, thus facilitating ATF6 translocation (Doroudgar and Glembotski 2013a). In other studies, it was shown that, in the mouse heart, acute activation of ATF6 induces about 400 genes in cardiac myocytes, some of which were not previously known to be ATF6 ER stress-inducible genes, and have subsequently been shown to participate in the adaptive gene program regulated by ATF6 in the mouse heart upon MI, I/R, or TAC (Belmont et al. 2008). Recently, it was shown that, compared to WT mice, ATF6 KO mouse hearts exhibit greater I/R damage, decreased function, and increased apoptosis and necrosis (Jin et al. 2017). In this study, it was shown that the expression levels of numerous oxidative stress response genes were decreased in ATF6 KO mouse hearts, identifying many of them for the first time as ATF6-inducible, ER stress response genes. Interestingly, the ATF6-dependent oxidative stress genes in the mouse heart encode antioxidant proteins that reside in outside of the ER, demonstrating, for the first time, that ATF6 links ER stress and oxidative stress response pathways, thus explaining a mechanism by which acute activation of ATF6 can mitigate I/R damage in the heart, which is caused mainly by ROS generation in mitochondria.

7 Conclusions

Secreted and membrane protein synthesis in the ER is of critical importance to normal cardiac myocyte contractile function; however, the processes that oversee and govern the quality of secreted and membrane protein synthesis in cardiac myocytes have been unappreciated and, therefore, unstudied until relatively recently. The ER-PQC and UPR in cardiac myocytes appear to be robust and able to meet the challenges of ER protein synthesis and folding during physiological growth of the heart. However, during pathologies that can also initiate myocardial growth, especially in the adult, the adaptive ER-PQC and UPR in cardiac myocytes is insufficient, leading to accumulation of proteotoxic, terminally misfolded proteins, aggregate formation and eventual impairment of cardiac myocyte contractile function, and heart failure. Emerging areas of interest in the field include determining what is different about physiological and pathological cardiac myocyte growth in terms of ER-PQC and the UPR, in hopes of designing therapeutic approaches to decrease myocyte death and loss of LV function during pathology. One clear area of interest has been spawned by the finding that the expression levels of essential components of the adaptive UPR machinery decrease as a function of age, while the propensity for developing cardiac pathology increases as a function of age. This finding has led to the exploration of therapeutic approaches aimed at enhancing the adaptive UPR in the aged, pathologic heart in hopes of improving the synthesis, folding and trafficking of secreted and membrane proteins in ways that are anticipated to enhance cardiac myocyte contractility and reduce the progression to heart failure.

References

Abarbanell AM, Herrmann JL, Weil BR, Wang Y, Tan J, Moberly SP, Fiege JW, Meldrum DR (2010) Animal models of myocardial and vascular injury. J Surg Res 162(2):239–249. doi:10.1016/j.jss.2009.06.021

Banerjee I, Fuseler JW, Price RL, Borg TK, Baudino TA (2007) Determination of cell types and numbers during cardiac development in the neonatal and adult rat and mouse. Am J Physiol Heart Circ Physiol 293(3):1883–1891. doi:10.1152/ajpheart.00514.2007

Belmont PJ, Tadimalla A, Chen WJ, Martindale JJ, Thuerauf DJ, Marcinko M, Gude N, Sussman MA, Glembotski CC (2008) Coordination of growth and endoplasmic reticulum stress signaling by regulator of calcineurin 1 (RCAN1), a novel ATF6-inducible gene. J Biol Chem 283(20):14012–14021. doi:10.1074/jbc.M709776200 [pii]

Bers DM (2002) Cardiac excitation-contraction coupling. Nature 415(6868):198–205. doi:10.1038/415198a

Bers DM (2008) Calcium cycling and signaling in cardiac myocytes. Annu Rev Physiol 70:23–49. doi:10.1146/annurev.physiol.70.113006.100455

Bezzina CR, Lahrouchi N, Priori SG (2015) Genetics of sudden cardiac death. Circ Res 116 (12):1919–1936. doi:10.1161/CIRCRESAHA.116.304030

Burke MA, Cook SA, Seidman JG, Seidman CE (2016) Clinical and mechanistic insights into the genetics of cardiomyopathy. J Am Coll Cardiol 68(25):2871–2886. doi:10.1016/j.jacc.2016.08.079

Catalucci D, Latronico MV, Ellingsen O, Condorelli G (2008) Physiological myocardial hypertrophy: how and why? Front Biosci 13:312–324

Chen YR, Zweier JL (2014) Cardiac mitochondria and reactive oxygen species generation. Circ Res 114(3):524–537. doi:10.1161/CIRCRESAHA.114.300559

Chung E, Leinwand LA (2014) Pregnancy as a cardiac stress model. Cardiovasc Res 101(4): 561–570. doi:10.1093/cvr/cvu013

Cooley CB, Ryno LM, Plate L, Morgan GJ, Hulleman JD, Kelly JW, Wiseman RL (2014) Unfolded protein response activation reduces secretion and extracellular aggregation of amyloidogenic immunoglobulin light chain. Proc Natl Acad Sci U S A 111(36):13046–13051. doi:10.1073/pnas.1406050111

Curran J, Mohler PJ (2015) Alternative paradigms for ion channelopathies: disorders of ion channel membrane trafficking and posttranslational modification. Annu Rev Physiol 77: 505–524. doi:10.1146/annurev-physiol-021014-071838

Dassanayaka S, Jones SP (2014) O-GlcNAc and the cardiovascular system. Pharmacol Ther 142 (1):62–71. doi:10.1016/j.pharmthera.2013.11.005

de Bold AJ (2011) Thirty years of research on atrial natriuretic factor: historical background and emerging concepts. Can J Physiol Pharmacol 89(8):527–531. doi:10.1139/Y11-019

DeFrances CJ, Podgornik MN (2006) 2004 National hospital discharge survey. Adv Data 371:1–19

Doroudgar S, Glembotski CC (2013a) ATF6 and thrombospondin 4: the dynamic duo of the adaptive endoplasmic reticulum stress response. Circ Res 112(1):9–12. doi:112/1/9 [pii] 10.1161/CIRCRESAHA.112.280560

Doroudgar S, Glembotski CC (2013b) New concepts of endoplasmic reticulum function in the heart: programmed to conserve. J Mol Cell Cardiol. doi:10.S0022-2828(12)00374-4 [pii] 10.1016/j.yjmcc.2012.10.006

Doroudgar S, Volkers M, Thuerauf DJ, Khan M, Mohsin S, Respress JL, Wang W, Gude N, Muller OJ, Wehrens XH, Sussman MA, Glembotski CC (2015a) Hrd1 and ER-associated protein degradation, ERAD, are critical elements of the adaptive er stress response in cardiac myocytes. Circ Res 117(6):536–546. doi:10.1161/CIRCRESAHA.115.306993 [pii]

Doroudgar S, Volkers M, Thuerauf DJ, Khan M, Mohsin S, Respress JL, Wang W, Gude N, Muller OJ, Wehrens XHT, Sussman MA, Glembotski CC (2015b) Hrd1 and ER-associated protein degradation, ERAD, are critical elements of the adaptive er stress response in cardiac myocytes. Circ Res 117(6):536–546. doi:10.1161/CIRCRESAHA.115.306993

Erickson JR, Pereira L, Wang L, Han G, Ferguson A, Dao K, Copeland RJ, Despa F, Hart GW, Ripplinger CM, Bers DM (2013) Diabetic hyperglycaemia activates CaMKII and arrhythmias by O-linked glycosylation. Nature 502(7471):372–376. doi:10.1038/nature12537

Foo RS, Mani K, Kitsis RN (2005) Death begets failure in the heart. J Clin Invest 115(3):565–571. doi:10.1172/JCI24569

Frangogiannis NG (2015) Pathophysiology of myocardial infarction. Compr Physiol 5(4): 1841–1875. doi:10.1002/cphy.c150006

Franke WW (1977) Structure and function of nuclear membranes. Biochem Soc Symp 42:125–135

Frey N, Katus HA, Olson EN, Hill JA (2004) Hypertrophy of the heart: a new therapeutic target? Circulation 109(13):1580–1589. doi:10.1161/01.CIR.0000120390.68287.BB

Fu HY, Okada K, Liao Y, Tsukamoto O, Isomura T, Asai M, Sawada T, Okuda K, Asano Y, Sanada S, Asanuma H, Asakura M, Takashima S, Komuro I, Kitakaze M, Minamino T (2010) Ablation of C/EBP homologous protein attenuates endoplasmic reticulum-mediated apoptosis and cardiac dysfunction induced by pressure overload. Circulation 122(4):361–369. doi:10.1161/CIRCULATIONAHA.109.917914

Fung G, Luo H, Qiu Y, Yang D, McManus B (2016) Myocarditis. Circ Res 118(3):496–514. doi:10.1161/CIRCRESAHA.115.306573

Gardner BM, Pincus D, Gotthardt K, Gallagher CM, Walter P (2013) Endoplasmic reticulum stress sensing in the unfolded protein response. Cold Spring Harb Perspect Biol 5(3):a013169. doi:10.1101/cshperspect.a013169

Genereux JC, Qu S, Zhou M, Ryno LM, Wang S, Shoulders MD, Kaufman RJ, Lasmezas CI, Kelly JW, Wiseman RL (2015) Unfolded protein response-induced ERdj3 secretion links ER stress to extracellular proteostasis. EMBO J 34(1):4–19. doi:10.15252/embj.201488896

Glembotski CC (2007) Endoplasmic reticulum stress in the heart. Circ Res 101(10):975–984. doi:101/10/975 [pii] 10.1161/CIRCRESAHA.107.161273

Glembotski CC (2008) The role of the unfolded protein response in the heart. J Mol Cell Cardiol 44(3):453–459. doi:10.S0022-2828(07)01283-7 [pii] 10.1016/j.yjmcc.2007.10.017

Glembotski CC (2012) Roles for the sarco-/endoplasmic reticulum in cardiac myocyte contraction, protein synthesis, and protein quality control. Physiology (Bethesda) 27(6):343–350. doi:27/6/343 [pii] 10.1152/physiol.00034.2012

Glembotski CC (2014) Roles for ATF6 and the sarco/endoplasmic reticulum protein quality control system in the heart. J Mol Cell Cardiol 71:11–15. doi:10.S0022-2828(13)00291-5 [pii] 10.1016/j.yjmcc.2013.09.018

Groenendyk J, Agellon LB, Michalak M (2013) Coping with endoplasmic reticulum stress in the cardiovascular system. Annu Rev Physiol 75:49–67. doi:10.1146/annurev-physiol-030212-183707

Heineke J, Molkentin JD (2006) Regulation of cardiac hypertrophy by intracellular signalling pathways. Nat Rev Mol Cell Biol 7(8):589–600. doi:10.1038/nrm1983 [pii]

Hetz C, Chevet E, Oakes SA (2015) Proteostasis control by the unfolded protein response. Nat Cell Biol 17(7):829–838. doi:10.1038/ncb3184

Hill JA, Olson EN (2008) Cardiac plasticity. N Engl J Med 358(13):1370–1380. doi:358/13/1370 [pii] 10.1056/NEJMra072139

Hodgson-Zingman DM, Karst ML, Zingman LV, Heublein DM, Darbar D, Herron KJ, Ballew JD, de Andrade M, Burnett JC Jr, Olson TM (2008) Atrial natriuretic peptide frameshift mutation in familial atrial fibrillation. N Engl J Med 359(2):158–165. doi:10.1056/NEJMoa0706300

Hoenig MR, Bianchi C, Rosenzweig A, Sellke FW (2008) The cardiac microvasculature in hypertension, cardiac hypertrophy and diastolic heart failure. Curr Vasc Pharmacol 6(4): 292–300

Ivanova EA, Orekhov AN (2016) The role of endoplasmic reticulum stress and unfolded protein response in atherosclerosis. Int J Mol Sci 17(2). doi:10.3390/ijms17020193

Jin JK, Blackwood EA, Azizi KM, Thuerauf DJ, Fahem AG, Hofmann C, Kaufman RJ, Doroudgar S, Glembotski CC (2017) ATF6 decreases myocardial ischemia/reperfusion damage and links ER stress and oxidative stress signaling pathways in the heart. Circ Res. doi:10.1161/CIRCRESAHA.116.310266

Kranias EG, Bers DM (2007) Calcium and cardiomyopathies. Subcell Biochem 45:523–537

Lazzeroni D, Rimoldi O, Camici PG (2016) From left ventricular hypertrophy to dysfunction and failure. Circ J 80(3):555–564. doi:10.1253/circj.CJ-16-0062

Lee CS (2015) Mechanisms of cardiotoxicity and the development of heart failure. Crit Care Nurs Clin North Am 27(4):469–481. doi:10.1016/j.cnc.2015.07.002

Lerchenmuller C, Rosenzweig A (2014) Mechanisms of exercise-induced cardiac growth. Drug Discov Today 19(7):1003–1009. doi:10.1016/j.drudis.2014.03.010

Li J, Umar S, Amjedi M, Iorga A, Sharma S, Nadadur RD, Regitz-Zagrosek V, Eghbali M (2012) New frontiers in heart hypertrophy during pregnancy. Am J Cardiovasc Dis 2(3):192–207

Li Y, Guo Y, Tang J, Jiang J, Chen Z (2014) New insights into the roles of CHOP-induced apoptosis in ER stress. Acta Biochim Biophys Sin (Shanghai) 46(8):629–640. doi:10.1093/abbs/gmu048

Liu M, Dudley SC Jr (2015) Role for the Unfolded Protein Response in Heart Disease and Cardiac Arrhythmias. Int J Mol Sci 17(1). doi:10.3390/ijms17010052

Liu X, Kwak D, Lu Z, Xu X, Fassett J, Wang H, Wei Y, Cavener DR, Hu X, Hall J, Bache RJ, Chen Y (2014) Endoplasmic reticulum stress sensor protein kinase R-like endoplasmic reticulum kinase (PERK) protects against pressure overload-induced heart failure and lung remodeling. Hypertension 64(4):738–744. doi:10.1161/HYPERTENSIONAHA.114.03811

Ljubojevic S, Bers DM (2015) Nuclear calcium in cardiac myocytes. J Cardiovasc Pharmacol 65 (3):211–217. doi:10.1097/FJC.0000000000000174

Louch WE, Sheehan KA, Wolska BM (2011) Methods in cardiomyocyte isolation, culture, and gene transfer. J Mol Cell Cardiol 51(3):288–298. doi:10.1016/j.yjmcc.2011.06.012

Luo G, Li Q, Zhang X, Shen L, Xie J, Zhang J, Kitakaze M, Huang X, Liao Y (2015) Ablation of C/EBP homologous protein increases the acute phase mortality and doesn't attenuate cardiac remodeling in mice with myocardial infarction. Biochem Biophys Res Commun 464(1):201–207. doi:10.1016/j.bbrc.2015.06.117

Lynch JM, Maillet M, Vanhoutte D, Schloemer A, Sargent MA, Blair NS, Lynch KA, Okada T, Aronow BJ, Osinska H, Prywes R, Lorenz JN, Mori K, Lawler J, Robbins J, Molkentin JD (2012) A thrombospondin-dependent pathway for a protective ER stress response. Cell 149 (6):1257–1268. doi:10.S0092-8674(12)00572-7 [pii] 10.1016/j.cell.2012.03.050

Maillet M, van Berlo JH, Molkentin JD (2013) Molecular basis of physiological heart growth: fundamental concepts and new players. Nat Rev Mol Cell Biol 14(1):38–48. doi:10.1038/nrm3495

Malhotra JD, Kaufman RJ (2007) The endoplasmic reticulum and the unfolded protein response. Semin Cell Dev Biol 18(6):716–731. doi:10.S1084-9521(07)00149-8 [pii] 10.1016/j.semcdb.2007.09.003

Martindale JJ, Fernandez R, Thuerauf D, Whittaker R, Gude N, Sussman MA, Glembotski CC (2006) Endoplasmic reticulum stress gene induction and protection from ischemia/reperfusion injury in the hearts of transgenic mice with a tamoxifen-regulated form of ATF6. Circ Res 98 (9):1186–1193. doi:10.1161/01.RES.0000220643.65941.8d [pii]

Meier-Ewert HK, Sanchorawala V, Berk JL, Ruberg FL (2011) Cardiac amyloidosis: evolving approach to diagnosis and management. Curr Treat Options Cardiovasc Med 13(6):528–542. doi:10.1007/s11936-011-0147-4

Millott R, Dudek E, Michalak M (2012) The endoplasmic reticulum in cardiovascular health and disease. Can J Physiol Pharmacol 90(9):1209–1217. doi:10.1139/y2012-058

Minamino T, Kitakaze M (2010) ER stress in cardiovascular disease. J Mol Cell Cardiol 48 (6):1105–1110. doi:10.S0022-2828(09)00469-6 [pii] 10.1016/j.yjmcc.2009.10.026

Miyazaki Y, Kaikita K, Endo M, Horio E, Miura M, Tsujita K, Hokimoto S, Yamamuro M, Iwawaki T, Gotoh T, Ogawa H, Oike Y (2011) C/EBP homologous protein deficiency attenuates myocardial reperfusion injury by inhibiting myocardial apoptosis and inflammation. Arterioscler Thromb Vasc Biol 31(5):1124–1132. doi:10.1161/ATVBAHA.111.224519

Molinaro M, Ameri P, Marone G, Petretta M, Abete P, Di Lisa F, De Placido S, Bonaduce D, Tocchetti CG (2015) Recent advances on pathophysiology, diagnostic and therapeutic insights in cardiac dysfunction induced by antineoplastic drugs. Biomed Res Int 2015:138148. doi:10.1155/2015/138148

Nakayama H, Bodi I, Maillet M, DeSantiago J, Domeier TL, Mikoshiba K, Lorenz JN, Blatter LA, Bers DM, Molkentin JD (2010) The IP3 receptor regulates cardiac hypertrophy in response to select stimuli. Circ Res 107(5):659–666. doi:10.1161/CIRCRESAHA.110.220038

Ngoh GA, Watson LJ, Facundo HT, Jones SP (2011) Augmented O-GlcNAc signaling attenuates oxidative stress and calcium overload in cardiomyocytes. Amino Acids 40(3):895–911. doi:10.1007/s00726-010-0728-7

O'Connell TD, Jensen BC, Baker AJ, Simpson PC (2014) Cardiac alpha1-adrenergic receptors: novel aspects of expression, signaling mechanisms, physiologic function, and clinical importance. Pharmacol Rev 66(1):308–333. doi:10.1124/pr.112.007203

Okada K, Minamino T, Tsukamoto Y, Liao Y, Tsukamoto O, Takashima S, Hirata A, Fujita M, Nagamachi Y, Nakatani T, Yutani C, Ozawa K, Ogawa S, Tomoike H, Hori M, Kitakaze M (2004) Prolonged endoplasmic reticulum stress in hypertrophic and failing heart after aortic constriction: possible contribution of endoplasmic reticulum stress to cardiac myocyte apoptosis. Circulation 110(6):705–712. doi:10.1161/01.CIR.0000137836.95625.D4 [pii]

Palade GE, Siekevitz P (1956) Liver microsomes; an integrated morphological and biochemical study. J Biophys Biochem Cytol 2(2):171–200

Pisoni GB, Molinari M (2016) Five questions (with their answers) on ER-associated degradation. Traffic 17(4):341–350. doi:10.1111/tra.12373

Plemper RK, Wolf DH (1999) Endoplasmic reticulum degradation. Reverse protein transport and its end in the proteasome. Mol Biol Rep 26(1–2):125–130

Reid DW, Nicchitta CV (2012) Primary role for endoplasmic reticulum-bound ribosomes in cellular translation identified by ribosome profiling. J Biol Chem 287(8):5518–5527. doi:10.1074/jbc.M111.312280 [pii]

Reid DW, Nicchitta CV (2015) Diversity and selectivity in mRNA translation on the endoplasmic reticulum. Nat Rev Mol Cell Biol 16(4):221–231. doi:10.1038/nrm3958

Roger VL, Go AS, Lloyd-Jones DM, Benjamin EJ, Berry JD, Borden WB, Bravata DM, Dai S, Ford ES, Fox CS, Fullerton HJ, Gillespie C, Hailpern SM, Heit JA, Howard VJ, Kissela BM, Kittner SJ, Lackland DT, Lichtman JH, Lisabeth LD, Makuc DM, Marcus GM, Marelli A, Matchar DB, Moy CS, Mozaffarian D, Mussolino ME, Nichol G, Paynter NP, Soliman EZ, Sorlie PD, Sotoodehnia N, Turan TN, Virani SS, Wong ND, Woo D, Turner MB, American Heart Association Statistics C, Stroke Statistics S (2012) Heart disease and stroke statistics—2012 update: a report from the American Heart Association. Circulation 125(1):e2–e220. doi:10.1161/CIR.0b013e31823ac046

Roh J, Rhee J, Chaudhari V, Rosenzweig A (2016) The role of exercise in cardiac aging: from physiology to molecular mechanisms. Circ Res 118(2):279–295. doi:10.1161/CIRCRESAHA.115.305250

Ron D, Walter P (2007) Signal integration in the endoplasmic reticulum unfolded protein response. Nat Rev Mol Cell Biol 8(7):519–529. doi:10.1038/nrm2199 [pii]

Rosenzweig A (2012) Medicine. Cardiac regeneration. Science 338(6114):1549–1550. doi:338/6114/1549 [pii] 10.1126/science.1228951

Ryoo HD (2016) Long and short (timeframe) of endoplasmic reticulum stress-induced cell death. FEBS J 283(20):3718–3722. doi:10.1111/febs.13755

Sano R, Reed JC (2013) ER stress-induced cell death mechanisms. Biochim Biophys Acta 1833 (12):3460–3470. doi:10.S0167-4889(13)00251-6 [pii] 10.1016/j.bbamcr.2013.06.028

Shen J, Chen X, Hendershot L, Prywes R (2002) ER stress regulation of ATF6 localization by dissociation of BiP/GRP78 binding and unmasking of Golgi localization signals. Dev Cell 3 (1):99–111. doi:10.S1534580702002034[pii]

Shi J, Guan J, Jiang B, Brenner DA, Del Monte F, Ward JE, Connors LH, Sawyer DB, Semigran MJ, Macgillivray TE, Seldin DC, Falk R, Liao R (2010) Amyloidogenic light chains induce cardiomyocyte contractile dysfunction and apoptosis via a non-canonical p38alpha MAPK pathway. Proc Natl Acad Sci U S A 107(9):4188–4193. doi:10.1073/pnas.0912263107

Shimizu I, Minamino T (2016) Physiological and pathological cardiac hypertrophy. J Mol Cell Cardiol 97:245–262. doi:10.1016/j.yjmcc.2016.06.001

Sobie EA, Lederer WJ (2012) Dynamic local changes in sarcoplasmic reticulum calcium: physiological and pathophysiological roles. J Mol Cell Cardiol 52(2):304–311. doi:10.1016/j.yjmcc.2011.06.024

Taylor RC (2016) Aging and the UPR(ER). Brain Res 1648(Pt B):588–593. doi:10.1016/j.brainres.2016.04.017

Thuerauf DJ, Marcinko M, Gude N, Rubio M, Sussman MA, Glembotski CC (2006) Activation of the unfolded protein response in infarcted mouse heart and hypoxic cultured cardiac myocytes. Circ Res 99(3):275–282. doi:10.1161/01.RES.0000233317.70421.03 [pii]

van Berlo JH, Maillet M, Molkentin JD (2013) Signaling effectors underlying pathologic growth and remodeling of the heart. J Clin Invest 123(1):37–45. doi:10.1172/JCI62839 [pii]

Vatta M, Ackerman MJ (2010) Genetics of heart failure and sudden death. Heart Fail Clin 6 (4):507–514, ix. doi:10.1016/j.hfc.2010.05.008

Wagner S, Maier LS, Bers DM (2015) Role of sodium and calcium dysregulation in tachyarrhythmias in sudden cardiac death. Circ Res 116(12):1956–1970. doi:10.1161/CIRCRESAHA.116.304678

Walter P, Ron D (2011) The unfolded protein response: from stress pathway to homeostatic regulation. Science 334(6059):1081–1086. doi:10.1126/science.1209038

Wang ZV, Deng Y, Gao N, Pedrozo Z, Li DL, Morales CR, Criollo A, Luo X, Tan W, Jiang N, Lehrman MA, Rothermel BA, Lee AH, Lavandero S, Mammen PP, Ferdous A, Gillette TG, Scherer PE, Hill JA (2014) Spliced X-box binding protein 1 couples the unfolded protein response to hexosamine biosynthetic pathway. Cell 156(6):1179–1192. doi:10.S0092-8674(14) 00025-7 [pii] 10.1016/j.cell.2014.01.014

Ward CW, Prosser BL, Lederer WJ (2014) Mechanical stretch-induced activation of ROS/RNS signaling in striated muscle. Antioxid Redox Signal 20(6):929–936. doi:10.1089/ars.2013.5517

Wasfy MM, Weiner RB (2015) Differentiating the athlete's heartPath6- from hypertrophic cardiomyopathy. Curr Opin Cardiol 30(5):500–505. doi:10.1097/HCO.0000000000000203

Webster KA, Discher DJ, Bishopric NH (1995) Cardioprotection in an in vitro model of hypoxic preconditioning. J Mol Cell Cardiol 27(1):453–458

Wilsbacher L, McNally EM (2016) Genetics of cardiac developmental disorders: cardiomyocyte proliferation and growth and relevance to heart failure. Annu Rev Pathol 11:395–419. doi:10.1146/annurev-pathol-012615-044336

Yajima T, Knowlton KU (2009) Viral myocarditis: from the perspective of the virus. Circulation 119(19):2615–2624. doi:10.1161/CIRCULATIONAHA.108.766022

Ye J, Rawson RB, Komuro R, Chen X, Dave UP, Prywes R, Brown MS, Goldstein JL (2000) ER stress induces cleavage of membrane-bound ATF6 by the same proteases that process SREBPs. Mol Cell 6(6):1355–1364. doi:10.S1097-2765(00)00133-7 [pii]

Zak R (1974) Development and proliferative capacity of cardiac muscle cells. Circ Res 35(2):suppl II:17–26

Zhou AX, Tabas I (2013) The UPR in atherosclerosis. Semin Immunopathol 35(3):321–332. doi:10.1007/s00281-013-0372-x

Printed by Printforce, the Netherlands